Financing Innovation in the United States, 1870 to the Present

Financing Innovation in the United States, 1870 to the Present

edited by Naomi R. Lamoreaux and Kenneth L. Sokoloff
with a foreword by William Janeway

The MIT Press
Cambridge, Massachusetts
London, England

MIT Press books may be purchased at special quantity discounts for business or sales promotional use. For information, please e-mail special_sales@mitpress.mit .edu or write to Special Sales Department, The MIT Press, 55 Hayward Street, Cambridge, MA 02142.

This book was set in Sabon on 3B2 by Asco Typesetters, Hong Kong, and was printed and bound in the United States.

Library of Congress Cataloging-in-Publication Data

Financing innovation in the United States, 1870 to the present / edited by Naomi R. Lamoreaux and Kenneth L. Sokoloff ; foreword by William Janeway.
    p. cm.
Includes bibliographical references and index.
ISBN-13: 978-0-262-12289-4 (hardcover : alk. paper)
1. High technological industries—United States—Finance—History.
2. Technological innovations—United States—Finance—History. I. Lamoreaux, Naomi R. II. Sokoloff, Kenneth Lee.

HC110.H53F56   2007
338.6'0410973—dc22   2006030095

10 9 8 7 6 5 4 3 2 1

A project organized by the Social Science Research Council

Contents

Foreword

William Janeway

The idea for the project that generated this book originated some twenty years ago. Having participated in the sale of the investment banking partnership F. Eberstadt & Co. to a global financial services group, I discovered the time to immerse myself in Fernand Braudel's great work, *Civilization and Capitalism* (1982). In the second volume, *The Wheels of Commerce*, Braudel identifies the unique attribute of the capitalist— that which distinguishes him decisively from participants in both the regulated markets of the traditional economy and the nascent "free" markets that were rising to challenge them: "The characteristic advantage of standing at the commanding heights of the economy ... consisted precisely of not having to confine oneself to a single choice, of being able, as today's businessmen would put it, to keep one's options open" (381).

My colleagues and I had been feeling our way into the emerging domain of venture capital. Braudel's insight delivered a shock of recognition, even as he amply documented that the sort of technological innovation that had become the focus of late-twentieth-century venture capitalists did not represent a relevant option for the preindustrial capitalists whose work he chronicled and placed in context. Braudel endorsed Simon Kuznets as "absolutely right when he says":

At the danger of exaggeration, one may ask whether there was *any* fixed, durable capital formation, except for the "monuments" in pre-modern times, whether there was any significant accumulation of capital goods with a long physical life that did not require current maintenance (or replacement) amounting to a high proportion of the original full value. If most equipment lasted no more than five or six years, if most land improvements had to be maintained by continuous rebuilding amounting to something like a fifth of the total value per years, and if most buildings were destroyed at a rate cumulating to fairly complete destruction over a period from 25 to 50 years, then there was little that could be classified as durable capital.... The whole concept of fixed capital may be a unique product of the modern economic epoch and of modern technology. (quoted in Braudel 1982, 158)

On the contrary, Braudel's capitalist found his apotheosis as a long-distance *arbitrageur*, financing trade and earning "super profits...based on the price difference between two markets very far apart, with supply and demand in complete ignorance of each other" (Braudel 1982, 405). Nonetheless, Braudel's transcendent image remained of the capitalist's unchanging goal: to escape from the "world of transparence and regularity," as he defines the "economy," where the possibility of profit is constrained and even eliminated by the regulations of the traditional market or the competition of the emerging free market. And so, it did not seem fanciful to imagine, the modern venture capitalist seeks the "super profits" that come from financing those innovation that disrupt old and define new markets.

Over the next fifteen years, as I became immersed in financing innovative technology as a partner of Warburg, Pincus, reading in the history of technology and the enterprises built to deploy technological innovation generated a growing sense of frustration. The rich literature on technological innovation is notable for the relatively modest degree of attention paid to the sources of capital that funded the deployment of new technologies. As Carlota Perez (2004) has written:

In Schumpeter's basic definition of capitalism as "that form of private property economy in which innovations are carried out by means of borrowed money", we find his characteristic separation of borrower and lender, entrepreneur and banker, as the two faces of the innovation coin. This is not, however, how his legacy has been interpreted and enriched by the great majority of Neo-Schumpeterians. The accent has almost invariably been on the entrepreneur to the neglect of the financial agent, no matter how obviously indispensable this agent may be to innovation.

The large and diverse body of scholarship on technology-driven industrial development—from Alfred Chandler (1977, 1990) and Richard Nelson (Nelson, Peck, and Kalachek, 1967) through Thomas Hughes (1983) and Leo Marx (Smith and Marx 1996), to Chris Freeman (Freeman and Soete 1997) and Nathan Rosenberg (1994)—provides limited, if any, insight into this critical nexus. The absence is all the more striking when considered in the still relevant light of the path-breaking work by Davis and North, *Institutional Change and American Economic Growth* (1971). Chapter 6, "Organization and Re-Organization in the Financial Markets: Savings and Investment in the American Economy, 1820–1950," offers a synoptic overview of the role of private and public finance in economic development. Davis and North thereby also define a research agenda that remains largely incomplete a generation on.

It was in this context that I began a conversation, initially informal, with Craig Calhoun, newly installed as president of the Social Science Research Council (SSRC). Craig engaged David Weiman, then a senior staff member of the SSRC, and the conversation began to take substantive shape: a research project whose purpose would be to generate a range of case studies in the financing of technological innovation. By 2001, Ashley Timmer had taken over responsibility at the SSRC, and—most important—Naomi Lamoreaux and Ken Sokoloff had agreed to lead the project. The result of their intellectual leadership, organizational focus, and (not least) original scholarship is this book.

This book contributes, individually and collectively, toward filling that missing dimension of economic history, where finance intersects invention to generate economically significant innovation. In the aftermath of the great dot-com/telecom bubble of 1998–2000, the relevance of the subject matter verges on the self-evident. But the chapters in this book, rooted in deep and often pioneering empirical excavation, do not only stand as exemplars of research methodology in a substantially unexplored domain. Their publication comes at a time when the dynamics of the capital markets and their role in economic evolution are once again a subject of theoretical as well as empirical study. On the one hand, a variety of imaginative approaches are being deployed to understand the empirical puzzles generated by the attempt to explain—or, rather, explain away—the functioning of the capital markets through the application of the rational expectations hypothesis (Mordecai, Jin, and Motolese 2005; Scheinkman and Xiong 2005; Weitzman 2005). Jointly and severally, this work offers the potential of a reintegration of theoretical finance into mainstream economic theory. On the other hand, these chapters provide specific context and content to inform and constrain renewed interest in the agenda implicitly established by Davis and

North and most recently redefined and renewed by Carlota Perez's work *Technological Revolutions and Financial Capital* (2002).

References

Braudel, Fernand. 1982. *The Wheels of Commerce*, Vol. 2 of *Civilization and Capitalism 15–18th Century*. New York: Harper & Row.

Chandler, Alfred D. 1977. *The Visible Hand: The Managerial Revolution in American Business*. With the assistance of Takashi Hikino. Cambridge, Mass.: Belknap Press of Harvard University Press.

Chandler, Alfred D. 1990. *Scale and Scope: The Dynamics of Industrial Capitalism*. Cambridge, Mass.: Belknap Press of Harvard University Press.

Davis, Lance E., and Douglas C. North. 1971. *Institutional Change and American Economic Growth*. Cambridge: Cambridge University Press.

Freeman, Chris, and Luc Soete. 1997. *The Economics of Industrial Innovation*. 3rd ed. Cambridge, Mass.: MIT Press.

Hughes, Tomas P. 1983. *Networks of Power: Electrification in Western Society, 1880–1930*. Baltimore: Johns Hopkins University Press.

Kurz, Mordecai, Hechui Jin, and Maurizio Motolese. 2005. "Determinants of Stock Market Volatility and Risk Premia." *Annals of Finance*, 1.

Nelson, Richard R., Merton J. Peck, and Edward D. Kalachek. 1967. *Technology, Economic Growth and Public Policy*. Washington, D.C.: Brookings Institution.

Perez, Carlota. 2002. *Technological Revolutions and Financial Capital: The Dynamics of Bubbles and Golden Ages*. Cheltenham: Edward Elgar Publishing.

Perez, Carlota. 2004. "Finance and Technological Change: A Neo-Schumpeterian Perspective," Working paper no. 14, Cambridge Endowment for Research in Finance, University of Cambridge.

Rosenberg, Nathan. 1994. *Exploring the Black Box: Techonology, Economics, and History*. Cambridge: Cambridge University Press.

Scheinkman, Jose A., and Wei Xiong. 2005. "Overconfidence and Speculative Bubbles." *Journal of Political Economy* 3.

Smith, Merritt Roe, and Leo Marx, eds. 1996. *Does Technology Drive History? The Dilemma of Technological Determinism*. Cambridge, Mass.: MIT Press.

Weitzman, Martin L. 2005. "A Unified Bayesian Theory of Equity 'Puzzles.'" Harvard University. ⟨http://econ-www.mit.edu/faculty/download_book.php?id=74.⟩

Acknowledgments

This book never would have been written if not for Bill Janeway's conviction that the methods by which finance is mobilized are crucial to the processes of innovation. Janeway's willingness to commit resources to support study of the subject, and his conversations with David Weiman and Ashley Timmer (both formerly of the Social Science Research Council), led to a series of informal discussions at the SSRC in New York that brought together a variety of scholars concerned with the record of technological development in the United States over the long run. Timmer and Craig Calhoun (president of the SSRC) encouraged us to organize a formal conference focusing on how innovation had been financed over American economic history, and the preliminary versions of these papers were presented at the Beckman Center of the National Academy of Sciences in Irvine, California, in March 2003. We are deeply grateful to Bill Janeway and the SSRC for their generous support of these meetings and the overall enterprise, as well as for their intellectual inspiration. We and the other authors of chapters in this book also benefited from the wise counsel of those who participated in the informal discussions at the SSRC or who served as discussants at the Irvine meeting: Iain Cockburn, Charles Calomiris, Linda Cohen, Michael Edelstein, Michael Fogarty, Bronwyn Hall, David Hounshell, Zorina Khan, Richard Nelson, Bhaven Sampat, and Scott Stern. Finally, we thank Elizabeth Murry of the MIT Press for her editorial suggestions and her unwavering belief in the project.

Introduction: The Organization and Finance of Innovation in American History

Naomi R. Lamoreaux and Kenneth L. Sokoloff

Recent decades have witnessed radical breakthroughs in technology but also major changes in our understanding of the processes of invention and innovation. Although many historians of technology traditionally attributed invention to serendipity, culture, or exogenous advances in science rather than the pursuit of material gain, scholars now widely consider invention to be a variety of entrepreneurial activity that is difficult to distinguish from innovation more generally. Inventors are thought to behave just like other entrepreneurs, investing their time and resources in generating new technological ideas with the goal of earning high returns.[1]

Some of this evolution of thinking has occurred as scholars, working with patent data and other related materials, have shown that both the rate and direction of inventive activity were systematically related to the growth of markets and other changes in economic conditions (Schmookler 1966, Sokoloff 1988, Sokoloff and Khan 1990). Another powerful contributor to the sea change in attitude has been the obvious role that small, entrepreneurial firms supported by venture capital have played in the technological dynamism of Silicon Valley and in other contemporary high-tech industries (Saxenian 1994; Gans and Stern 2003; Arora, Fosfuri, and Gambardella 2001).

No longer does anyone seriously question the notion that the production of new technological knowledge increases, much like that of any other commodity, as more resources are committed to it. There is less consensus, however, about the factors that account for variation over time, place, industry, or other circumstances in the commitment of resources to invention and innovation. One approach treats investments in these activities as governed by the same considerations as other business projects—that is, whether to undertake an investment turns on the likelihood of developing a good new product or a way of producing an existing product at lower cost, as well as on the size of the market for the product or technology. Proponents of this view attribute the high rates of invention in the United States today to the potential for new technological discoveries created by recent foundational advances in science, the comparative advantage of this country in research and development (R&D), and the high returns to new technology available in a context of vast international product markets. Their implicit assumption is that promising projects will somehow always manage to be recognized, pursued, and funded, and these scholars are not particularly concerned with who carries them out or how the enterprise is organized.[2]

By contrast, another perspective emphasizes the importance of the institutional environment for determining whether efforts to develop new technologies will go forward and which inventor or entity will do the work. The resolution of such issues frequently turns on the extent to which the institutional environment permits the difficulties raised by the generally uncertain nature of investment in new technologies to be resolved. According to this view, eras of dramatic or accelerated technological change are likely to be triggered by the emergence of new organizations or instruments that make it easier for technologically creative individuals or firms to attract backing for their efforts.[3]

Yet a third view emphasizes the importance of the government's adoption of a more active role in promoting the development of new technologies, especially over the twentieth century, by making direct allocations of resources to R&D or by contracting for R&D-intensive products (Mowery and Rosenberg 1989). Increased government spending is seen as critical to ongoing technological progress because many socially worthy projects would otherwise fail to obtain funding. Either private investors would judge the expected private (as opposed to social) returns inadequate (as in the case of so-called basic research), or the scale and risk of the project in question would be too daunting.

Because technological change is so vital for long-run economic growth, it is of fundamental importance to understand how technologically creative individuals and firms obtain the resources needed to undertake their investments in invention and innovation. It is also important to understand how the availability of such resources, including the manner in which they are accessed as well as the amounts that can be raised, influences the rate, direction, and organization of technological development. Conversely, one should also inquire whether the nature of the technology itself (as reflected, say, in the cost of conducting inventive activity or in the ability to enforce property rights to new discoveries) influences the way in which finance is mobilized. Achieving this level of comprehension requires detailed knowledge of the ways in which the financing of innovation and patterns of technical progress have changed and interacted over time. Although scholars have some sense of the ways in which innovative firms mobilize capital today, only recently have they devoted much attention to understanding how innovation was financed in the past. This book is intended to help remedy this deficiency in our knowledge.

Early Institutional Framework

The extraordinary record of the United States at invention and innovation cannot be understood without appreciating the fundamental importance of the patent system. When this democratic republic emerged on the world scene, it was thinly populated, overwhelmingly rural, and seen as something of an economic and technological backwater by European observers. Few outside its borders considered it to be a prospective rival for industrial leadership. The United States lacked the great universities and scientists that Europe enjoyed, and the scale and sophistication of its financial and manufacturing enterprises paled in comparison. Yet it would not be long before the young nation shocked the world with its display of technological creativity at the 1851 Crystal Palace exhibition. How did this change happen? Observers attributed much of the country's rapid technological progress to its distinctive patent system, and it is no coincidence that Britain and many other European countries over the next decade began to set up or modify their patent institutions to make them more like those of the Americans (Rosenberg 1969, Khan and Sokoloff 2004).

The framers of the U.S. Constitution and its early laws were bold, ambitious, and optimistic, and they quite self-consciously set up a patent system that represented a radical break from the precedents of the Old World. In their view, providing broad access to a well-specified and enforceable property right to new technology would stimulate technical progress, and nearly all of the innovations they made in the design of patent institutions aimed to strengthen and extend incentives and opportunities for inventive activity to a much broader range of the population than would have enjoyed them under traditional intellectual property institutions. European patent systems of this era required extremely high fees (if not political influence as well) to obtain patent rights, awarded monopoly rights to those who were first to file (even if not responsible for the invention), and eschewed positive efforts to promote the diffusion of the details of technical advances. In direct contrast, the United States chose to set the fee for obtaining a patent at a level far lower than anywhere else (less than 5 percent of the level in Britain), reserve the right to a patent to "the first and true" inventor anywhere in the world, and require that the specifications of patented inventions be made public immediately (Khan and Sokoloff 1998).

The design of the patent system reflected a strong belief in the utility of defining tradable assets in new technological knowledge.[4] Such assets provided material incentives for investment in inventive activity, helped inventors mobilize capital to support their efforts, and encouraged the flow of new technological knowledge from the inventor to the enterprises that would exploit it commercially. A key innovation in the design of the system, albeit one that was not permanent until 1836, was the practice of examining applications for novelty and conformity with the statutes before granting patents.[5] This feature was of fundamental significance because approval from technical experts reduced uncertainty about the validity (or value) of patents. Inventors could more readily sell or license patents and realize a return to their ideas in that way, use the patent rights (or the prospects thereof) to raise funds to continue developing or commercializing the inventions, or accomplish both ends simultaneously. Private parties could always, as they did under the registration systems prevailing in Europe, expend the resources needed to make the same determination as the examiners, but there were scale economies and positive externalities when the government absorbed the cost of certifying a patent grant as legitimate and made the information public.[6] Trade in patented technologies, as a result, was much more extensive—both in absolute terms and relative to the volume of patenting—in the United States than in countries such as Britain and France where the registration system endured into the twentieth century (Khan and Sokoloff 2004).

The structure of the U.S. system was based on the idea that people from all walks of life were capable of making significant contributions to the advance of technological knowledge, but that in order to realize their potential, they needed secure property rights to their knowledge. Those who came from humble backgrounds were particularly sensitive to the provision of such rights. Without clear title to their inventions (that is, without the ability to obtain a patent that would have a high likelihood of withstanding legal challenge), they would have been plagued by problems of asymmetric information and other high transactions costs in their attempts to attract investors. Similarly, employers would have had difficulty working out arrangements to encourage workers to develop ideas about how processes or products might be improved and to offer them up.

The patent system seems to have worked as intended and well. Encouraged by the low costs of filing for patents and the relatively rapid development of mechanisms for enforcing them, Americans from the outset were enthusiastic about establishing their claims to intellectual property.[7] By 1810, despite its lag in industrial development, the United States far surpassed Britain in patenting per capita. Moreover, a much broader socioeconomic range of the population responded to the opportunities made available to them by obtaining patents on their inventions (Khan and Sokoloff 1998). The generation of new technological knowledge appears to have been highly responsive to potential returns during this early stage of industrialization. Patenting rates were strongly procyclical. They were also markedly greater, even after controlling for region, urbanization, and the local composition of the labor force, in areas that obtained low-cost access to markets through extensions of navigable waterways. The activities of inventors whose discoveries have been considered by subsequent writers to be especially significant displayed similar or even stronger patterns (Sokoloff 1988, Khan and Sokoloff 1993).

The patent system allowed inventors to realize income from their inventions by employing them in the production of goods for market, selling or licensing the rights to other producers, or pursuing both avenues simultaneously. During the first decades of the nineteenth century, many, if not most, inventors sought to profit directly by establishing manufacturing or other businesses.[8] With few enterprises of this era incorporated and trading in equity shares extremely limited, the bulk of their capital came, as it did for most businesses, from the inventors' own savings, funds provided by partners (which the value of patent assets likely helped to mobilize), retained earnings, and other informal sources. Local banks provided some support as well.[9] By midcentury, however, inventors were increasingly operating as independent entrepreneurs who specialized at invention and extracted the returns from their discoveries by selling off or licensing the patent rights to a variety of different firms to which they had no long-term attachments. This business strategy appears to have evolved from the practice of inventors in the early industrializing Northeast (even those who had their own manufacturing enterprises) selling geographically delimited patent rights in different areas of the country in order to take advantage of opportunities to realize additional returns from the segmented markets of the time. A very high volume of such arm's-length trades is evident from the late 1830s well into the 1860s, by which decade the development of a national market was diminishing the value of such limited rights.[10] As technologically creative individuals learned how much income could be realized by transferring their patent rights to those better positioned to exploit the inventions commercially, they were drawn toward specialization at what they did best (Lamoreaux and Sokoloff 1996, 1999a, 1999b, 1999c, 2001).

The ability to trade patent rights also promoted specialization by providing a means for the technologically creative to mobilize capital to support the development of their ideas. Many of the assignments registered with the U.S. Patent Office during the 1840s, 1850s, 1860s, and 1870s, for example, involved patentees' transferring partial shares of their rights to one or two other individuals while maintaining a share for themselves. No doubt some of these agreements were partnerships whose purpose was the direct commercial exploitation of the invention in question. But inventors with many patents often made partial assignments of different patents to different individuals. Moreover, the patentee and the assignee often subsequently sold off their patent rights to a third party. Hence, it seems likely that many partial assignments of patents were tendered in return for an infusion of funds.

This move toward greater specialization at invention was likely further promoted by the increasing value of specific human capital, which in turn encouraged those who possessed the requisite skills and knowledge to exploit their comparative advantage. Inventive activity surged during the middle of the nineteenth century as the mechanization of previously hand-powered methods of manufacture and the rise of new-technology industries, such as railroads and telegraphy, alerted and attracted individuals to the high returns available to talented inventors. To be effective at making contributions at the frontiers of these new and more complex technologies, however, increasingly required knowledge obtained through long experience or instruction. This development, together with an enhanced ability to trade in patented technologies, helped spur the rise of a class of highly specialized inventors (Lamoreaux and Sokoloff 1999c, 2001; Khan and Sokoloff 2004).

The design of the patent system, and in particular the 1836 reform restoring the practice of examining applications for novelty, facilitated the expansion of commerce in patented technologies, but the broad and active market in patented technologies that had emerged by the middle of the nineteenth century required institutions beyond those that specify and enforce property rights. Major support for the evolution of such a market came rather quickly with a proliferation of patent lawyers and agents in cities and other locations with high rates of patenting that was already well under way by the 1840s. In addition to assisting inventors in preparing their patent applications, these professionals soon began to take on the role of intermediaries in a market for technology. Some projected themselves nationally, with offices or correspondents throughout the country and periodicals that reached a broad readership. One of the most important was Munn and Company, the largest patent agency of the nineteenth century and publisher of *Scientific American*. Journals like *Scientific American* featured articles about important new inventions, printed lists of patents recently issued, and offered to provide readers with copies of full patent specifications for a small fee. They also included pages of advertisements placed by patent agents and lawyers soliciting clients, detective agencies specializing in patent issues, inventors seeking partners with capital to invest, patentees hoping to sell or license rights to their technologies, and producers of patented products trying to increase their sales.[11] In short, they reflected a deepening and broadening market for technology. Some inventors chose to market their inventions themselves, perhaps consulting one of the advice manuals written for the how-to-do-it-yourself audience, but many turned to the big national agencies.[12] The manuscript assignment records indicate that Munn and Company alone handled (mainly by mail) roughly 15 percent of all assignment contracts recorded by the Patent Office in 1866. By the mid-1870s, however, Munn's share of the market had dropped below 5 percent as more and more of the intermediation work came to be done by local patent lawyers and agents who were able to provide more customized services.[13]

The rapid emergence of institutions supporting a market for technology, and the high volume of trade in patents, are but two reflections of the profound impact that the enhanced ability to trade in patent rights had on the incentive to invest in inventive activity and on the way in which inventive activity was organized during the mid to late nineteenth century. Other evidence for this impact comes from the radical acceleration in patenting rates that began in the 1840s, not long after the 1836 reform, and that resulted in more than a 900 percent increase in per capita patenting rates by 1870 (see figure I.1).[14] Although far from conclusive, the surge in inventive activity, which coincided with a sharp acceleration in the rate of manufacturing labor productivity growth (Sokoloff 1986), lends further support to the idea that the existence of a market for technology based on a patent system with broad access and examination was an important stimulus to technical progress. Moreover, consistent with what theory would predict as a response to an expanding market, invention became an increasingly specialized activity. The dramatic change in who was responsible for producing new technological knowledge is evident in the substantial rise in the proportion of all patents that were awarded to inventors who were relatively specialized at invention. From roughly the first third of the nineteenth century to the last third, the proportion of patents awarded to inventors who received

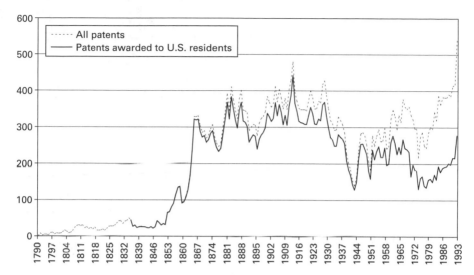

Figure I.1
Rate of patenting per million residents in the United States, 1790–1998

ten or more patents over their entire career jumped from less than 5 percent to between 25 and 36 percent (see table I.1). As would be expected, it was these relatively specialized or productive inventors who were most active in trading away the rights to their patents. Among the cohort of inventors active during the last third of the nineteenth century, patentees with more than twenty career patents assigned away nearly 60 percent of their patents at issue, while those with five or fewer career patents assigned less than 20 percent.

A skeptic might question whether these assignments by productive inventors were really arm's-length transactions or transfers of patent rights from employees to employers. Although specialization at invention across individuals can occur within a single firm, and indeed is often a celebrated feature of many large twentieth-century firms with R&D laboratories, these assignments appear to have taken place across enterprises—between patentees and assignees who had no long-term attachment to each other. In previous work, we traced the careers of a random sample of patentees over fifty years and found that in general, the most productive inventors exhibited a remarkable degree of contractual (as well as geographic) mobility. Roughly half of them, as weighted by

Table I.1
Distribution of patents by patentee commitment to patenting, 1790–1911

	Number of "career" patents by patentee					
	One patent	Two patents	Three patents	Four to five patents	Six to nine patents	Ten or more patents
1790–1811	51.0%	19.0%	12.0%	7.6%	7.0%	3.5%
1812–1829	57.5	17.4	7.1	7.6	5.5	4.9
1830–1842	57.4	16.5	8.1	8.0	5.6	4.4
1870–1871	21.1	12.5	9.9	15.8	11.8	28.9
1890–1891	19.5	10.3	10.3	10.3	13.8	35.9
1910–1911	33.2	14.3	8.2	9.8	9.4	25.0

Source: The figures from 1790 to 1842 are from Sokoloff and Khan (1990). The figures for the latter years were computed from the longitudinal B data set we constructed. See Lamoreaux and Sokoloff (1999c).

the number of their career patents, assigned their patents to four or more distinct assignees over their careers (Lamoreaux and Sokoloff 1999c). Overall, the evidence suggests that the patent system, and the ease of trading in new technological knowledge that secure property rights made possible, contributed to the rise of highly specialized inventors who were able to operate without any substantial involvement in the direct commercial exploitation of their inventions or long-term attachments with any single firm.

Another illustration of the fundamental importance of the market for technology comes from the regional patterns of its development. Two different gauges of the market are presented in table I.2, and each indicates that trade in patents evolved alongside increases in, or higher levels of, patenting activity. The geographic correspondence arose out of two mutually reinforcing phenomena. Institutional supports for trade in technology, such as patent lawyers and agents, tended to emerge, or locate, in areas where there were high levels of inventive activity, and proximity to those institutional supports in turn encouraged greater investment in invention and innovation. Whether one measures the extent of the market for technology by examining the location of patent agents and lawyers, who were crucial intermediaries in the market, or by calculating the proportion of patents that patentees had already assigned before the

Table I.2
Regional shares of total patents, great inventor patents, patent attorneys, and population, as well as assignment rates at issue, 1790–1930

Region	1790–1829	1830–1845	1846–1865	1866–1885	1886–1905	1906–1930
New England						
Patents	34.4%	30.1%	24.7%	19.7%	16.7%	11.4%
Great inventor patents	55.1	34.1	29.6	29.1	29.1	18.3
Patent attorneys	—	—	—	—	20.5	14.4
Population	21.0	13.2	10.1	9.1	7.6	7.2
% assigned at issue	—	—	—	26.5	40.8	50.0
Middle Atlantic						
Patents	54.5	52.3	48.3	40.6	37.6	30.8
Great inventor patents	35.5	57.7	55.7	51.5	41.1	62.0
Patent attorneys	—	—	—	—	45.1	42.2
Population	34.4	30.0	26.5	23.1	20.5	21.1
% assigned at issue	—	—	—	20.6	29.1	36.1
Midwest						
Patents	3.0	8.3	20.8	30.3	34.5	36.8
Great inventor patents	1.9	3.2	13.3	13.6	22.9	14.5
Patent attorneys	—	—	—	—	28.9	28.9
Population	3.3	17.3	29.2	34.0	36.0	32.6
% assigned at issue	—	—	—	12.4	26.2	28.1
South						
Patents	8.1	9.2	5.1	6.0	6.8	10.8
Great inventor patents	7.5	5.0	1.4	1.5	2.3	3.6
Patent attorneys	—	—	—	—	3.2	6.3
Population	41.3	39.7	32.9	31.9	31.5	31.7
% assigned at issue	—	—	—	6.4	25.0	22.7
West						
Patents	—	—	1.0	3.4	4.6	10.2
Great inventor patents	—	—	0.0	2.9	2.7	1.6
Patent attorneys	—	—	—	—	2.3	8.2
Population	—	—	1.4	1.9	4.5	7.5
% assigned at issue	—	—	—	0.0	25.4	21.4

Note: The Midwest region combines figures for the East North Central and West North Central. The geographic distribution of patent agents and attorneys was computed as the geographic distribution of attorneys registered with the patent office who were located outside the District of Columbia.
Sources: Lamoreaux and Sokoloff (1999a, 1999b, 2003); Khan and Sokoloff (2004).

patent was granted, it is clear that trade in patent rights was concentrated in New England and the Middle Atlantic—the regions with the highest rates of patenting per capita throughout the nineteenth century and into the twentieth. Although generally less than a third of the nation's population resided in these two regions, they were home to more than two-thirds of the patent attorneys in the United States outside Washington D.C. Indeed, patent agents were overrepresented in these regions relative to the regional shares of patents. Inventors generally found locations where the market for technology was centered very favorable to their activity, but it is telling that the attraction was most powerful for those making the most important discoveries. The share of the U.S. population residing in New England and the Middle Atlantic never exceeded 40 percent between 1850 and 1900 (it declined from roughly 38 to 28 percent over these years), but the two regions accounted for between 45 and 70 percent of all patents and consistently more than 70 percent (ranging as high as 85 percent) of patents awarded to the nation's great inventors. Moreover, the geographic concentration of great inventors in areas where the market for technology was especially active is even more pronounced at the level of counties (Lamoreaux and Sokoloff 2001, Khan and Sokoloff 1993).

Evolving Links between Inventors and Commercializing Firms

The highly developed market for technology that emerged in the United States during the half-century following the 1836 revision of the patent law gave technologically creative individuals remarkable freedom to specialize at inventive activity and maintain their independence from the firms that bought or licensed the rights to their inventions. In the words of the noted historian of technology Thomas Hughes (1989), the late nineteenth century was a "golden age for independent inventors." Gail Borden, Charles Brush, Thomas Edison, Sidney Short, and Elmer Sperry are but a few of the pioneering inventors and innovators who took advantage of the opportunities that the ability to trade in property rights to new technological knowledge allowed them.[15] By the beginning of the twentieth century, however, this golden age was coming to an end. The contractual mobility of even the most productive inventors declined as they increasingly established long-term attachments with particular firms, joining companies as principals or employees and assigning a high proportion of their patents to those enterprises (Lamoreaux and Sokoloff 2005, chapter 1 this volume).

A variety of factors likely contributed to this development, but the crucial one may have been the growing difficulty that inventors and innovators faced in obtaining outside sources of finance during the Second Industrial Revolution. Although rapidly expanding product markets were raising the incentives to invest in inventive activity, the costs of carrying out R&D were also rising with the growing complexity and capital intensity of technology.[16] Under such circumstances, talented inventors seeking to work on the frontiers of technology were more inclined, if not compelled, to make long-term commitments to firms in order to gain access to the resources necessary to move forward on their projects. Firms also had important reasons for forging closer relationships with the inventors they sponsored. First, if they were to supply the growing amounts of capital required for the development of new capital-intensive technologies, they naturally wanted to protect their investments by securing the attention and support of the key inventors for extended periods. Effective exploitation of a complex technology in a dynamic and competitive environment often involved adaptations or incremental improvements that such figures were best qualified to undertake. In addition, firms wanted to preclude the possibility that the inventor might undercut the value of the technological discovery with a subsequent invention. The cost of such a development would of course be greater with more capital-intensive technologies and larger scales of exploitation, and thus the desire for firms to control the future course of technical advance likely increased over time.[17]

By the late nineteenth century, it was clear to observers that technological change was a permanent feature of the industrial economy and that substantial returns could be obtained through investing in the development of frontier technologies. Railroads and telegraphy were perhaps the first grand-scale examples of industries created or revolutionized by important inventions, but others such as electricity, telephones, steel, chemicals, and automobiles soon followed.[18] As interest in these sorts of opportunities grew, technologically creative entrepreneurs increasingly sought out investors (and vice versa) because the greater technical complexity and capital intensity of the new technologies meant that effective programs of inventive activity and commercial exploitation required more financial backing than before. Ventures to exploit these new technologies also needed more funds than their counterparts from previous generations because they now operated in national or international markets and thus had to attain a larger scale. Moreover, although the returns to technological leadership generally increased with the size of markets, firms felt compelled to invest more to keep up with the cutting edge. By the early 1900s, as a result, many firms in new-technology industries were beginning to set up and expand internal research laboratories, not only to monitor the latest advances in technology, but also to carry out their own programs of R&D.[19]

Most of the firms that invested heavily in R&D facilities in the early twentieth century originated as entrepreneurial companies formed to exploit the discoveries of particular inventors. Perhaps the most famous example is General Electric, formed from a merger of two core enterprises that had been organized by investors with the aim of commercializing the inventions of Thomas Edison and Elihu Thomson (Passer 1953, Carlson 1991). Chapter 1 by Naomi Lamoreaux, Margaret Levenstein, and Kenneth Sokoloff and chapter 2 by Steven Klepper show that this phenomenon of organizing companies around prominent inventors was quite important during the Second Industrial Revolution. Indeed, although the studies are structured very differently, both find that start-ups played a critical role in the development of cutting-edge technologies and that the initial financing for these enterprises typically was raised informally from local backers, many of whom were personally acquainted with the inventors involved. More formal financial institutions, such as banks or organized securities markets, would not become major sources of long-term capital for these sorts of firms until much later.

Lamoreaux, Levenstein, and Sokoloff focus their study on one of the key sites of the Second Industrial Revolution, Cleveland, a city that spawned a number of highly innovative firms across a wide range of industries such as electricity, steel, automobiles, chemicals, precision tools and machinery, and scientific instruments. Drawing on a variety of sources ranging from patent records and city directories to company archives, they examine the activities of, and relationships between, the inventors and firms that were most involved in pioneering new technologies. The bulk of the capital, and much of the impetus, to establish these enterprises came from local investors who were optimistic—sometimes wildly so—about the prospects for new technologies and wanted to help spur the growth of the Cleveland economy. By plugging into the networks of inventors that formed around pioneering enterprises such as the Brush Electric Company, they were able to obtain information about the technologies that were most likely to pay off.

The willingness of Cleveland's leading citizens to invest in firms focused on the development of new technologies was instrumental in attracting talented inventors such as Brush, Short, and Sperry to locate in the city. Entrepreneurial and independent minded as these gifted inventors may have been, in order to secure a return to, and obtain support for, their inventive activity, they were willing to move and make the longer-term, if not career-long, commitments that those who were putting up the capital increasingly demanded. Rather than maintaining their independent status, these inventors usually became principals—not employees—of the firms established to exploit their technological discoveries, and they committed themselves (generally on quite favorable terms) to stay with the new firm for a period of years and to transfer their patents to it. Sometimes they took leadership roles in directing their firms, but more often they seem to have held positions that were more advisory or directly related to programs of R&D. Not surprisingly, the inventors and other principals in these companies did not always see eye to eye. Individuals with extraordinary technological creativity could be reluctant to give up control to the professional managers that investors preferred, and it was not uncommon for them to move on from the original company and organize additional start-ups.

Through meticulous examination of the histories of a rather complete sample of early automobile companies, Klepper draws a similar picture for Detroit. Just as Lamoreaux, Levenstein, and Sokoloff find that the most productive inventors in Cleveland were those who had distinguished themselves at invention before becoming principals in newly organized firms, Klepper finds that the most successful start-ups in automobiles (in terms of growth or survival over time) were organized around individuals whose experience in the industry allowed them to secure backing to develop their ideas by establishing new firms. By contrast, firms that were started by entrepreneurs without track records in the industry did not generally fare well. As was the case for Cleveland, moreover, the main sources of finance for such enterprises were informal. The major backers tended to be local businessmen who had personal knowledge of the lead inventor or entrepreneur. It was not really until the 1920s that these enterprises turned to banks or public offerings of securities for additional infusions of capital.

Although a small number of major new-technology companies, such as General Electric, were able to raise substantial amounts of capital on the New York and other stock exchanges by the beginning of the twentieth century, it is striking how few of the dynamic start-ups featured in these accounts of Cleveland and Detroit tapped these sources of finance, even years after their establishment and apparent success. One possible explanation is that retained earnings may have been sufficient to support the growth of such firms. Another possibility is that the poor quality of the information available to outsiders about such firms, or the higher levels of risk associated with such endeavors, may have constrained the market for their securities. As Larry Neal and Lance Davis show in chapter 3, the limited role that stock markets played in mobilizing capital for new, innovative companies was not unique to the United States. Around the turn of the century, however, the New York Stock Exchange (NYSE) reformed its rules in ways that expanded its role in the economy compared to securities markets in Britain, France, and Germany.

The reforms were precipitated by the long deflation of the late nineteenth century. Steadily falling prices had slowly but surely increased financial pressure on companies, such as railroads, that were major debtors, creating by the mid-1890s a severe financial crisis that sent about a quarter of the nation's railroad mileage into receivership. The crisis stimulated a number of important innovations in the nation's financial markets that went beyond the railroad reorganizations spearheaded by J. P. Morgan and other major private bankers. In particular, NYSE members responded to the declining profitability of their brokerage businesses by instituting some important rule changes that made listing on the exchange an imprimatur of quality. Although small start-ups, such as those in Cleveland and Detroit, would continue to depend on local networks and financial institutions for capital until the 1920s, these rule changes put the NYSE in a position to take on the task of financing the large-scale enterprises needed to exploit Second Industrial Revolution technologies intensively and systematically—for example, the utilities that would provide cheap access to electricity to the entire economy.

The impact of the rule changes can be seen in data that Neal and Davis provide on the prices of seats on the NYSE. It can also be seen in the information Mary O'Sullivan has collected on stock issues for cash from 1897 until the present. Her data, set out in chapter 4, show that, beginning during the first decade of the twentieth century, the value of new issues soared, achieving levels relative to GDP that have never been seen again since the 1929 crash, not even during the great bull market of the late twentieth century. Also consistent with Neal and Davis's argument is her finding that utilities were responsible for a larger proportion of new issues by far than any other sector of the economy, accounting for 46 percent of the total in the period 1921 to 1925 (the first period for which it is possible to breakdown the data by industry), 29 percent during the second half of the 1920s, and 34 percent over the entire period 1921 though 1955.

Although Sullivan finds no clear statistical relationship between investment in the stock market and capital formation in the economy as a whole, she shows that the decade of the 1920s represented something of a watershed, with the stock market playing for the first time an important role in channeling finance to new firms. She documents this new role by examining three industries in detail: automobiles, aviation, and radio. In the case of autos, most new firms entered the industry before the 1920s, and O'Sullivan confirms the patterns found by Lamoreaux, Levenstein, and Sokoloff for Cleveland and by Klepper for Detroit. Securities markets did not play an important role in the finance of new firms, which instead obtained capital directly from local investors. Only later, when the automobile industry was in its consolidation phase, did entrepreneurs tap markets to finance mergers. Before the late 1920s, the situation in the aviation industry was much the same. However, Charles Lindbergh's transatlantic flight attracted investors' interest, and young firms ranging from aircraft producers to parts suppliers to airline operators were consequently able to raise sums on the stock market that were enormous relative to the amount of capital already invested in the industry. In radio, the role of the stock market was even more dramatic. The beginning of commercial broadcasting both attracted investors and sparked a wave of entry during the early 1920s. The result was a surge in stock issues so large that commentators talked about "a new radio stock a day" and estimated that more shares had been sold than radio receivers.

What accounted for the change in the role of the markets? O'Sullivan underscores the importance of the institutional changes that Neal and Davis outlined in their chapter, but argues that the resulting dominance of the NYSE had the important, if unintended, consequence of spurring other exchanges to develop businesses that were complementary to rather than competitive with the NYSE. For example, the New York Curb Exchange became a testing ground for firms whose issues could move to the Big Board once they had proved to be good investments. Regional exchanges in Boston, Philadelphia, Chicago, Cleveland, and other cities played a similar role for new firms in their vicinities. O'Sullivan also believes that there were significant changes on the demand side during the 1920s, particularly a dramatic increase in the number of investors who were willing to risk buying stock in new firms in order to profit from capital gains.

Thomas Nicholas, in chapter 5, also finds a change in investors' behavior during the 1920s, using a novel data set he constructed that includes balance sheet information, measures of the quantity and quality of intellectual property, and stock market prices for a large sample of publicly traded companies for the twenty years preceding the 1929 crash. Before the 1920s, Nicholas finds, investors' valuations of companies were largely determined by their investments in physical capital. During the 1920s, however, investors appear to have increased their appreciation for firms' investments in intellectual capital as well. Using the number and significance (gauged by citations) of firms' patented inventions as a measure, Nicholas estimates that intellectual capital rose from a minuscule fraction of the stock market valuation of firms to roughly 40 percent during the 1920s. Through a careful analysis of the relationship between firms' share prices and their portfolios of patents, Nicholas concludes that the stock market boom of the 1920s was not a bubble but rather was driven to a large extent by investors' increasing ability to evaluate new technological knowledge. This growing ability in turn led to a significant reduction in the cost of capital for, and a powerful stimulus to, investment in programs of R&D.

The growing prominence of R&D-intensive firms on the stock exchanges reflects in part the rise over the first decades of the twentieth century of large companies that carried out major programs of inventive activity through in-house research laboratories. General Electric, Du Pont, Westinghouse, General Motors, and IBM are among the best-known examples of firms that grew to dominate their industries by relying on a strategy of developing new technologies in the lab and then extracting returns from them directly through the production of goods and services (Passer 1953, Hounshell and Smith 1988, Wise 1985, Sloan 1964). Their success at innovation, many have contended, owed to their superior ability to mobilize resources for the support of R&D—facilitating specialization by the technologically creative at what they did best—and to their expertise at selecting which projects were most worth pursuing.[20] By contrast, small, innovative firms and entrepreneurially minded inventors who wanted to be independent faced greater challenges in obtaining the funds needed to support their efforts. Even when they had a good idea, the higher levels of risk associated with investment in any single R&D program made the prospect of sinking funds into it forbidding. Large firms with extensive resources had the advantage of being able to lower the overall risk of investing in inventive activity by backing a broad and diversified set of projects.

Although the story that Margaret Graham tells in chapter 6 about how Corning came to be the innovator and central player in optical fiber is one of triumph, the case study highlights the many obstacles that small and medium-size firms faced. Established during the nineteenth century, Corning Glass Works had a long history of innovation and success as a specialty glass producer. Closely held and controlled by the founding family into the second half of the twentieth century, it had flourished by focusing attention on unrecognized technological opportunities ("uncluttered paths") and by shrewdly and frequently entering into joint ventures with other firms to leverage its resources and gain access to specialized

knowledge or capabilities. As Graham recounts, however, federal antitrust policy had already constrained Corning from cooperating with
other glass producers when, in the 1960s, the firm came to realize that
optical transmission could revolutionize the technology of communications. The potential returns were enormous, but the amount of resources
Corning would require to carry out the R&D, defend its rights to its discoveries, and build the manufacturing and marketing capacity to exploit
the technology in the marketplace were daunting for a small company
that was virtually a new entrant in an industry dominated by the behe-
moth AT&T. Indeed, at times the development of optical fiber was
nearly called off as executives questioned the wisdom of sinking so much
capital into one long-term and risky project. Nevertheless, as Graham
makes clear, Corning persevered where many larger firms would have
given up. In this period, large, managerially directed enterprises were imposing financial controls that evaluated the costs of new research projects against the potential gains. Corning adopted these practices as well,
but although the numbers for optical fiber did not satisfy the criteria
normally used by the firm to evaluate whether to go forward, the management of this closely held firm was in a position to overrule the conventional analysis and effectively bet the company on the success of the
project. AT&T proceeded much more slowly and cautiously with its research and as a result lost out on this important innovation.

For most of U.S. industrial history, innovative firms raised money by
exploiting networks of personal connections. Only in the 1920s did formal institutions such as securities markets come to play a role in the finance of new firms. Growing excitement about new technologies such
as aviation and radio spurred interest in the issues of new high-tech
firms, and investors began to price stocks by assessing firms' portfolios
of patents. By that time, however, much inventive activity was already
moving into the R&D laboratories of the country's largest firms. Although these labs could lower the risk of investing in R&D by taking
on a diverse range of projects, corporate managers attempted further to
reduce uncertainty by imposing financial tests that had to be passed before resources could be devoted to a new one. Not surprisingly, these
efforts by managers to rationalize their research budgets had the effect
of eroding the innovativeness of their labs. As a result, by the late twen-

tieth century, some companies would decide to cut their research budgets, spin off their labs into separate companies, or even shut down their research facilities entirely. Independent firms would again become an important source of new technological discoveries.

The Role of Government

The increasing cost of conducting R&D clearly had powerful effects on the organization and financing of inventive activity. In addition to fostering a shift in the location of privately funded research, away from small enterprises toward large firms with in-house research labs, it also likely contributed to a major increase in the role of government over the course of the twentieth century. Part of the dramatic growth in government funding can certainly be understood in terms of the traditional view that there should be public support for those R&D projects (so-called basic research) that were expected to yield high social returns but whose private returns would be insufficient to induce businesses or other private agents to underwrite the investment (Arrow 1962, Nelson 1959). The government's commitment to medical research through the National Institutes of Health (NIH) and other federal funding agencies is an obvious example. When one considers, however, that the big change in policy took place during the 1930s and 1940s, with the federal government funding more than half of the total annual national investment in R&D by the early 1950s, more would seem to have been involved. Concern with ensuring that the country was the leader in military technology and nuclear energy was no doubt a major driver, and indeed most government backing for R&D at midcentury came through the Defense Department. But the wide range of research programs that the government supported suggests that another impetus may have been the greater ease of procuring finance from public than from private sources, and not that the returns would have been difficult for private agents to appropriate. Regardless of the reasons, once the government became the key patron of R&D, it exercised enormous influence on the direction and the level of inventive activity, as well as on where R&D was conducted and how the resulting knowledge was diffused (Mowery and Rosenberg 1989).

Chapter 7 by Kira Fabrizio and David Mowery examines the federal government's role in financing innovation in the information technology (IT) industry (semiconductors, computer hardware, computer software, and the Internet) after World War II. It is beyond the scope of their study to consider the counterfactual of what would have happened to the industry if there had been no significant federal funding, but the authors lay out the details of government involvement in the industry and argue that its impact went "well beyond the amelioration of the 'market failures.'" Although they acknowledge that the IT sector was a major beneficiary of private investment in R&D at least as far back as the 1960s, they argue that government concern for promoting the generation of new technologies in this area was virtually indispensable to the creation of the networks of researchers that shared information across universities, government agencies, and private firms and spawned a remarkably broad array of important innovations. Government financing of innovation had several distinct features that made it so important. First, federal research dollars simultaneously supported work on a broad array of technological alternatives. Second, both federal policy and the relatively weak intellectual property rights that accrued to developments produced with the aid of government funds encouraged the rapid diffusion of new technological knowledge. Third, government procurement expenditures provided firms in the IT industry with a profitable market for their products and also the opportunity to work out the practical difficulties they faced in scaling up production.

The authors recognize that the case of IT may have been a special and unrepresentative episode in the annals of government funding of R&D. That the government happened to intervene at a crucial early stage in the development of the industry and that IT was a *general purpose technology* with many applications and externalities helps to account for the long-term significance of federal funding. Nevertheless, because government involvement meant much higher levels of support for R&D than private sources would otherwise have provided, and because there was more rapid diffusion of new technological knowledge than would typically occur with a strict regime of private property (patent) rights, this case provides some support for those who argue for a more active government role in fostering technological development.

Steven Usselman does not directly dispute the idea that government can, and often does, play a major and beneficial role in promoting innovation and technological progress, but he contends that proponents of federal support for R&D are sometimes inclined to claim too much. In chapter 8, he traces the history of the IBM model 360, a major innovation in mainframe computer systems that has been considered a prime example of the benefits of government spending. Based on an examination of IBM's internal documents, as well as interviews with many central figures in the firm, especially those involved in overseeing the program, Usselman contends that the development of the 360 system was neither stimulated nor funded by federal procurement for the military and the space programs. His persuasive analysis concludes instead that the model 360 was quite consciously designed for the private market. According to Usselman, IBM's top executives had become convinced that military projects were leading the company into technological "cul-de-sacs whose arcane lessons" were difficult to translate into commercial applications. With the 360 program, they deliberately used the company's own resources to develop a computer technology targeted at the kinds of private sector customers that IBM had traditionally served. Although Usselman does not deny that military procurement played an important role in financing technological development more generally at IBM, his study is a reminder that the potential for extracting returns though commercial exploitation remains a powerful incentive for investment in inventive activity. Government support for R&D can have large impacts, but the idea that market failures inhibit the private provision of technological knowledge may be exaggerated.

The Return of Market Forces

For most of the twentieth century, the role of the market for technology was much more circumscribed than it had been during the late nineteenth century. As large companies and the federal government provided a greater share of the funding for technological discovery, the conduct of R&D came increasingly to be concentrated in private research labs and universities. Large firms primarily relied on direct commercial exploitation to realize the returns from their investments in R&D. In combination with the rather weak protections for intellectual property extended to government-supported research, that reliance meant that trade in patented technological knowledge had become a rather secondary activity by the middle of the twentieth century.

Given the decline in such trade, it is perhaps not surprising that economists and other observers worked out plausible theoretical rationales for why it should not be important. One common explanation was that it was generally more efficient to integrate R&D and commercial development within a single firm because the expertise acquired in marketing new products could be used to focus inventive activity in the most profitable directions and because difficulties in enforcing patent rights meant that firms could more easily realize their returns if the details of a new technology could be kept secret and the speed at which it was brought into production increased. A related view was that there were high costs to transacting in technological knowledge in most industries and that these costs limited the ability of independent inventors or firms specialized at invention to obtain the finance they needed to undertake investments in R&D or to earn a return to their efforts by selling off or licensing the rights to their discoveries. Implicitly, if not explicitly, the dominant opinion came to be that outside of a very few highly unusual industries, independent researchers either had to depend primarily on government support for their creative activities or associate themselves with businesses that were directly involved in commercializing inventions.[21]

In recent decades, however, the rise of venture capital firms as well as other developments in financial institutions and in the operation of the global economy have triggered a substantial change in the organization of inventive activity and the resurgence of an active market for new technological knowledge. Start-up enterprises that specialize in R&D and plan to realize their returns by selling off or licensing the rights to their discoveries have proliferated throughout the economy. At the same time, firms in traditional R&D-intensive industries like pharmaceuticals have downsized their in-house research capacities. Patenting activity has accelerated, and a higher volume of trade in patented technologies is also evident (Kortum and Lerner 1998, 2000; Lerner 2001). These developments have given further impetus to a reexamination of, and debate about, the role of patent systems and the feasibility or effectiveness of reliance on patents and markets in allocating resources to R&D.

In chapter 9, Ashish Arora, Marco Ceccagnoli, and Wesley Cohen seek to improve our understanding of the circumstances conducive to trade in technology by analyzing data collected by the Carnegie Mellon Survey on the Nature and Determinants of Industrial R&D for the years 1991–1993. The authors posit that such trade can yield important social welfare benefits by facilitating a division of labor between entities that are best suited to engage in technological discovery and those best positioned to exploit particular inventions commercially. They further posit that such trade would facilitate a more rapid and larger-scale exploitation of new technological knowledge. They recognize, however, that there might be barriers to the realization of these gains. Some of these barriers might result from efficiencies in carrying out both R&D and the commercial exploitation of new technologies within the same firm, but it is also possible that welfare-enhancing trades might not occur because parties cannot adequately protect themselves against opportunistic behavior.

In their multivariate analysis, Arora, Ceccagnoli, and Cohen find that complementarities between marketing and invention are indeed associated with the vertical integration of R&D within the firm and, as a result, with less extensive trade in patented technologies. But they also find that in industries where patent protection is considered stronger (that is, patent rights are more easily enforced), there is more licensing of new technological knowledge. Because patents are not an available means of protection when writing contracts for the generation of knowledge in the future, firms entering into such ventures must rely on other means of safeguarding their intellectual property, for example, maintaining secrecy. But patents do play an important role in making it possible for firms subsequently to exploit the resulting technology by licensing it to other parties. Although not the focus of their study, the systematic patterns that Arora, Ceccagnoli, and Cohen found in the organization of inventive activity and trade in technological knowledge are consistent with the idea that the early-twentieth-century rise of large, vertically integrated firms with in-house R&D facilities contributed to the secular decline in patenting rates.

The number of small firms engaging in inventive activity has grown enormously over the past several decades, fueled in large measure by the growing availability of venture capital. This development in turn has contributed to a marked tendency to disintegrate R&D and the commercial exploitation of new technologies. Large firms are cutting back resources for in-house R&D and instead are increasingly focusing on marketing and commercializing technologies developed by independent entities. In chapter 10, Joshua Lerner explores the problems that small R&D firms face when they try to raise funds from venture capitalists or form alliances with larger enterprises. These problems result from uncertainty about the value of their technologies, asymmetric information between the firms' entrepreneurs and the parties with whom they are contracting, difficulties in evaluating intangible assets in the firms' balance sheets, and changing conditions in financial markets. Some of these problems are beyond the ability of the parties to resolve—for example, capital is always easier to raise when financial markets are booming. But other difficulties can be significantly alleviated. Just as in the late nineteenth century, the emergence of intermediaries has helped resolve problems of asymmetric information between sellers and buyers of new technologies. Moreover, such problems have also been mitigated by contractual provisions affecting the allocation of control rights. Lerner reports on a study of biotechnology alliances he conducted with Robert Merges showing that the extent to which small R&D firms had to give up control rights to their larger partners varied systematically with the magnitude of their own financial resources and with the strength of their patent positions (Lerner and Merges 1998). He also reports on a study (coauthored with Antoinette Schoar) of venture capital contracts in Asia. Lerner and Schoar (2005) find that as Asian venture capitalists gained experience (and encountered problems), they became more sophisticated about the allocation of control rights and, as a result, the provisions they included in their contracts converged rapidly toward U.S. practice.

At the same time as venture capitalists were learning how to structure contracts to resolve problems associated with investing in new high-tech enterprises, investors were developing the ability to evaluate firms' intangible assets. As Michael Darby and Lynne Zucker demonstrate in chapter 11, investors during the 1990s appear to have become remarkably sophisticated in their assessments of firms' technological capabilities. Working with a data set that includes virtually all of the nonpublic firms established before 1990 in the segment of the biotech industry based on recombinant DNA technology, they identify the factors that predict which firms ultimately go public and how much they are able to raise in their initial public offerings (IPOs). Like Lerner, they find that firms have an easier time raising capital when market conditions are hot. But like Nicholas, they show that investors are nonetheless quite discriminating in what they are willing to finance. Among the variables that increase the probability of going public in any given year are the quality of the firm's science base (as gauged by the publication and patenting records of scientists associated with the company) and how many rounds of venture capital the firm had already obtained. The same factors also predict the amount that can be raised through an IPO, though here the quality of the firm's science base plays an even stronger role. Just as in the late nineteenth century, today's high-tech enterprises find it easier to mobilize capital if they can provide objective evidence of the extraordinary technological capabilities of their principals.

Conclusion

The U.S. economy has realized impressive, and relatively continuous, rates of technological change since the early 1800s.[22] Because this achievement has occurred over such a long period of time and under quite varied circumstances, it is difficult to identify a single factor that can account for entrepreneurs' ongoing ability to secure financing for their ventures. Indeed, perhaps the most striking aspect of the record of innovation over American economic history is the flexibility that technologically creative entrepreneurs have exhibited in adjusting their business and career plans so as to obtain financing for, and extract the returns from, their projects. As the costs of engaging in inventive activity and the relative availability of finance for independent ventures changed over time, entrepreneurs moved in and out of large firms and tapped into alternative sources of funding such as the federal government.

During early industrialization, when neither production technology nor inventive activity required large amounts of physical capital, technologically creative individuals seem to have been able to fund their relatively small-scale R&D projects with infusions from friends and family and from retained earnings. To the extent that they needed to raise additional funds by attracting partners, they were greatly aided by a patent system that provided broad access to well-defined and easily enforced property rights in new technological knowledge. Secure property rights in patented technology also enabled creative inventors to extract their returns by selling off and licensing their inventions to firms and individuals who were better able to exploit them commercially. The emergence of intermediaries to facilitate these transactions in the market for technology encouraged a division of labor that enabled those with a talent for invention to specialize in what they did best.

The amounts of capital required to develop new technologies increased dramatically during the Second Industrial Revolution, however, and pioneering inventors and innovators often had to be flexible to secure the backing they needed. Some, such as Thomson and Sperry, opted to migrate to cities where groups of investors offered infusions of capital. Others chose to commit for extended periods to work for nationally oriented companies. These companies had the ability, especially from the 1920s on, to tap the securities markets for funds and hence could raise the resources needed to support the inventors' work. Although they generally integrated R&D and the commercialization of new technologies within their bounds, they preserved a productive division of labor inside the firm between those who invented and those who were responsible for marketing and other commercial activities. As a result, though the ranks of independent inventors were much depleted during the early twentieth century, the continued rapid pace of technical change at the economy level suggests that the rate of invention was not adversely affected to any significant degree. Nonetheless, because large firms focused on commercially exploiting the returns to the discoveries they made in-house, they had less reason to make tradable assets out of the new technological knowledge they generated. The movement toward vertical integration may thus help explain the secular decline in patenting rates from the late nineteenth century to roughly 1980 that has long been a puzzle to scholars of technological change.

There may well have been, and may still be, complementarities in having R&D and the commercialization of new technologies carried out by the same firm. However, the disintegration of these activities that occurred in many high-tech industries following the rise of venture capital institutions during the late twentieth century highlights the possibility that what was most important about large firms with extensive in-house R&D was their access to capital and ability to diversify investments across different projects. The enormous growth and evident success of government support for R&D during the middle of the century also points to there having been a problem in raising sufficient amounts of capital for big programs of research in new fields of technology. The intriguing implication is that the development of financial institutions, and sources of support for R&D more generally, had a profound impact on the organization of inventive activity, if not on the pace of productivity growth. As scholars continue to reconstruct what has happened in the past—how the mobilization of resources for technological advance has varied over time and across industries—it appears increasingly likely that they will find that there is no globally superior way to organize invention. They are, however, likely to discover that the way in which inventive activity is financed has profound consequences for both the direction of technological change and the competitive dynamism of the economy.

Notes

1. Perhaps the most traditional view is that significant inventions were largely the work of individuals of extraordinary genius or insight whose discoveries were not much influenced by the desire for material gain, whereas innovators, by definition responsible for the first commercial applications of the new technological knowledge (that is, of inventions), were motivated by profit. See Schumpeter (1934) and Mokyr (1990). In recent years, scholars have increasingly deemphasized the distinction between inventors and innovators and have implicitly or explicitly treated decisions about investment in inventive activity as akin to other entrepreneurial investments. Indeed, as the chapters in this volume suggest, many scholars have come to employ the terms *inventor* and *innovator*

interchangeably. See also Khan and Sokoloff (1993), Baumol (2002), and Lamoreaux and Sokoloff (2005).

2. For an example of this approach, see Jorgenson, Gollop, and Fraumeni (1987).

3. For discussion of how the growth of venture capital might have spurred the rate of innovation in the 1980s and 1990s, see Kortum and Lerner (1998, 2000). For a discussion of how special features of the U.S. patent system stimulated technologically creative individuals to increase their rate of invention during the nineteenth century, see Khan (2005). For a more general treatment, see Scotchmer (2004).

4. The early framers of the patent system could not have foreseen the precise nature and extent of trade in patented inventions that would develop, but from the very first patent law of 1790, they included provisions explicitly designed to support trade in patent rights, and both the courts and the U.S. Patent Office acted to facilitate such transfers. For example, the Patent Office served as a registry that anyone could consult to identify the owner of the rights to a particular technology. To ensure that the registry remained up to date, the law specified that any contract transferring the property rights to a patent from one party to another would not be legally binding unless a copy was deposited with the Patent Office within a few months. The framers of the early patent institutions seem to have conceived of a market for technology that would work in a manner not unlike those for other valuable assets, with well-specified and tradable property rights both encouraging production and improving allocation. Of course, the ability of patentees to find buyers or licensees for their patents depended on the security of these property rights. Responsibility for enforcing patent rights was left to the federal courts, and judges quickly developed an effective set of principles for protecting the rights of patentees and also of those who purchased or licensed patented technologies. See Khan (1995, 2005).

5. The 1836 law actually represented something of a return to the original conception, as the 1790 law had stipulated that all applications for patents be assessed by a committee consisting of the secretary of state, the secretary of war, and the attorney general. Thomas Jefferson, the first secretary of state, was regarded by his fellow cabinet members as more qualified for the task and appears to have borne most of the responsibility for reviewing the applications. When he expressed concern about how much time this duty consumed, Congress passed a law in 1793 that shifted to a straightforward system of granting patents to all applications that met the administrative requirements (a fee, detailed specifications, and a model). A serious problem with this so-called registration system, however, was that it left any questions that arose about the validity of a patent to be settled by the judiciary. Inventors argued that this system substantially raised the cost of enforcing their patent rights and lowered their returns. Congress began to hold hearings about revising the system in the early 1820s, but it was not until 1836 that action was taken. See Khan and Sokoloff (2004).

6. The design of the patent system in Britain discouraged trade in patents in other ways as well. For example, the high fees encouraged many inventors to extract returns by keeping their discoveries secret rather than by filing for patents and marketing their inventions to those better positioned to directly exploit them. This inclination toward secrecy was encouraged by the ability of those with wealth to patent the inventions of others who lacked the resources to pay the fee. See Dutton (1984) and MacLeod (1988, 1999).

7. See Sokoloff and Khan (1990). An examination of the careers of 160 great inventors active during the antebellum period indicates that all but a few energetically patented their inventions as a means of pursuing the material returns. Khan and Sokoloff (1993). Also see Hounshell (1984).

8. An analysis of the biographies and records at patenting of over 400 great inventors in the United States who were born between 1735 and 1885 indicates there was substantial change over time in how they extracted their returns. Those active during the first half of the nineteenth century were more often than not engaged in direct exploitation of their inventions, but many also earned income from selling off or licensing patent rights—for example, to producers in geographically segmented markets. By the second half of the nineteenth century, an increasing share realized their income from their patents by selling off the rights. For those born after 1865, however, there was a marked increase in the share who worked as employees, especially in capital-intensive sectors such as transportation and electricity. See Khan and Sokoloff (1993, 2004), as well as Lamoreaux and Sokoloff (2005).

9. For evidence on how early nineteenth-century manufacturing enterprises mobilized their capital, see Lamoreaux (1994) and Davis (1957, 1958, 1960, 1966). Also see Hounshell (1984).

10. See Cooper (1991) for a discussion of how Thomas Blanchard extracted returns from his inventions by licensing and selling off patent rights, as well as litigating to enforce his patents. By unfortunate coincidence, the building housing the Patent Office was destroyed by fire in 1836, and with it were lost the sales (assignment) contracts for patents recorded prior to that date. It is impossible, therefore, to be certain about the actual level of trade in patents before the 1836 law. A variety of indirect indicators suggest, however, that there was indeed a major increase in sales of patent rights about this time. Based on an examination of the assignment records for a number of years during the 1840s and 1850s, we estimate that three to eight times as many times assignment contracts for patents were filed than the number of patents granted. We benefit here from the legal requirement that all patent assignments had to be filed with the U.S. Patent Office within three months in order to be legally binding. "Geographic" assignments accounted for 80 to 90 percent of the contracts registered with the Patent Office during the 1840s. They accounted for just under a quarter of all assignments in 1870—that is, after improvements in transportation led to the emergence of national product markets and a corresponding movement toward national assignments.

11. The patent agencies that published these kinds of journals were themselves in the business of buying and selling of patents, as well as helping inventors obtain patents for their inventions, and they undoubtedly saw their publications as a means to attract a larger customer base. For a fascinating account of *Scientific American* and Munn and Company, see Borut (1977). Other journals similar to *Scientific American* included the *American Artisan*, published by Brown, Coombs & Company; the *American Inventor*, by the American Patent Agency; and the *Patent Right Gazette*, by the U.S. Patent Right Association (which, despite its name, functioned as a general patent agency). Lamoreaux and Sokoloff (1999c, 2003).

12. For an example of one of many advice manuals, see Simonds (1871).

13. For an extended treatment of the evolution of intermediaries in this market for technology, see Lamoreaux and Sokoloff (2003).

14. Given that the lag between the 1836 reform and the onset of the acceleration is nearly a decade, one might seriously question whether the latter can be attributed to the former. Although the lag is troubling, it is worth noting that the persistent and severe contraction triggered by the panic of 1837 and lasting through 1843 would likely have depressed rates of inventive activity.

15. For discussions of the careers of these great inventors, see Khan and Sokoloff (1993), Hughes (1971), Taylor (1978), and Israel (1998). See also chapter 1, this volume.

16. For estimates of changes in capital intensity over time by industry, see Kendrick (1961).

17. See Usselman (2002). As firms committed more resources to support inventive activity internal to the firm, they became increasingly concerned with obtaining the property rights to any discoveries their employees made. During the early twentieth century, the judiciary increasingly sided with employers in disputes with employees involving patents and trade secrets. The logic seemed to be that the firm's claim was more deserving when it provided substantial inputs to the inventive activity. See Fisk (1998, 2001).

18. For discussions of developments in some of these industries, see Usselman (2002), Adams and Butler (1999), Passer (1953), and Hughes (1983).

19. Mowery (1983, 1995). As Mowery shows, one also sees during this period the emergence of large companies such as Arthur D Little, that conducted R&D on contract in a broad range of technologies.

20. For example, Alfred Sloan argued that the abundant resources large corporations could offer technologically creative individuals expanded their range of possibilities, and cited the career of Charles Kettering, who was instrumental in developing the self-starter, quick-drying paints, improved blends of gasoline, improved designs of engines, as well as Freon for General Motors. See Sloan (1964, esp. chap. 14), as well as the discussion of Sloan's career in Livesay (1979, chap. 8).

21. For recent discussions and critiques of these views, see Zeckhauser (1996), Teece (1986, 1988), Arora, Fosfuri, and Gambardella (2001), and Gans and Stern (2003).

22. For estimates of remarkably stable rates of manufacturing productivity growth over 150 years, see Kendrick (1961) and Sokoloff (1986).

References

Adams, Stephen B., and Orville R. Butler. 1999. *Manufacturing the Future: A History of Western Electric.* Cambridge: Cambridge University Press.

Arora, Ashish, Andrea Fosfuri, and Alfonso Gambardella. 2001. *Markets for Technology: The Economics of Innovation and Corporate Strategy.* Cambridge, Mass.: MIT Press.

Arrow, Kenneth J. 1962. "Economic Welfare and the Allocation of Resources for Invention." In Universities–National Bureau Committee for Economic Research and the Committee on Economic Growth of the Social Science Research Council, *The Rate and Direction of Inventive Activity: Economic and Social Factors* (609–625). Princeton, N.J.: Princeton University Press.

Baumol, William J. 2002. *The Free-Market Innovation Machine: Analyzing the Growth Miracle of Capitalism.* Princeton, N.J.: Princeton University Press.

Borut, Michael. 1977. "The *Scientific American* in Nineteenth Century America." Ph.D. dissertation, New York University.

Carlson, W. Bernard. 1991. *Innovation as a Social Process: Elihu Thomson and the Rise of General Electric, 1870–1900.* Cambridge: Cambridge University Press.

Cooper, Carolyn C. 1991. *Shaping Invention: Thomas Blanchard's Machinery and Patent Management in Nineteenth-Century America.* New York: Columbia University Press.

Davis, Lance E. 1957. "Sources of Industrial Finance: The American Textile Industry, A Case Study." *Explorations in Entrepreneurial History* 9, 189–203.

Davis, Lance E. 1958. "Stock Ownership in the Early New England Textile Industry." *Business History Review* 32, 204–222.

Davis, Lance E. 1960. "The New England Textile Mills and the Capital Markets: A Study of Industrial Borrowing, 1840–1860." *Journal of Economic History* 20, 1–30.

Davis, Lance E. 1966. "The Capital Markets and Industrial Concentration: The U.S. and the U.K.: A Comparative Study." *Economic History Review* 19, 255–272.

Dutton, H. I. 1984. *The Patent System and Inventive Activity during the Industrial Revolution, 1750–1852.* Dover, N.H.: Manchester University Press.

Fisk, Catherine L. 1998. "Removing the 'Fuel of Interest' from the 'Fire of Genius': Law and the Employee-Inventor, 1830–1930." *University of Chicago Law Review* 65, 1127–1198.

Fisk, Catherine L. 2001. "Working Knowledge: Trade Secrets, Restrictive Covenants in Employment, and the Rise of Corporate Intellectual Property, 1800–1920." *Hastings Law Journal* 52, 441–535.

Gans, Joshua S., and Scott Stern. 2003. "The Product Market and the Market for 'Ideas': Commercialization Strategies for Technology Entrepreneurs." *Research Policy* 32, 333–350.

Hounshell, David. 1984. *From the American System to Mass Production, 1800–1932: The Development of Manufacturing Technology in the United States*. Baltimore: Johns Hopkins University Press.

Hounshell, David A., and John Kenley Smith Jr. 1988. *Science and Corporate Strategy: Du Pont R&D, 1902–1980*. Cambridge: Cambridge University Press.

Hughes, Thomas Parke. 1971. *Elmer Sperry: Inventor and Engineer*. Baltimore: Johns Hopkins University Press.

Hughes, Thomas Parke. 1983. *Networks of Power: Electrification in Western Society, 1880–1930*. Baltimore: Johns Hopkins University Press.

Hughes, Thomas Parke. 1989. *American Genesis: A Century of Invention and Technological Enthusiasm, 1870–1970*. New York: Viking.

Israel, Paul. 1998. *Edison: A Life of Invention*. New York: Wiley.

Jorgenson, Dale W., Frank Gollop, and Barbara Fraumeni. 1987. *Productivity and U.S. Economic Growth*. Cambridge, Mass.: Harvard University Press.

Kendrick, John W. 1961. *Productivity Trends in the United States*. Princeton, N.J.: Princeton University Press.

Khan, B. Zorina. 1995. "Property Rights and Patent Litigation in Early Nineteenth-Century America." *Journal of Economic History* 55, 58–97.

Khan, B. Zorina. 2005. *The Democratization of Invention: Patents and Copyrights in American Economic Development, 1790–1920*. Cambridge: Cambridge University Press.

Khan, B. Zorina, and Kenneth L. Sokoloff. 1993. "'Schemes of Practical Utility': Entrepreneurship and Innovation among 'Great Inventors' in the United States, 1790–1865." *Journal of Economic History* 53, 289–307.

Khan, B. Zorina, and Kenneth L. Sokoloff. 1998. "Patent Institutions, Industrial Organization and Early Technological Change: Britain and the United States, 1790–1850." In Maxine Berg and Kristine Bruland, eds., *Technological Revolutions in Europe: Historical Perspectives* (292–313). Cheltenham: Edward Elgar.

Khan, B. Zorina, and Kenneth L. Sokoloff. 2004. "Institutions and Technological Innovation during Early Economic Growth: Evidence from 'Great Inventors,' 1790–1930." In Theo Eicher and Cecilia Penalosa Garcia, eds., *Institutions and Growth* (123–158). Cambridge, Mass.: MIT Press.

Kortum, Samuel, and Josh Lerner. 1998. "Stronger Protection or Technological Revolution: What Is behind the Recent Surge in Patenting." *Carnegie-Rochester Conference Series on Public Policy* 48, 247–304.

Kortum, Samuel, and Josh Lerner. 2000. "Assessing the Contribution of Venture Capital to Innovation." *RAND Journal of Economics* 31, 674–692.

Lamoreaux, Naomi R. 1994. *Insider Lending: Banks, Personal Connections, and Economic Development in Industrial New England.* Cambridge: Cambridge University Press.

Lamoreaux, Naomi R., and Kenneth L. Sokoloff. 1996. "Long-Term Change in the Organization of Inventive Activity." *Proceedings of the National Academy of Sciences* 93, 12686–12692.

Lamoreaux, Naomi R., and Kenneth L. Sokoloff. 1999a. "The Geography of the Market for Technology in the Late-Nineteenth- and Early-Twentieth Century United States." In Gary D. Libecap, ed., *Advances in the Study of Entrepreneurship, Innovation, and Economic Growth* (Vol. 11, 67–121). Greenwich, Conn.: JAI Press.

Lamoreaux, Naomi R., and Kenneth L. Sokoloff. 1999b. "Inventive Activity and the Market for Technology in the United States, 1840–1920." Working paper 7107, NBER, Cambridge, Mass.

Lamoreaux, Naomi R., and Kenneth L. Sokoloff. 1999c. "Inventors, Firms, and the Market for Technology in the Late Nineteenth and Early Twentieth Centuries." In Naomi R. Lamoreaux, Daniel M. G. Raff, and Peter Temin, eds., *Learning by Doing in Markets, Firms, and Countries* (19–57). Chicago: University of Chicago Press.

Lamoreaux, Naomi R., and Kenneth L. Sokoloff. 2001. "Market Trade in Patents and the Rise of a Class of Specialized Inventors in the Nineteenth-Century United States." *American Economic Review* 91, 39–44.

Lamoreaux, Naomi R., and Kenneth L. Sokoloff. 2003. "Intermediaries in the U.S. Market for Technology, 1870–1920." In Stanley L. Engerman et al., eds., *Finance, Intermediaries, and Economic Development* (209–246). Cambridge: Cambridge University Press.

Lamoreaux, Naomi R., and Kenneth L. Sokoloff. 2005. "The Decline of the Independent Inventor: A Schumpeterian Story?" Working paper W11654, NBER, Cambridge, Mass.

Lerner, Josh. 2001. *Venture Capital and Private Equity: A Casebook.* 2nd edition. New York: John Wiley & Sons.

Lerner, Josh, and Robert P. Merges. 1998. "The Control of Technology Alliances: An Empirical Analysis of the Biotechnology Industry." *Journal of Industrial Economics* 46, 125–156.

Lerner, Josh, and Antoinette Schoar. 2005. "Does Legal Enforcement Affect Financial Transactions: The Contractual Channel in Private Equity." *Quarterly Journal of Economics* 120, 223–246.

Livesay, Harold C. 1979. *American Made: Men Who Shaped the American Economy*. Boston: Little, Brown.

MacLeod, Christine. 1988. *Inventing the Industrial Revolution: The English Patent System, 1660–1800*. Cambridge: Cambridge University Press.

MacLeod, Christine. 1999. "Negotiating the Rewards of Invention: The Shop-Floor Inventor in Victorian Britain." *Business History* 41, 17–36.

Mokyr, Joel. 1990. *The Lever of Riches: Technological Creativity and Economic Progress*. New York: Oxford University Press.

Mowery, David C. 1983. "The Relationship between Intrafirm and Contractual Forms of Industrial Research in American Manufacturing, 1900–1940." *Explorations in Economic History* 20, 351–374.

Mowery, David C. 1995. "The Boundaries of the U.S. Firm in R&D." In Naomi R. Lamoreaux and Daniel M. G. Raff, eds., *Coordination and Information: Historical Perspectives on the Organization of Enterprise* (147–176). Chicago: University of Chicago Press.

Mowery, David C., and Nathan Rosenberg. 1989. *Technology and the Pursuit of Economic Growth*. Cambridge: Cambridge University Press.

Nelson, Richard R. 1959. "The Simple Economics of Basic Scientific Research." *Journal of Political Economy* 67, 297–306.

Passer, Harold C. 1953. *The Electrical Manufacturers, 1875–1900: A Study in Competition, Entrepreneurship, Technical Change and Economic Growth*. Cambridge, Mass.: Harvard University Press.

Rosenberg, Nathan, ed. 1969. *The American System of Manufactures: The Report of the Committee on the Machinery of the United States 1855, and the Special Reports of George Wallis and Joseph Whitworth 1854*. Edinburgh: Edinburgh University Press.

Saxenian, AnnaLee. 1994. *Regional Advantage: Culture and Competition in Silicon Valley and Route 128*. Cambridge, Mass.: Harvard University Press.

Schmookler, Jacob. 1966. *Invention and Economic Growth*. Cambridge, Mass.: Harvard University Press.

Schumpeter, Joseph A. 1934. *The Theory of Economic Development: An Inquiry into Profits, Capital, Credit, Interest, and the Business Cycle*. Trans. Redvers Opie. Cambridge, Mass.: Harvard University Press.

Scotchmer, Suzanne. 2004. *Innovation and Incentives*. Cambridge, Mass.: MIT Press.

Simonds, William Edgar. 1871. *Practical Suggestions on the Sale of Patents*. Hartford: Privately printed.

Sloan, Alfred P. 1964. *My Years with General Motors*. Garden City, N.Y.: Doubleday.

Sokoloff, Kenneth L. 1986. "Productivity Growth in Manufacturing during Early Industrialization: Evidence from the American Northeast, 1820 to 1860." In

Stanley L. Engerman and Robert E. Gallman, eds., *Long-Term Factors in American Economic Growth* (679–736). Chicago: University of Chicago Press.

Sokoloff, Kenneth L. 1988. "Inventive Activity in Early Industrial America: Evidence from Patent Records, 1790–1846." *Journal of Economic History* 48, 813–850.

Sokoloff, Kenneth L., and B. Zorina Khan. 1990. "The Democratization of Invention during Early Industrialization: Evidence from the United States, 1790–1846." *Journal of Economic History* 50, 363–378.

Taylor, Jocelyn Pierson. 1978. *Mr. Edison's Lawyer: Launching the Electric Light*. New York: Topp-Litho.

Teece, David J. 1986. "Profiting from Technological Innovation: Implications for Integration, Collaboration, Licensing, and Public Policy." *Research Policy* 15, 285–305.

Teece, David J. 1988. "Technological Change and the Nature of the Firm." In Giovanni Dosi et al., eds., *Technical Change and Economic Theory* (256–281). London: Pinter.

Usselman, Steven W. 2002. *Regulating Railroad Innovation: Business, Technology, and Politics in America, 1840–1920*. Cambridge: Cambridge University Press.

Wise, George. 1985. *Willis R. Whitney, General Electric, and the Origins of U.S. Industrial Research*. New York: Columbia University Press.

Zeckhauser, Richard. 1996. "The Challenge of Contracting for Technological Information." *Proceedings of the National Academy of Sciences* 93, 12743–12748.

1

Financing Invention during the Second Industrial Revolution: Cleveland, Ohio, 1870–1920

Naomi R. Lamoreaux, Margaret Levenstein, and Kenneth L. Sokoloff

Technological change was so rapid and transformative during the late nineteenth and early twentieth centuries that the period is commonly known as the Second Industrial Revolution. Although the progress of this era is conventionally attributed to the rise of big business and the beginnings of the shift toward in-house R&D, this was also a time (like the Third Industrial Revolution of the late twentieth century) when large numbers of new firms were being formed in the high-tech sectors of the economy. Based on recent experience, it might be assumed that financing for such enterprises depended on the preexistence of formal financial institutions—for example, well-functioning and highly liquid securities markets that enabled early-stage investors to make money by taking enterprises public. Although a variety of different kinds of exchanges had emerged in the United States over the course of the nineteenth century, none were as yet well suited to play this kind of a role. How then was venture capital mobilized? The question is an important one because by examining the formation of new firms in an institutional context very different from today's, we can develop a more basic understanding of the difficulties in financing cutting-edge ventures and also how these problems can be overcome.

Questions of this sort are often best answered through detailed local studies that make it possible to track the financial sources tapped by individual inventors and firms. We chose to focus our study on Cleveland, Ohio, a center of inventive activity in a remarkable number of important industries, including electric light and power, steel, petroleum, chemicals, and automobiles, and began our research by collecting samples of patentees residing in Cleveland at several different times. We then compiled the histories of the most productive of these patentees, gathering information on their activities and how they were financed from archival and manuscript collections, city directories, newspapers, and a variety of secondary sources. We searched the extant records of formal financial institutions such as the Cleveland Stock Exchange, the Cleveland Trust Company, and other banks or banklike institutions for evidence of their connections with patentees or with enterprises that exploited the inventors' discoveries.

Perhaps not surprisingly, we found that neither banks nor securities markets played a direct role in financing these new ventures. This is not to say that such institutions were unimportant. Once new firms were established, local banks and trust companies helped to support their ongoing activities in a variety of ways—for example, by financing their working capital and underwriting bond issues. Moreover, some of the largest new enterprises eventually were listed on the Cleveland Stock Exchange (and even later on the New York Stock Exchange). But formal financial institutions did not play a leading role in the creation of new enterprises. Instead, entrepreneurs seeking to exploit new technologies typically raised capital directly from wealthy individuals. These investors bought substantial shares in the equity of new firms, held on to their investments for long periods of time, and often played an important and ongoing role in management.

All too often the willingness of such individuals to provide early-stage backing is ascribed simply to personal connections, as if it were self-evident that entrepreneurs would be able to command the savings of their families, friends, or other people with whom they had close personal associations. Although entrepreneurs may often have found it easier to raise funds from people who knew them well than from strangers, even the closest members of their families were often reluctant to put their money into ventures exploiting untried new inventions unless they had some way of determining that the technology was likely to work and have a market.[1] Such a determination required access to expertise. The contribution of this study is to put some structure on what are normally dismissed as informal channels of finance by describing the mechanisms that enabled entrepreneurs in high-tech industries to tap both personal and impersonal sources of funds.[2] In particular we show how a small number of successful enterprises could stimulate investment, first by demonstrating the wealth-creation possibilities of cutting-edge technologies and then by becoming hubs of overlapping networks of inventors and financiers. Our findings thus parallel those of Steven Klepper in his study of spin-offs in the Detroit automobile industry (see chapter 3). Rather than formal financial institutions, a few key firms served as channels for directing funds to new innovative enterprises.

After providing a detailed look at the mechanisms by which the city's most important hub enterprise, the Brush Electric Company, spawned a whole host of spin-off and start-up enterprises, we offer a quantitative picture of inventive activity in Cleveland and of the ways in which inventors profited from their discoveries. We find that many productive inventors were principals in firms formed to exploit their inventions, though by the turn of the twentieth century, there was also a significant group of inventors who were either employees of large firms or had some other kind of long-term attachment with a large firm. These parallel developments, we show, replicate patterns apparent in the national data. By the turn of the century, the increasing complexity of technology was raising the costs of engaging effectively in inventive activity, and inventors were finding it necessary to associate themselves with firms. Intriguingly, however, there was a strong regional divergence in the ways inventors formed such associations. Whereas in the Northeast, particularly in the Middle Atlantic, the most productive inventors were likely to be employed in large, integrated firms, in the Midwest they were more likely to become principals in firms organized to exploit their inventions. We speculate that differences in the depth or organization of capital markets in the two regions may help to explain this regional pattern. In the Northeast, the nation's main capital markets were well suited to the finance of large firms and may have pulled local capital into the exchanges for this purpose. At a greater distance, however, local venture capital markets seem to have been quite effective in furnishing funds for the formation of new enterprises.

The Rise of Cleveland's Manufacturing and Financial Sectors

Located on Lake Erie at the terminus of the Ohio Canal, Cleveland had long been the commercial center of northeastern Ohio. The city's first heavy industrial enterprise, a firm that produced steam furnaces, was founded in the 1830s, and its first iron-rolling mills were built in the 1850s (Miller and Wheeler 1990). But Cleveland's rise as an industrial powerhouse was largely a post–Civil War phenomenon. As late as 1870, Cuyahoga County, where Cleveland is located, ranked only number twenty-two in manufacturing output among counties nationwide. By 1920, however, it had risen to fourth place (U.S. Census Office 1870–1920). Moreover, in the interim, Cleveland had become a hotbed of patenting activity. In 1900 it ranked eighth out of all U.S. cities in the total number of patents granted to residents, and if the calculation is limited to patents deemed by official examiners to have made significant contributions to the industrial art of the period, Cleveland was the fifth most technologically important city in the country (Fogarty, Garofalo, and Hammack n.d.).

Much of the growth in Cleveland's manufacturing sector occurred in industries associated with the Second Industrial Revolution. Cleveland's location gave it convenient access to Lake Superior iron ore, so it is not surprising that iron and steel was the city's leading industry in terms of value of output throughout the nineteenth century, falling to second place in 1910 and to third in 1920 (see table 1.1). The machine tool industry was also persistently among the city's top three. By 1910, however, automobiles had become the third largest industry, and it would climb by 1920 to number one. During the same decade, electrical machinery rose to fourth place, so that the city's top industries were now automobiles, machine tools, iron and steel, and electrical machinery. Chemical products such as paints and varnishes also had a major presence in the city and its surrounding areas.[3]

Cleveland's financial sector grew alongside its manufacturing industries over the half-century between 1870 and 1920. In 1870 the city was home to five national banks, most with origins dating to the 1840s, and one substantial savings institution, the Society for Savings, founded in 1849. By 1920 the city had thirty-eight banks, savings institutions, and trust companies, with total deposits amounting to more than $800 million. More than a dozen national banks were founded during these decades, though because of mergers, the net gain in number was only two. After the Ohio legislature passed enabling legislation, a dozen trust companies were organized, eleven of which were still active at the end of

Table 1.1
Cleveland's largest industries, 1870–1920

Industry rank	1870[a]	1880	1890	1900	1910	1920
1	Coal, rectified	Iron and steel	Iron and steel	Iron and steel	Iron and steel, steel works, and rolling mills	Automobiles
2	Iron, forged and rolled	Slaughtering and meatpacking	Foundry and machine shop products	Foundry and machine shop products	Foundry and machine shop products	Foundry and machine tools
3	Flour mill products	Foundry and machine shop products	Petroleum Refining	Slaughtering and meatpacking, wholesale	Automobiles	Iron and steel
4	Meat, packed pork	Clothing, men's	Slaughtering and meatpacking, wholesale	Clothing, women's factory product	Slaughtering and meatpacking	Electrical machinery
5	Iron, castings (not specified)	Liquors, malt	Carpentering	Liquors, malt	Clothing, women's	Clothing, women's

[a] 1870 data are for Cuyahoga County. All other years are for the city of Cleveland.
Source: U.S. Census of Manufactures (1870–1920).

the period. Similarly, an 1868 law permitting the formation of building-and-loan and savings-and-loan associations stimulated a rash of entry, though many of these institutions were short-lived.[4]

The city's leading businessmen had been active in founding financial institutions since the formation of the Commercial Bank of Lake Erie in 1816, so it is not surprising that industrialists were involved in organizing and running many of the banks and trust companies formed after 1870.[5] Robert Hanna, for example, helped to organize the Ohio National Bank in 1876, eight years after founding the Cleveland Malleable Iron Company. His nephew and partner, Marcus Hanna, organized the Union National Bank in 1884, served as its president for at least a decade, and later sat on the boards of the Commercial National Bank, the Guardian Trust Company, and the People's Savings and Loan Association.[6] C. A. Grasselli, a leading chemical manufacturer, was a founder and then president of the city's first trust company, Broadway Savings and Trust (1884), and also of its second trust company, Woodland Avenue Savings and Trust (1886).[7] Prominent among the founders of the Cleveland Trust Company (1895) were Fayette Brown, president of the Brown Hoisting Machine Company, and Jacob D. Cox, owner of the Cleveland Twist Drill Company.[8] Brown's son, Harvey H. Brown, who replaced him as president of the Hoisting Company, served on the board of the Bank of Commerce (Cleveland Stock Exchange Handbook 1903). Many savings institutions also had industrialists among their organizers. For example, the founding board of directors of the Detroit Street Savings and Loan included Theodor Kundtz, an inventor and manufacturer of sewing machine cabinets and auto bodies, and E. R. Edson, an inventor and manufacturer of machinery for extracting oil and other products from fish.[9]

Undoubtedly the industrialists who founded these banks, trust companies, and savings institutions aimed to benefit their own enterprises, but it seems that they were primarily concerned with increasing their access to working capital. In any event, the institutions they created mainly supported the city's industries by providing firms with short-term commercial credit. Although they sometimes assisted new firms in other ways, for example, by handling their bond issues or accepting their securities as collateral for individual loans, the extant records suggest that such aid was relatively rare. The bottom line is that these institutions were not directly involved in founding new companies.[10]

Cleveland's equity markets also developed during this period. Daily quotations of government bond prices first appeared in the *Cleveland Plain Dealer* in 1880, and by 1886 the daily lists included prices for the stocks of local banks, street railroad companies, iron mines, and "miscellaneous" securities, including a small number of industrials. The continued growth of both local brokerage houses and trading in local securities resulted in 1900 in the formal organization of the Cleveland Stock Exchange (CSE) Although the city's brokers led in forming the exchange, its founding members included prominent Cleveland industrialists such as Harvey Brown, Jeptha Wade, and Daniel R. Hanna.[11]

As was the case for other exchanges at that time, railroads initially dominated the listings. Compared to the New York Stock Exchange (NYSE), however, the Cleveland market from the beginning handled the securities of a more diverse set of firms. For example, 52 percent of the firms listed on the NYSE in 1900 were railroads, and in 1910 the figure was still 48 percent. By contrast, railroads accounted for only 40 percent of the listings on the Cleveland exchange in 1903, and by 1910 the share had fallen to 15 percent (Cleveland Stock Exchange Handbook 1903, Cull and Davis 1994). This decline in the relative position of railroads on the CSE owed mainly to the listing of new banks, trust companies, and utilities, including several local electric light companies and nine local telephone companies. Between 1910 and 1914, however, the number of manufacturing firms on the CSE more than doubled. The newly listed manufacturers included some of the most successful of the innovative firms formed over the previous several decades (American Multigraph, the Bishop-Babcock-Becker Company, Brown Hoisting Machine, National Carbon, Wellman-Seaver-Morgan, and the White Company). But these firms did not turn to the CSE to raise capital, nor did their largest shareholders use the market to increase the liquidity of their investments. Trading in the equities of these manufacturing firms was at best light, and it seems that the listings were mainly useful to local brokers who from time to time had small lots of these securities to offer the public.[12]

Although venture capitalists today often make their profits by taking firms public and then cashing out their investments, that does not seem to have been the practice in the early twentieth century. To the contrary, investors in start-up enterprises appear to have taken their profits over the long run in the form of dividends on their shareholdings. The key question thus becomes to understand how it came to be that so many wealthy Clevelanders were induced to invest their savings in firms formed to exploit cutting-edge technology.

Clusters of Inventors as Sources of Information for Financiers

Before they would be willing to risk their assets in new technological ventures, those with wealth to invest had to be convinced that it was possible to earn high rates of return in such enterprises. They also had to be convinced that the specific technologies in which they were being asked to invest were especially promising. The first condition could be relatively easily satisfied by some enticing examples of ventures that had made their backers rich. The second was more difficult, because evaluating the merits of alternative projects required considerable technological expertise.

There were a variety of ways in which investors could obtain assessments of new technologies coming on the market, but one of the simplest was to tap into the discussions that inventors themselves were having about exciting new discoveries.[13] In the late nineteenth century, certain kinds of enterprises were particularly well placed to become focal points for such conversations. Perhaps the most important example was the telegraph. Because operators were responsible for repairing and improving the equipment they used, local telegraph offices developed intense "shop cultures," to use Paul Israel's (1992) phrase, that encouraged employees to keep abreast of advances in electrical technology and experiment with new applications. Thomas Edison was only the most famous of the many inventors who got their starts in this way. Western Union provided financial assistance to a number of new ventures that came out of its offices—Edison's early company to exploit an improved stock ticker is a good example—but local capitalists also tapped into these clusters of inventive activity in order to learn about promising new companies (Israel 1992, 1998, Adams and Butler 1999).

One such enterprise was the Telegraph Supply Company of Cleveland, Ohio, organized to exploit the inventions of George B. Hicks, developer of the Hicks telegraph repeater. In 1872, a young businessman named George W. Stockly bought a large interest in the company and took responsibility for the commercial side of its affairs, assuming the title of vice president and general manager. Shortly after Stockly joined the company, Hicks suddenly died. Stockly had no technical training and, to save his investment, took two steps. He brought in a new set of officers who were well connected and technologically expert. These included the patent solicitor Mortimer D. Leggett, who had previously served as the U.S. commissioner of patents, and the banker James J. Tracy, who belonged to the "Ark," a natural science club whose leading members were engaged, around the same time, in founding the Case School of Applied Science. He also began to form relationships with promising young inventors and invite them to come work in the Telegraph Supply Company's shops.[14]

The Brush Electric Company and Its Demonstration Effect

Among the young men whom Stockly encouraged was Charles F. Brush, whose invention of an arc-lighting system would spark Cleveland's technology boom.[15] Brush had been interested in electricity from an early age and had built his first arc light while still a student at Cleveland High School. During the late 1860s, he attended the University of Michigan, majoring in mining engineering because the school did not yet have a program in electricity. He then returned to Cleveland and attempted to earn a living as an analytical chemist. When he found he could not make ends meet, he joined an iron-dealing partnership with his childhood friend, Charles Bingham. All the while, he continued to experiment in his spare time with electric lighting (Kennedy 1885, Brush 1905, Gorman 1961, Eisenman 1967).

Stockly, another long-time friend, hired Brush to do some consulting work for the Telegraph Supply Company. The two men got to talking about the future of electricity, and Stockly, impressed by Brush's ideas, offered him the use of the company's shop to develop his arc-lighting system. When Brush successfully demonstrated a new dynamo, Stockly negotiated a contract that gave Telegraph Supply exclusive rights to market the device in exchange for royalties. Brush's reputation as an inventor got a boost in 1878 when his dynamo won a competition at the Franklin Institute in Philadelphia, but it was the backing of Stockly and the other well-connected officers of the Telegraph Supply Company that translated Brush's technical triumph into a commercial success.[16]

Looking for a dramatic way to publicize Brush's invention, Stockly and his associates negotiated a contract with the city of Cleveland to light Monumental Park (now Public Square). Advance publicity brought out a large crowd the evening of April 29, 1879, when Cleveland officials threw a switch, and twelve strategically placed arc lamps flooded the park with light. News of the event spread quickly throughout the country, generating a rush of interest in the new type of street lighting, followed by orders for installations. The successful demonstration also helped the officers of Telegraph Supply line up investors, and the next year the firm was reorganized as the Brush Electric Company with an authorized capital of $3 million, an enormous amount for the time.[17]

Brush Electric installed about 80 percent of the nation's arc-lighting systems during the early 1880s and made the businesspeople who initially bought its stock rich. As Jacob D. Cox (1951) founder of the Cleveland Twist Drill Company, later regretfully noted: "The original holders made immense sums of money but, as I had no funds to invest, I missed this rare opportunity" (90–91). Brush himself became a wealthy man, earning royalties in excess of $200,000 a year on his patents during 1882 and 1883. Indeed, his royalty account accumulated so quickly that the company fell behind on its payments, and to settle the debt, Brush agreed in 1886 to take $500,000 in stock.[18] By the second half of the decade, however, the company was losing ground to new competitors, and Brush, Stockly, and the other major shareholders sold out to the Thomson Houston Electric Company at what appears to have been a handsome price. According to a report in the *New York Times*, the controlling shareholders (Stockly, Tracy, Leggett, Brush, and Stockly's sister) owned 30,000 of the company's 40,000 outstanding shares and sold them for $75 each. The par value of the stock was $50, and its market price was estimated to be $35.[19]

The success of Brush and other early electric lighting companies grabbed investors' attention.[20] But investors who lacked technological expertise could easily make mistakes and put their funds in poorly conceived or even fraudulent enterprises. Con artists quickly took advantage of the heightened interest in electrical innovations by organizing wildcat companies "whose only purpose seemed to be to foist a lot of worthless stock upon a gullible public." Henry I. Hoyt, president of the Gramme Electrical Company, claimed that the wildcatters owned no patents and displayed other companies' machines to gullible investors as if they were their own. "'Why, Sir,' exclaimed Mr. Hoyt, indignantly, 'I have even had men come to me and ask to borrow dynamo-electric generators to exhibit in the offices of the so-called 'electric light companies.'" Hoyt estimated that between forty and fifty such "speculative enterprises" were out raising capital.[21]

One way in which investors could reduce their risk of being taken was to invest in companies promoted by businesspeople who had a proven record of success. Not surprisingly, the men most visibly associated with the Brush Electric Company had a comparatively easy time raising capital for subsequent ventures. A good example is Washington Lawrence, a major early investor in Brush Electric and for a time the company's general manager. In 1882 Lawrence took the unusual step of selling his interest and investing the proceeds in real estate, but he returned to the industry in 1886 to buy a controlling share of the Boulton Carbon Company, a spin-off venture that supplied carbons for arc lights to Brush and other firms. Lawrence reorganized Boulton as the National Carbon Company, bringing in wealthy investors such as Myron T. Herrick, a local lawyer who had founded a hardware company and who built the Society for Savings into a major financial institution, and Webb C. Hayes, the son of former President Rutherford B. Hayes. He then used the firm as a vehicle to acquire competing enterprises and expand into batteries and other components of electrical systems, creating in the process one of the country's earliest industrial research laboratories.[22]

Another good example of the ease with which men associated with Brush Electric could raise capital for innovative projects is Brush's own promotion of the Linde Air Products Company. Brush became aware during the 1890s of the work of Carl von Linde, a German scientist who had developed a process for liquefying air. Linde wanted to market his invention in the United States but was prevented by a conflicting patent. Believing that Linde's invention had priority, Brush bought a one-third share in the patent and financed the necessary litigation. The case dragged on for several years, but Brush ultimately won and, after a brief period during which he and Linde disagreed about terms, set about organizing the Linde Air Products Company. In early 1907 he held a meeting in his office to present a prospectus for the company to a small group of prominent Cleveland businessmen. Virtually all of those present immediately agreed to invest, and the company was launched with a capital of $250,000 and with Brush as president, the chemical entrepreneur C. A. Grasselli as vice president, and J. L. Severance of Standard Oil as secretary-treasurer.[23] After some initial technical problems, which Brush himself resolved, the enterprise grew rapidly. By 1910 R. G. Dun & Company gave it a credit rating of "excellent," despite having no information with which to measure its pecuniary strength, and by 1917 its authorized capital had increased to $15 million. In that year, the firm merged with National Carbon, Union Carbide, and several other firms to form Union Carbide and Carbon Company, with Brush and the other investors exchanging their Linde stock for twice the number of shares in the new combine.[24]

Important enterprises like Brush Electric could solve the problem of information in another way as well—by becoming gathering places for inventors. From early on, the Brush factory was a magnet for ambitious young men who came to work in its shops, network with other technologically creative people, and catch the eye of investors eager to finance the next Charles Brush. The inventors who worked in such close proximity shared ideas and cooperated in solving problems, but they also critiqued each other's inventions. Local capitalists could plug into the conversations and use the collective judgments of the group to inform their investment decisions.

A Gathering Place for Inventors and a Spawning Ground for New Companies

The inventors who flocked to the Brush Electric Company came there for many different reasons and in many different ways. Some already had developed promising ideas and were invited to use the company's shops to work out the details. In 1883, for example, Edward M. Bentley and Walter H. Knight left their jobs at the Patent Office in Washington to seek backers for a plan they had devised for electrically powered streetcars. Stockly was among the first capitalists they contacted, and he agreed to invest in the firm, brought the young men to Cleveland, and put Brush employees to work building the necessary equipment, including an experimental track. The very visible position that Bentley and Knight occupied at Brush enabled them to convince the East Cleveland Railway Company to lay a trial line and also to attract additional local backers to invest in the newly formed Bentley-Knight Electric Railway Company. Although the initial trials were successful and generated orders for streetcar systems in other cities, the company subsequently ran into technical and financial difficulties. In 1889 Thomson-Houston bought out most of its shareholders in order to gain control of the valuable Bentley-Knight patents.[25]

The Brush firm also incubated the Cowles Electric Smelting and Aluminum Company, the brainchild of Eugene and Alfred H. Cowles, sons of a prominent Cleveland newspaper publisher. Eugene started out as a journalist but began to study electricity after covering an early exhibit of Brush lighting for his father's newspaper. In 1880 he left the news business to organize and manage the Brush Electric Light and Power Company, the Brush-affiliated utility in Cleveland. His brother Alfred was at that time studying physics and chemistry at Cornell, but the two brothers soon joined forces in New Mexico, where Eugene, who had tuberculosis, had gone for health reasons and to inspect mining properties. The proximity of the mines they visited to good sources of water power inspired the brothers to investigate the possibility of smelting ores electrically. Lacking the equipment they needed to experiment, they returned to Cleveland (and to Brush), built an experimental furnace, applied for a patent, and in 1885 organized their company.[26]

Although some observers were initially skeptical about whether the process would be economical (Brush scoffed that it was simply an expensive way to burn coal), their successful operation at the Brush facility silenced doubters. By 1886 Brush himself was using Cowles aluminum to make hubs for armatures. Moreover, financiers could visit the Brush works, see the Cowles' furnaces in operation, and calculate production costs. With capital supplied by their father and other investors (most notably zinc smelter Frederick William Matthiessen, who became the company's president), they built a larger plant in Lockport, New York, where they could obtain cheap power from Niagara Falls (the local Brush Electric Light and Power Company had generated the first hydro-electricity there just a few years earlier).[27]

The company did well until 1893, when it was shut down by court or-der in a patent dispute. Although in 1903 the Cowles Company emerged victorious from the litigation with a hefty financial settlement, ten years of enforced idleness had done its aluminum business irreparable harm, and it licensed its patents to the defeated party, the Pittsburgh Reduction Company, later ALCOA. Meanwhile, a group of investors had organized the Electro Gas Company to use the Cowles's technology for the produc-tion of calcium carbide, the main component of acetylene gas. The appli-cation was discovered and patented by Thomas L. Willson, who had worked for a short time at Brush and presumably had there become familiar with the Cowles' smelting methods. The Cowles Company received 12.5 percent of the capital stock of the new firm in exchange for a license to use its furnace patents. In 1898 Electro Gas was reorga-nized as Union Carbide.[28]

Sidney H. Short, another early inventor of electric streetcars, also ran his company inside the Brush factory. Growing up in Columbus, Ohio, Short (like Brush) had long been interested in electricity, and he had amused himself as a child by equipping his parents' home with burglar alarms and other electrical devices. While attending Ohio State during the late 1870s, he patented and sold a transmitter for telephones. Gradu-ating in 1880, he accepted a professorial position at the University of Denver, where he taught physics and chemistry and pursued his research in electrical applications. Within several years he had demonstrated his "Joseph Henry," a trolley car driven by an electric motor. In his own words, "so impressed were the capitalists then interested in my experi-ments that the Denver Tramway Company was at once organized" to build an electric streetcar in that city. Obtaining a contract for a system in St. Louis, he secured financial backing from an Ohio investor who arranged to have the necessary dynamos custom-made at the Brush Electric Company. Short then moved to Cleveland to supervise the work and experiment in the company's shops. Brush encouraged Short's efforts and helped to finance the resulting Short Electric Railway Company, which operated out of a Brush building (Short 1989, Smith 1955, *Dictionary of American Biography*, 1928–1936, Moley 1962).

Another fertile source of new technological ideas and companies was Brush Electric's workforce. Of course, as a general rule, the employees of technologically cutting-edge companies are well positioned to start such ventures; not only are they up on the latest advances, but their jobs often give them the opportunity to demonstrate their skills and knowledge to potential backers and customers.[29] Brush, however, facilitated the transition from employee to inventor-entrepreneur by instituting a training course designed to remedy the company's acute difficulties in hiring people with the expertise needed to install and maintain electrical lighting and power systems. The men who took the course gained familiarity with all aspects of the company's electrical technology and also the opportunity to hobnob with the inventors who had gravitated to its shops.

The career of John C. Lincoln, founder of the Lincoln Electric Company, provides insight into the connections and opportunities the Brush training course offered young would-be inventors. The son of an impoverished minister, Lincoln developed an interest in electricity during his high school years and pursued it at Ohio State by taking all of the relevant courses the university offered. He also worked during his spare time for the company that installed Columbus's first electric streetcar line. With the help of a relative, he secured a position at Brush Electric in 1888, enrolling in the training course that Brush had created for his employees. About a year later, Brush introduced Lincoln to Short, who promptly hired him to assist in demonstrating and installing his electric streetcar system. While traveling in Short's employ, Lincoln obtained his first patent, an electric brake for streetcars, which he sold for $500 to the great inventor Elmer Sperry (who was then also working on streetcars at the Brush facility).

Lincoln returned to Cleveland and to Brush Electric as Short's superintendent of construction in 1892, but the two men had a falling out when Short blamed Lincoln for the failure of a component. Shortly thereafter,

Lincoln joined forces with Samuel K. Elliott and his brothers, W. H. and Emmett, in an enterprise to manufacture electric motors. This venture too originated in a Brush connection, for Samuel was a fellow student in the training course. Lincoln designed a motor that quickly gained a respected place in the market, and in 1895 the four men organized the Elliott-Lincoln Electric Company with financial backing from members of a local family named Crawford. With the economy in depression, the firm experienced hard times, and Lincoln fought with his associates over the future direction of the business. Forced out of the company in 1896, he started his own venture to produce electric motors, the Lincoln Electric Company. The firm's initial capital came from his own meager savings, but Lincoln was able to build up the business rapidly by custom-designing motors for local firms. By the end of the next decade, Lincoln had spun off the production of motors to the Lincoln Motor Works (later Reliance Electric) with capital supplied by the Cleveland industrialist Peter M. Hitchcock (a relative who was an early supporter of Brush and who had originally gotten Lincoln the job there), and was increasingly devoting his energies to the development of the arc-welding equipment for which the firm would become famous. In the meantime, he wrote a popular handbook on electricity that sold 40,000 copies, invented an automobile powered by a variable-speed electric motor that he considered his greatest invention, and designed generators for electric vehicles that allowed owners to recharge their car's batteries in their own garages.[30]

The spate of mergers and buyouts that followed the formation of General Electric in 1892 also spurred a number of Brush-connected inventors to form independent enterprises. For example, employees of what had been Brush's lamp affiliate, the Swan Lamp Manufacturing Company, formed the Adams-Bagnall Electric Company in 1895 (Covington 1999). Similarly, after selling his streetcar company to General Electric in 1893, Short almost immediately reentered the business by joining forces with the Walker Manufacturing Company, a cable and machinery builder, to develop an electric traction system. By this time, his abilities as an inventor in this promising new field of technology were so well appreciated that he was able to mobilize financial support on what seem to be remarkably favorable terms. For example, in a contract dated July

1, 1896, he agreed in return for generous compensation and the title of vice president to work for the company and to assign it the rights to all of his patents that he then owned and controlled, as well as to inventions he would patent in the future, in the areas of "Dynamo electric machinery, Street Railway motors and car equipment, Arclighting machinery, and Alternating machinery." Although we do not know what salary he received, his contract specified that he would be paid an additional royalty of "20 cents per horse power upon all of the electrical apparatus sold and delivered to customers" and that the assignments of his patents would be revoked if the royalties due were not paid within three months of the delivery of the apparatus, or if the company failed to sell and deliver electrical equipment totaling at least $300,000 in any calendar year. Most telling was the contract's "reversionary clause," which specified that if Short were to leave the employment of the Walker Company, the rights to all of his patents would revert back to him. The implication is that Short's participation in the company was so desirable that he was able to find people willing to invest in developing his technology, even though he had the right to withdraw at any time with all of the assets he had brought to the firm.[31] Indeed, the next year, a group of New York capitalists who were "largely interested in street railway and electric lighting systems in many of the principal cities" (they included ex-Governor Roswell P. Flower, Anthony N. Bradey, James W. Hinkley, and Perry Belmont, brother of August Belmont) bought a controlling interest in Walker and negotiated with Short and the company's other officers for the sale of the firm and its assets to Westinghouse in 1898.[32] Short then left for England to help set up a company that would exploit his traction patents in Europe. When he died suddenly of appendicitis in London in 1902, he left an estate valued at about $2.5 million (Smith 1955).

By the late 1880s, the Brush factory had become to such an extent the location of choice for inventors working on electrical projects that technologically creative people continued to gravitate to the site even after Brush and his associates sold out to Thomson-Houston. For example, Walter C. Baker invented his electric automobile there, organizing the Baker Motor Vehicle Company in 1898 with the assistance of his father-in-law, Rollin C. White, one of the founders of the White Sewing

Machine and Cleveland Machine Screw Companies.[33] Around the same time, Elmer Sperry developed his own electric vehicle at the Brush facility. Sperry had originally been enticed to Cleveland and to Brush by Washington Lawrence and other major investors associated with National Carbon. Collectively known as the Sperry Syndicate, the group contracted with Sperry in 1890 to develop a prototype for an electric streetcar, promising that if the prototype proved workable, the syndicate would either form its own company to build the cars or sell or license the patents to another company that would. The arrangement was very early-stage financing. Although Sperry already had some patents in the area, he had not yet developed a working model. Sperry developed his streetcar over the next couple of years, and in 1892 the syndicate arranged to exploit the invention in a joint venture with the Thomson-Houston Electric Company (which a few months later became General Electric). The resulting Sperry Electric Railway Company contracted to pay Sperry a lucrative salary as consultant in addition to a share in the company's profits.[34]

When Sperry got interested in the idea of an electric automobile a few years later, he again turned to the syndicate for financial help. Their backing provided the support he needed to develop his vehicle (at Brush), which was then licensed to the Cleveland Machine Screw Company (Sperry received stock in the company and the position of electrical engineer). In 1900, the American Bicycle Company bought this business, along with Sperry's patents, and the next year assigned Sperry's electric storage battery inventions to yet another new company, the National Battery Company. Sperry helped to get this enterprise up and running and served for a short time as its general manager (Hughes 1971, Hritsko 1988, Wager 1986).

The Limited Role of Formal Financial Institutions

Completely missing from the foregoing narrative is any role for formal financial institutions in the founding of the original Brush Electric Company or the many start-ups and spin-offs that came out of Brush cluster. The entrepreneurs who organized and promoted these new ventures secured investment capital largely by relying on personal connections. These could be familial, as when the father of Eugene and Alfred Cowles provided much of the initial capital for the Cowles Electric Smelting and Aluminum Company; they could result from friendships, as when George Stockly agreed to support Brush's initial work in electrical lighting; or they could be based on the recommendations of men who had established their expertise in the community, as when Brush secured backing for the Linde Air Products Company simply by assuring local businessmen of the merits of the technology. Association with a hub enterprise such as Brush could in and of itself be a means of attracting both attention and funds. Thus, Bentley and Knight, as well as Short, were able to use their very visible association with Brush to raise local capital for their streetcar companies.

The wealthy Clevelanders who bought shares in these new high-tech enterprises seem to have been motivated by the returns they expected to earn from owning and holding them rather than the profits they could reap by selling them off after an initial run-up in price. Although a few investors cashed out their investments relatively early (as Lawrence did when he sold off his Brush stock), the practice seems to have been uncommon. A search of Cleveland newspapers reveals, for example, that from the time of the formation of the Brush Electric Company until the late 1880s, when the idea of selling or merging the firm was beginning to be discussed, the only mention of Brush shares available on the market occurred around the time Lawrence was selling out.[35] Before the formation of the CSE in 1900, the only firms associated with the Brush network for which share prices were quoted in the Cleveland papers were Brush Electric itself and the Walker Manufacturing Company. Even after the formation of the exchange, we do not see much trading in the equities of concerns associated with this hub. The one major exception, National Carbon, was listed on the exchange from the very beginning, but by that time it was a consolidation of a large number of previously competing firms.

This is not to say that formal financial institutions played no role at all in the financing of these firms. Once enterprises associated with the hub got off to a good start, they were undoubtedly able to tap into other sources of finance, particularly trade credit from suppliers but also short-term commercial loans from banks and other financial intermediaries. Here the network of financiers, which overlapped with that of inventors, was likely to have proved important. Some of the men who invested their savings in the new enterprises were also officers and directors of banks. For example, James J. Tracy, one of the original incorporators of Brush Electric, became vice president of the Society for Savings after a long career in various other Cleveland financial institutions.[36] Similarly, Myron T. Herrick, a member of the Sperry Syndicate and one of the initial investors in National Carbon, was secretary-treasurer and then president of the Society for Savings, a founder of the Euclid Avenue National Bank, a director of the American Exchange National Bank, and a director of the Garfield Savings Bank.[37] Some of the inventors and other businessmen involved in such start-ups and spin-offs also helped to organize financial institutions during this period—Brush himself was a founder and vice president of the Euclid Avenue National Bank (Eisenman 1967)—and it is likely that they did so because they thought their companies would benefit. But it is important to emphasize the secondary role of such formal financial institutions. The primary role was played by local businessmen who hoped to replicate the experience of the investors who had made so much money from their initial stake in Brush Electric. Moreover, the information networks that formed around the Brush enterprise helped to convince them that they could invest in these cutting-edge enterprises with some assurance of success.

Beyond Brush

In Cleveland the Brush Electric Company was one of the earliest and most important examples of an enterprise around which extensive overlapping networks of inventors and financiers formed. But it was not the only such enterprise. In the machine tool sector, for example, the White Sewing Machine Company played a similar role, assisting complementary enterprises and spinning off a host of new firms. The founder, Thomas H. White, had moved his small sewing machine business to Cleveland in the late 1860s, and in combination with Howard W. White (his half-brother) and Rollin C. White (no relation) formed what later became known as the White Sewing Machine Company.[38] Once the firm was well established, White began to encourage vertically related ventures. For example, he convinced the fledgling precision machine tool firm of Warner and Swasey to move from Chicago to Cleveland. White helped the two partners to find a suitable location for their shop, fed them information about potential customers, and assured them that "if anyone undertakes to squeeze you, let me know, and I will see that they don't." He also stepped forward when they needed a respected business-man to guarantee a contract.[39] White played an even more important role in the success of Theodor Kundtz's furniture company, providing financial backing as well as buying the major part of his output. At that time, sewing machine furniture consisted of little more than tables on which the machines were bolted. With White's support, Kundtz inno-vated by designing and patenting convertible tables that, when not in use for sewing, closed up to become attractive pieces of furniture.[40]

Like many other enterprises using machine tool technology, the White Sewing Machine Company made a number of products over the years besides sewing machines, including kerosene street lamps, roller skates, phonographs, bicycles, and precision tools. In 1890 the Whites spun off their machine tool business as a separate concern, the Cleveland Ma-chine Screw Company, headed by Rollin C. White. This firm also diver-sified its output (Rose 1950). Thomas White's son, Rollin H. White, was a gifted inventor who had double-majored in mechanical and electrical engineering at Cornell. Rollin and his brother Windsor (who also had engineering training) developed a type of safety bicycle while working in the Screw Company's employ. The company had acquired a local bicycle stamping concern, the A. L. Moore Company, and it produced the bicycles until it sold off this part of its business to the American Bicycle consolidation in 1898.[41] In the late 1890s, the company also moved into automobiles when Elmer Sperry arranged for it to produce the electric car he had designed at Brush, assigning Cleveland Machine Screw his patents in exchange for shares of its stock and agreeing to assume the po-sition of electrical engineer. This business, along with Sperry's patents, was sold to American Bicycle in 1900 (Hughes 1971, Hritsko 1988, Wager 1986).

Around the same time, Walter C. Baker founded his own company to produce the electric vehicles he had designed in a Brush shop. Financial backing for the company came from his father-in-law, Rollin C. White, president of Cleveland Machine Screw Company. Baker's father, George W. Baker, had been an inventor and long-time employee of the White

Sewing Machine Company. Walter attended the Case School of Applied Science and after a brief stint as a civil engineer returned to work at the Screw Company. There he invented a revolutionary antifriction ball bearing that could be used for bicycles, carriages, and automobiles, and with the assistance of his father-in-law and several other men, he organized the American Ball Bearing Company in 1895, the same year he received his patent.[42]

When Thomas White's son, Rollin H., developed a new kind of flash boiler for steam vehicles in 1899, the White Sewing Machine Company added the production of automobiles to its already diversified product line. The vehicles proved so successful that the Whites spun off production into a separate automobile concern, the White Company, in 1906.[43] Bowing to trends in popular demand, in 1909 the company began producing gasoline vehicles, the main components of which were designed by other companies, and began to phase out the production of steam cars in 1911. Forced to spend more of his time simply managing production, Rollin had comparatively little outlet for his creativity and was stimulated by a visit to a Hawaiian plantation owned by another brother, Clarence, to turn his energies toward designing agricultural equipment. He invented the first crawler-type tractor and, with Clarence's help, founded the Cleveland Motor Plow Company in 1916 (later renamed the Cleveland Tractor Company and shortened to Cletrac). Once Cletrac's success was ensured, he founded another car company, the Rollin Motor Company, in 1923. That venture lasted only a few years, though the cars it produced embodied notable technological advances.[44]

By the time the White Company was organized, the formal financial institutions that Cleveland's industrialists had helped to create were more accessible to new high-tech start-ups and spin-offs. Hence the Whites' automobile venture was listed on the CSE in 1912, just six years after its formation. Nonetheless, the role played by such institutions in the creation and promotion of new firms was still relatively minor. As in the case of the firms associated with the Brush hub, early investment still came primarily through local informal channels. Family and friends played a prominent role, as did upstream or downstream enterprises that had special reasons to encourage the development of complementary businesses. Other significant amounts came from businesspeople in the local community who were eager to follow the example of those who had gotten rich from investing in cutting-edge technologies. Here the networks that formed around firms like White played a critical role. Because they were collecting points for technological expertise, they served an important vetting function. Inventors seeking validation for their ideas gravitated to these hubs. So did businesspeople in search of profitable investments.

Quantitative Patterns in Cleveland and Beyond

In order to provide a more systematic picture of the ways in which Cleveland inventors financed the creation and exploitation of new technological knowledge, we collected data from the *Annual Reports of the Commissioner of Patents* on patents awarded to the city's inventors over three three-year periods: 1884, 1885, and 1886; 1898, 1900, and 1902; and 1910, 1911, and 1912. Focusing our attention on the most productive of the patentees, we selected from the full set of Cleveland inventors those who patented a minimum number of patents during each set of years. We then tracked these productive inventors through city directories and other sources to learn their occupations. We collected all of their patents for specific sets of years,[45] categorizing them according to whether they were assigned at issue and to whom. Assignments to companies were classified in the following way. We first checked to see whether the assignee was a company in which the patentee was an officer or proprietor. If not, we classified as "national companies" all assignees for which financial information was reported in the *Commercial and Financial Chronicle* or in *Poor's* or *Moody's Manual of Industrial Securities* (indicating that the company was important enough to tap the national capital markets), that had the word *National* in the name of the firm, or that were listed in an early-1920s National Research Council directory of companies with research laboratories. The remaining companies were divided into two groups: other "local companies," if the firm was located in Cleveland (or elsewhere in Cuyahoga County), and "other companies" for the rest.

As we have shown, once key enterprises like the Brush Electric Company or White Sewing Machine proved successful, they attracted inventors and investors in a fertile mix that led to the formation of additional enterprises. One would expect, therefore, that once the resulting technology boom got under way, it would have been easier for talented inventors to obtain the financial backing they needed to exploit their ideas commercially and continue their inventive activities. The data on assignments at issue in table 1.2 are consistent with this expectation. By the turn of the century, productive Cleveland patentees were assigning more than 50 percent of their patents at issue. Moreover, a third of the patents they assigned went to firms in which they were principals. The proportions for 1910–1912 were slightly higher, suggesting that the technology boom was continuing apace, at least through the pre–World War I period.

The table also suggests, however, that an increasing proportion of the patents granted to productive Cleveland inventors were being assigned to other kinds of firms, particularly to the large-scale enterprises that were rising to national prominence during this period. We can gain insight into how the inventors who assigned their patents to large national firms may have differed from those who assigned to companies in which they were principals by breaking the patentees down according to the total number of patents they obtained in the sample years and whether they were officers or proprietors of firms.[46] As Tables 1.3A and 1.3B show, the two groups of inventors were quite distinct. Patentees who were principals were unlikely to assign their patents to large national firms, whereas the overwhelming majority of patents assigned to such firms (over 90 percent for the 1898–1902 sample) came from inventors with more than fifteen patents who were not principals in firms.[47] This last group of patentees differed from other similarly productive inventors in another striking way as well. Those who were not principals assigned a much larger proportion of their patents at issue. If we focus our attention only on patentees from the 1898–1902 group who obtained more than fifteen patents in the years sampled, those who were not principals assigned fully three-quarters of their patents at issue, whereas the figure for principals was less than 40 percent. In other words, members of the

Table 1.2
Distribution of Cleveland patents by assignee type

Type of assignee, if any	1884–1886 sample	1898–1902 sample	1910–1912 sample
Not assigned at issue			
Number	306	395	271
Percent	(77.7)	(47.1)	(44.8)
Assigned to individual			
Number	33	30	27.5
Percent	(8.4)	(3.6)	(4.5)
Assigned to company where patentee is principal			
Number	6	148	118.5
Percent	(1.5)	(17.6)	(19.6)
Assigned to national company			
Number	5	95	121
Percent	(1.3)	(11.3)	(20.0)
Assigned to local company			
Number	25	77	58
Percent	(6.3)	(9.2)	(9.6)
Assigned to other company			
Number	19	90	10
Percent	(4.8)	(10.7)	(1.7)
Total number of patents	394	839	606

Notes and sources: The 1884–1886 sample (42 patentees) comprises inventors who were Cleveland residents and received three or more patents during 1884, 1885, and 1886 (except for John Walker whose name was too common for us to make precise matches). It includes the patents they were awarded in those years, as well as in 1881, 1882, 1888, and 1889. The 1898–1902 sample (36 patentees) consists of inventors who were Cleveland residents, obtained a patent in 1900, and had a total of at least three patents in 1898, 1900, 1902, plus several inventors resident in Cleveland and prominent enough to be profiled in the *Dictionary of American Biography*. The patent record for this sample consists of all patents the inventors were awarded in 1892 through 1912, except for the years 1895, 1901, and 1904. The 1910–1912 sample (107 patentees) consists of inventors resident in Cleveland who received a patent in 1912 and at least three patents during 1910, 1911, and 1912. The patent record for this sample consists of their total patents for these three years.

Table 1.3
Distribution of patents by assignee type, patentee productivity, and relationship
to assignee

	Category of patentee				
	15 or fewer patents and principal	15 or fewer patents and not a principal	More than 15 patents and principal	More than 15 patents and not a principal	Total
A. 1884–1886 Cleveland sample[a]					
Number of patentees	14	21	5	2	42
Type of assignee					
Not assigned					
Number	76	87	131	12	306
Percent	(90.5)	(63.0)	(97.0)	(32.4)	(77.2)
Assigned to individual					
Number	4	26	0	3	35
Percent	(4.8)	(18.8)	(0.0)	(8.1)	(8.8)
Assigned to company where patentee is principal					
Number	3	0	3	0	6
Percent	(3.6)	(0.0)	(2.2)	(0.0)	(1.5)
Assigned to national company					
Number	1	4	0	0	5
Percent	(1.2)	(2.9)	(0.0)	(0.0)	(1.3)
Assigned to local company					
Number	0	18	1	6	25
Percent	(0.0)	(13.0)	(0.7)	(16.2)	(6.3)
Assigned to other company					
Number	0	3	0	16	19
Percent	(0.0)	(2.2)	(0.0)	(43.2)	(4.8)
Total number of patents	*84*	*138*	*135*	*37*	*394*
Percentage of total patents in sample	*(21.3)*	*(35.0)*	*(34.3)*	*(9.4)*	*(100.0)*

Table 1.3
(continued)

	Category of patentee				
	1–5 patents	6–15 patents	More than 15 patents and principal	More than 15 patents and not a principal	Total
B. 1898–1902 Cleveland sample[b]					
Number of patentees and number who are principals	6 patentees, 1 principal	9 patentees, 5 principals	13 patentees, all principals	7 patentees, no principals	35 patentees, 19 principals
Type of assignee					
Not assigned					
Number	9	41	269	76	395
Percent	(60.0)	(49.4)	(61.1)	(25.3)	(47.1)
Assigned to individual					
Number	2	6	14	8	30
Percent	(13.3)	(7.2)	(3.2)	(2.7)	(3.6)
Assigned to company where patentee is principal					
Number	0	21	119	11	151
Percent	(0.0)	(25.3)	(27.0)	(3.7)	(18.0)
Assigned to national company					
Number	0	0	9	86	95
Percent	(0.0)	(0.0)	(2.0)	(28.7)	(11.3)
Assigned to local company					
Number	4	14	27	34	79
Percent	(26.7)	(16.9)	(6.1)	(11.3)	(9.4)
Assigned to other company					
Number	0	1	2	85	88
Percent	(0.0)	(1.2)	(0.5)	(28.3)	(10.5)
Total number of patents	*15*	*83*	*440*	*300*	*838*
Percentage of total patents in sample	*(1.8)*	*(9.9)*	*(52.5)*	*(35.8)*	*(100.0)*

Table 1.3
(continued)

---

[a] Many patents are credited to John Walker during the years 1884, 1885, and 1886. Because *John Walker* was a common name, however, and a number of different individuals with that name appear in the Cleveland city directories, we were not able to reliably identify which individuals received the various patents. Hence we excluded the patents credited to Walker in computing the estimates in this table.

[b] The small number of assignments made by patentees classified as nonprincipals to firms in which the patentee was a principal involve cases where the patentee's status as principal was brief.

Source: See table 1.2.

latter group seem (like Sidney Short) to have been able to retain a striking degree of independence from the investors who financed their firms.

The same pattern of increasing assignments to firms in which the patentee was a principal and to large national companies is apparent in a nationwide sample of more than five hundred patentees whose surnames began with the letter B.[48] The sample was generated by first drawing three random cross-sectional samples from the lists of patents reported in the *Annual Reports of the Commissioner of Patents* for the years 1870–1871, 1890–1891, and 1910–1911, and then, for those patentees whose last names began with B, collecting data on all the patents they received during the twenty-five years before and after they appeared in our samples, including information on whether and to whom the patents were assigned at issue. Table 1.4 divides the inventors into three subsamples (according to the cross-section from which they were originally drawn) and also by the number of patents they obtained over their careers. It then categorizes each of their patents according to whether it was assigned at issue, and if it was, whether the assignment was in whole or part to an individual or to a company. Assignments to companies were classified in the same way as for the Cleveland sample, with the exception of the category for assignments to companies in which the patentee was a principal. Because we had less personal information about the patentees in the national sample, we restricted this category to companies that bore the patentee's surname.

Perhaps the most striking feature of the estimates presented in table 1.4 is the widening contrast between the assignment behavior of the

Table 1.4
Distribution of patents by assignee type, patentee subsample, and number of career patents

Type of assignee	Categories of patentees by career patents			
	1–2 patents (col. %)	3–5 patents (col. %)	6–9 patents (col. %)	10 or more patents (col. %)
Not assigned				
1870–1871 subsample	82.4	88.6	87.7	75.3
1890–1891 subsample	72.9	70.5	60.6	45.6
1910–1911 subsample	85.0	78.1	57.5	39.3
Assigned to individual				
1870–1871 subsample	13.2	8.6	6.6	14.3
1890–1891 subsample	12.9	20.1	19.2	13.5
1910–1911 subsample	9.0	9.7	7.5	6.0
Assigned to company with same name				
1870–1871 subsample	0.0	0.0	0.0	1.7
1890–1891 subsample	0.0	1.6	3.7	6.1
1910–1911 subsample	0.0	0.0	5.8	24.6
Assigned to national company				
1870–1871 subsample	0.0	0.0	0.8	1.2
1890–1891 subsample	1.4	0.0	0.5	9.9
1910–1911 subsample	0.0	1.9	0.0	14.8
Assigned to other local company				
1870–1871 subsample	1.5	0.7	2.5	4.5
1890–1891 subsample	10.0	3.9	5.3	15.9
1910–1911 subsample	1.5	3.9	15.8	8.4
Assigned to other company				
1870–1871 subsample	2.9	2.1	2.5	2.9
1890–1891 subsample	4.3	3.9	10.6	9.0
1910–1911 subsample	3.9	6.5	13.3	7.0
Number of patents				
1870–1871 subsample	*68*	*140*	*122*	*749*
1890–1891 subsample	*80*	*129*	*188*	*2,060*
1910–1911 subsample	*133*	*155*	*120*	*1,777*

Table 1.4
(continued)

Source: The table is based on a longitudinal data set constructed by selecting all of the inventors whose family names began with the letter *B* from three random cross-sectional samples drawn from the *Annual Reports of the Commissioner of Patents* for the years 1870–1871, 1890–1891, and 1910–1911. We then collected from the *Annual Reports* all of the patents received by these inventors during the twenty-five years before and after they appeared in the samples, including information on whether and to whom the patents were assigned at issue. Companies to which the patentees assigned their inventions were classified as follows. We first checked to see whether the assignee was a company with the same name as the patentee. If not, we classified as "large integrated companies" all assignees for which financial information was reported in the *Commercial and Financial Chronicle* or in *Poor's* or *Moody's Manual of Industrial Securities* (indicating that the company was important enough to tap the national capital markets) or, alternatively, that were listed in an early 1920s National Research Council directory of companies with research laboratories. The remaining companies were divided into two groups: "other local companies," if the assignee was located in the same city as the patentee, and "other companies" for all the rest.

most specialized or productive patentees (those with ten or more patents over their careers) and inventors who obtained only a small number of patents over the fifty years we followed them. By the 1910–1911 subsample, the specialized inventors were assigning more than 60 percent of their patents at issue, while inventors with five or fewer career patents were assigning only about a fifth. The identities of the assignees to whom these two groups of inventors transferred their patents also diverged increasingly over time. Whereas inventors with five or fewer patents continued to assign mainly to individuals, often maintaining shares of the patents themselves, inventors with ten or more patents were overwhelmingly assigning their patents to companies. Moreover, by the third subsample, 24.4 percent of the patent assignments made by this group of productive inventors went to large, integrated companies and 40.5 percent to firms that bore the name of the patentee. Indeed, virtually all of the patents assigned to large integrated companies came from inventors with ten or more career patents, and the share of this productive group in patents assigned to companies that bore the same name as the inventor was almost as high.

As table 1.5 shows, these patterns had a pronounced regional character. Inventors, especially highly productive ones, in the Midwest (East North Central states) were disproportionately likely to assign their patents at issue to companies that bore their names: 56.7 percent of their assignments went to such firms and only 7.4 percent to large, integrated enterprises.[49] The pattern in the Middle Atlantic was just the opposite, with 36.2 percent of assignments going to the large national firms and only 4.4 percent to companies that shared the name of the inventor. New England was an intermediate case, with 35.4 percent of assignments going to the larger enterprises and an equivalent number to firms with the same name as the inventor. Although regional differences in industrial composition might in principle account for these disparities, the same qualitative pattern holds when we control for the sectoral classification of the patents.[50]

The national data show more clearly than the Cleveland samples that the increased tendency to assign to large firms and to firms in which the inventor was likely to be a principal represented a significant break in trend. As Lamoreaux and Sokoloff have shown in previous work (1999a, 1999b, 2005, forthcoming), during the middle decades of the nineteenth century, trade in patent rights boomed, and productive inventors responded by specializing in the generation of new technological ideas, extracting their returns by selling off the patent rights to individuals or firms better positioned to exploit them commercially. As a result, they exhibited a high degree of contractual mobility, assigning their patents to many different parties over the course of their careers. By the turn of the century, however, ongoing technological change had raised the amount of capital required for technological discovery, and inventors found it increasingly difficult to maintain such a high degree of independence. As table 1.6 indicates, after rising between the subsamples of 1870–1871 and 1890–1891, the proportion of patentees who assigned their patents at issue to four or more distinct assignees over their careers decreased between the 1890–1891 and 1910–1911 groups. The fall was greatest in the Middle Atlantic, where the proportion of assignments going to large, national enterprises was highest. Intriguingly, contractual mobility actually increased in the Midwest, where inventors were most likely to assign their patents to firms that bore their names. As we have

Table 1.5
Distribution of patents by assignee type, subsample, and region

Type of assignee	New England (col. %)	Middle Atlantic (col. %)	East North Central (col. %)
Not assigned			
1870–1871 subsample	76.1	75.6	83.0
1890–1891 subsample	24.7	58.1	51.3
1910–1911 subsample	35.0	38.1	44.6
Assigned to individual			
1870–1871 subsample	14.3	13.8	10.6
1890–1891 subsample	11.6	9.8	23.1
1910–1911 subsample	8.9	5.2	5.2
Assigned to company with same name			
1870–1871 subsample	0.6	2.3	0.5
1890–1891 subsample	3.4	5.0	6.8
1910–1911 subsample	23.0	2.7	31.4
Assigned to national company			
1870–1871 subsample	0.0	0.0	0.0
1890–1891 subsample	15.5	9.4	3.8
1910–1911 subsample	23.0	22.1	4.1
Assigned to other local company			
1870–1871 subsample	7.5	3.9	1.0
1890–1891 subsample	30.8	9.5	10.6
1910–1911 subsample	3.7	8.2	8.4
Assigned to other company			
1870–1871 subsample	1.6	4.4	0.0
1890–1891 subsample	14.1	8.2	4.4
1910–1911 subsample	6.5	23.8	6.4
Number of patents			
1870–1871 subsample	*322*	*434*	*218*
1890–1891 subsample	*555*	*947*	*707*
1910–1911 subsample	*383*	*601*	*1,050*

Notes and sources: See table 1.4.

Table 1.6
Change over subsamples in patentees' contractual mobility, by region

Number of different assignees	New England (col. %)	Middle Atlantic (col. %)	East North Central (col. %)	Other U.S. (col. %)	n
No assignees					
1870–1871 subsample	51.2	64.8	60.5	57.1	87
1890–1891 subsample	25.7	42.4	36.7	41.7	69
1910–1911 subsample	34.8	48.4	55.0	60.4	111
One assignee					
1870–1871 subsample	39.0	18.5	26.3	14.3	38
1890–1891 subsample	28.6	12.1	38.3	33.3	49
1910–1911 subsample	17.4	29.0	13.8	27.1	46
Two to three different assignees					
1870–1871 subsample	2.4	9.3	5.3	21.4	11
1890–1891 subsample	25.7	25.8	16.7	12.5	39
1910–1911 subsample	34.8	12.9	18.8	10.4	36
Four or more different assignees					
1870–1871 subsample	7.3	7.4	7.9	7.1	11
1890–1891 subsample	20.0	19.7	8.3	12.5	28
1910–1911 subsample	13.0	9.7	12.5	2.1	20
Number of patentees					
1870–1871 subsample	*41*	*54*	*38*	*14*	*147*
1890–1891 subsample	*35*	*66*	*60*	*24*	*185*
1910–1911 subsample	*23*	*62*	*80*	*48*	*213*

Note: Each patentee is represented in the table by one patent that was randomly selected from his list of career patents.
Source: See table 1.4.

already seen for Cleveland patentees, inventors who were able to attract investment in enterprises formed to exploit their inventions seem to have been able to maintain a greater degree of independence than other patentees in that they continued to control after issue a higher proportion of the patents they obtained.

As might be expected, inventors found it easier to obtain financial backing for new enterprises if they could demonstrate early in their careers that they had the "right stuff"—that is, the ability to generate economically valuable ideas. In table 1.7 we look at how the patenting and assignment behavior of inventors with ten or more patents evolved over their careers, using the date of the earliest patent to mark the beginning of each inventor's career.[51] Evidence that it was becoming more difficult for patentees to establish themselves and secure sufficient capital to support their inventive activity is provided by the dramatic change that occurred over time in the proportion of patents obtained at different stages of their careers. For specialized inventors in the 1870–1871 subsample, the patents were spread fairly evenly over time, with about a third obtained in the early stage of their careers, a third in the intermediate stage, and a third later on. By contrast, productive members of the 1910–1911 subsample obtained only about 15 percent of their patents in the first stage of their careers and more than 55 percent in the last.[52] Indeed, what is most striking about the evidence in the table is the disproportionate extent to which assignments to companies bearing the name of the inventor occurred in the later phase of the patentees' careers. For the last (1910–1911) subsample, only 1.3 percent of the patents obtained by inventors in the first five years of their careers went to companies that bore the name of the patentee. During the next ten years of their careers, the figure increased to 17.1 percent and, after fifteen years had elapsed, to 35.4 percent. The pattern was similar for assignments to large, integrated companies, though there is evidence that inventors who obtained credentials in the form of college degrees in science or engineering were able to secure this kind of employment earlier in their careers (Lamoreaux and Sokoloff 2005).

The findings on patenting trends in the nation as a whole shed additional light on the important role played by firms like the Brush Electric Company. In an environment in which it was increasingly difficult for inventors to raise the capital they needed to pursue new discoveries, such hub enterprises made it possible for those with promising ideas to stand out from the crowd and attract the financial backing they needed to carry out their inventive activity. The networks of inventors that formed around such firms not only performed an important vetting function for capitalists, they also made it possible for talented inventors to establish their credentials. It is no wonder that so many technologically creative individuals gravitated to such facilities.

Table 1.7
Distribution of patents by assignee type and stage of career

Type of assignee	5 years or less since first patent (col. %)	More than 5 years and 15 years or less since first patent (col. %)	More than 15 years since first patent (col. %)
Not assigned			
1870–1871 subsample	81.9	75.3	68.9
1890–1891 subsample	62.0	52.7	36.6
1910–1911 subsample	45.6	50.3	32.1
Assigned to individual			
1870–1871 subsample	10.3	18.1	14.4
1890–1891 subsample	16.1	16.5	11.0
1910–1911 subsample	14.1	7.1	2.8
Assigned to company with same name			
1870–1871 subsample	0.4	0.0	4.8
1890–1891 subsample	2.2	4.2	8.4
1910–1911 subsample	1.3	17.1	35.4
Assigned to national company			
1870–1871 subsample	0.0	0.0	0.0
1890–1891 subsample	7.1	6.3	12.9
1910–1911 subsample	12.1	7.3	19.2
Assigned to other local company			
1870–1871 subsample	6.6	3.1	7.6
1890–1891 subsample	8.4	15.1	18.6
1910–1911 subsample	17.1	11.7	4.1
Assigned to other company			
1870–1871 subsample	0.8	3.5	4.4
1890–1891 subsample	4.3	5.4	12.6
1910–1911 subsample	9.8	6.5	6.3
Number of patents			
1870–1871 subsample	*243*	*255*	*251*
1890–1891 subsample	*323*	*651*	*1,086*
1910–1911 subsample	*305*	*479*	*993*

Note: Only patents awarded to patentees with ten or more career patents are included in this table.
Source: See table 1.4.

Conclusion

By the turn of the century, inventors were finding it increasingly necessary to form long-term attachments with firms in order to continue their inventive activity. The conventional story is that they responded to the difficulties they faced in maintaining their independence by moving into employment positions in the R&D facilities that large-scale enterprises were beginning to build. Our data suggest that the conventional story may indeed be correct for inventors in the northeastern region of the country, especially the Middle Atlantic states, but that in the Midwest, the prevailing pattern was for inventors to become principals in companies formed to exploit their inventions. Why this regional pattern developed is beyond the scope of this chapter to explain. One obvious hypothesis is that the well-developed financial markets of the Northeast efficiently funneled the bulk of that region's savings into the large-scale enterprises increasingly headquartered there. The midwestern economy, however, was not yet fully integrated into national financial system (Davis 1965). It may well be that capital markets in this region, because they were more confined to local investment projects, offered greater opportunities to entrepreneurs seeking funds for start-ups to exploit new technologies. It may also be that boosters committed to building up their local economy (and their own enterprises in the process) helped to create an environment that encouraged wealthy midwesterners to invest locally.

Regardless, the main contribution of our study is to detail the channels through which these local pools of capital flowed into the large numbers of high-tech enterprises formed in cities like Cleveland during the late nineteenth and early twentieth centuries. We find that formal financial institutions played a supporting or secondary role—that venture capital was mainly mobilized more informally. We are able to go further, moreover, and show how such informal finance worked in actual practice. We thus highlight the dual role played by hub enterprises such as the Brush Electric Company. On the one hand, the money they made for their backers had an important demonstration effect, enticing local savers to risk their money in cutting-edge enterprises so that they would not miss out on such rich opportunities again. On the other hand, successful enterprises like Brush also had an information effect. They became gathering places for talented inventors, who not only stimulated each other's technological creativity but also served a useful vetting function for investors. The buzz they generated about which new technologies were most likely to work conveyed an enormous amount of useful information to the financiers who plugged into these networks. In the process, Brush and other similar enterprises sparked a technology boom in Cleveland that helped to propel it from a medium-size city specializing in resource handling and processing to one of the largest and most prosperous industrial cities in the nation.

Notes

We express our appreciation to Marigee Bacolod, Charles Balck, Lisa Boehmer, Nancy Cole, Homan Dayani, Yael Elad, Gina Franco, Svjetlana Gacinovic, Charles Kaljian, Kristen Kam, Paul Kim, Anna Maria Lagiss, Stephen Lamoreaux, Huagang Li, Catherine Truong Ly, Matthew Purdy, Iane Saenam, Yolanda McDonough Summerhill, Dhanoos Sutthiphisal, Jicky Thantrong, and Hong Tran for their outstanding research assistance, and to Sue Hanson, head of special collections at Case Western Reserve University's Kelvin Smith Library, and Ann Sindelar and other staff members at the Western Reserve Historical Society for their help and guidance. We also benefited greatly from discussions with Richard Baznik, Deidre Cooper, Virginia Dawson, Michael S. Fogarty, Gasper S. Garofalo, Stephen Haber, David Hammack, Carol Heim, Susan Helper, Monty Hindman, Thomas P. Hughes, Margaret Jacob, William Janeway, Max Kent, Zorina Khan, Steven Klepper, Timur Kuran, Josh Lerner, Andrea Maestrejuan, Deirdre McCloskey, Gary Previtz, Daniel Raff, Jean-Laurent Rosenthal, Joshua Stein, Richard Sylla, Ross Thomson, Manuel Trajtenberg, David Weiman, Gavin Wright, Mir Yarfitz, Luigi Zingales, and participants in presentations at Case Western Reserve University, New York University, Stanford University, Universidad Carlos III, the University of Massachusetts, the University of Pennsylvania, Utrecht University, the Business History Conference, the NBER, and the SSRC Conference on the Finance of Innovation. Finally, we gratefully acknowledge the financial support we have received for this research from the Social Science Research Council, the National Science Foundation, and the Academic Senate at the University of California, Los Angeles.

1. A good example is Alexander E. Brown, son of Fayette Brown, a prominent Cleveland merchant banker, iron dealer, and manufacturer. Born in 1852, Alexander attended Brooklyn Polytechnical Institute in the early 1870s and then took a job from 1873 to 1874 as chief engineer with the Massillon (Ohio) Bridge Company. While in the company's employ, he invented a method of using scrap iron and steel to build bridge columns. Returning to Cleveland, he attempted to

pursue a career as an inventor, but found himself strapped for funds. As he complained to his older brother Harvey H. Brown in 1880, "I have spent so much time and money on this case, in what was necessary, but which . . . is only a loss or expense to me." He begged Harvey, an iron ore dealer, to help him defray the cost of obtaining the patent and also of acquiring the Canadian rights, promising him in exchange a quarter interest in the patent. As he explained, "I have my Electric Lamp patents to get yet, and they will cost like 'sin' for I will have to get English and other patents for them." Apparently family members were not convinced of the value of the inventions and were not willing to provide much support for Alexander's endeavors in electrical lighting. His invention of a hoisting machine was another story, however. His father, himself an accomplished inventor, recognized its potential and stepped in to organize the Brown Hoisting & Conveying Machine Company with a capital stock of $100,000. Letter from Alexander E. Brown to Harvey H. Brown, July 30, 1880, Container 1, Folder 1, Harvey Huntington Brown Papers, 1848–1923, Mss. 3342, Western Reserve Historical Society Manuscript Collections; "Brown, Alexander Ephraim" and "Brown, Fayette," *Encyclopedia of Cleveland History*; "Brown, Fayette," *Dictionary of American Biography*; and Rose (1950).

2. There is an enormous literature on the role of familial and other personal connections (or more generally networks) in finance. The basic idea underpinning this literature is that people are better able to judge the creditworthiness of those with whom they are well acquainted or connected than they are strangers. Scholars, however, have not taken the further step that we take here of asking how those with funds make decisions about whether to commit resources (and how much to commit) to a friend or relative who is otherwise trustworthy but is setting up a venture at the technological cutting edge.

3. Cleveland was also the home of Standard Oil, and petroleum refining was the city's third largest manufacturing industry in 1890. As crude oil production shifted to other regions, however, the city's refining capacity declined.

4. "Banks and Savings and Loans," *Encyclopedia of Cleveland History*; "Banks and Finance: The Solid Institutions of a Prosperous City," *Cleveland Plain Dealer*, July 14, 1892, 13.

5. *Cleveland Plain Dealer*, July 14, 1892, 13.

6. *Cleveland Plain Dealer*, July 14, 1892, 1, 6; "Eberhard Manufacturing Company," "M. A. Hanna Company," and "Marcus Alonzo Hanna," *Encyclopedia of Cleveland History*; Croly (1965).

7. "Our Prominent Men: Leading Spirits of the Younger Financial Institutions," *Cleveland Plain Dealer*, March 22, 1887, 1. See also July 7, 1892, 13.

8. Minutes of the Board of Directors of the Cleveland Trust Company, 1903–1906, Folder 125, Container 12, Ameritrust Corporation Records, Ms. No. 4750, Western Reserve Historical Society Manuscript Collections.

9. *Cleveland Stock Exchange Handbook*, 1903; Minutes of the Detroit Street Savings & Loan, 1895–1901, Folder 1, Container 1, Ameritrust Corporation Records.

10. For example, in May 1904, Cleveland Trust underwrote $15,000 in bonds for the Addresso Printograph Company and $10,000 in bonds for the Brilliant Electric Company. In March 1905, it underwrote $30,000 in bonds for the Whiting Electric Company. Cleveland Trust regularly underwrote much larger bond issues for more established local enterprises, however. For example, for the Wellman-Seaver-Morgan Company, it underwrote $800,000 in bonds in June 1903, $400,000 in October 1903, and $1.1 million in October 1905. Reports of the Cleveland Trust Company's Trust Department, 1903–1906, Folder 125, Container 12, Ameritrust Corporation Records.

We have found one example of an institution (the Fairmount Savings Bank) accepting shares of new firms as collateral for loans to individuals. Of the nearly 150 loans approved by the savings bank from July 1903 to November 1904, approximately 20 percent were backed by equities. Of these, about 50 percent were issued by small manufacturing enterprises. Journal of the Finance Committee of the Fairmount Savings Bank (1903), Folder 12, Ameritrust Corporation Records.

11. *Cleveland Plain Dealer*, October 5, 1900, 5.

12. *Cleveland Stock Exchange Handbook*, 1914. Daily trades on the exchange were reported once a week in the Cleveland journal *Finance*.

13. A more expensive method was to hire a patent lawyer to investigate the merits of new technologies offered for sale. See Lamoreaux and Sokoloff (2003).

14. Stockly (1901); letter from W. S. Culver to W. D. Stockley, April 24, 1928, Box 4, Folder 5, Charles F. Brush Collection, Kelvin Smith Library, Case Western Reserve University; Kennedy (1885); Orth (1990); Rose (1950); Cooper and Schmitz (1993); "Leggett, Mortimer Dormer," and "Tracy, James Jared, Sr.," *Encyclopedia of Cleveland History*.

15. Another inventor who worked on electrical devices at the shop was Alexander E. Brown, whose hoisting machine invention revolutionized the handling of cargo on the Great Lakes. Rose (1950).

16. Typescript of address by Charles F. Brush to the Franklin Institute, April 18, 1929, Box 9, Folder 11, Charles F. Brush Collection; Stockly (1901); Kennedy (1885); Brush (1905); Gorman (1961); Eisenman (1967); and "Brush, Charles Francis," "The Brush Electric Co.," and "Electrical and Electronics Industries," *Encyclopedia of Cleveland History*. See also the following documents: "Agreement between Charles F. Brush and Telegraph Supply Company," June 7, 1876, "Agreement between Charles F. Brush and Telegraph Supply Company," March 24, 1877, and "Memorandum of Agreement between Telegraph Supply Company and Charles F. Brush," March 24, 1877, all in Box 21, Folder 12, Charles F. Brush Collection.

Intriguingly, other boyhood friendships did not have similar benefits. For example, Brush lived during his high school years at the same boarding house as John D. Rockefeller, who despite his great wealth does not seem to have been a supporter of technological innovation. (In 1898 Elmer Sperry took Rockefeller for a ride in the electric automobile he had recently invented. Rockefeller refused to take over the controls and advised Sperry afterward to move on to other things.) Later, however, Brush would join Frank Rockefeller in a real estate venture. Eisenman (1967); Hughes (1971).

17. Brush (1905); Gorman (1961); Cox (1951). Subsequent negotiations between Brush and the company suggest that only half of the authorized capital was ever paid in. Brush himself did not take stock in the company until a number of years later. Instead, he agreed to assign all of his existing and future lighting system patents to the enterprise in exchange for royalties. See "Memorandum of Supplementary Agreement between Brush and the Brush Electric Co. (formerly Telegraph Supply)," September 1, 1880, and "Agreement between Brush and the Brush Electric Company," July 27, 1886, Box 21, Folder 12, Charles F. Brush Collection.

18. As part of the deal, Brush Electric's capital stock was reduced from $3 million to $1.5 million fully paid-in shares. The company then issued an additional $0.5 million in stock for Brush. In addition, the company promised that before making any dividends, it would pay Brush "an amount not less than one fourth part of the whole sum proposed to be divided, ... such payments to continue until [the company's] indebtedness shall be fully paid." "Agreement between Brush and the Brush Electric Company," July 27, 1886, Box 21, Folder 12, Charles F. Brush Collection. The royalty statements are in Box 15. Brush also sold his British and other foreign patents to the Brush Electric Light Corporation, Limited, in England for a "very large price." According to a report published in *Scientific American* (April 2, 1881, 211), "The sums paid for these foreign patents are ... greater than have ever been paid for any other foreign patents obtained by an American."

19. *New York Times*, January 21, 1890, 1; Eisenman (1967); Rose (1950).

20. See, for example, the testimony of Edward M. Bentley, a pioneer inventor of electric street railways: "My experience and observation has been that all electrical inventions since 1876 have excited unusual public interest and since Siemen's road in 1879 I think electric railways have had their full share of public interest in electrical matters. ... We had no particular difficulty [in interesting capitalists]. The men we most wanted were readily secured" (5). "Testimony Taken in Behalf of Walter H. Knight," Rudolph M. Hunter v. Walter H. Knight, Case 10,553, Interference Case Files, 1836–1905, Records of the Patent Office, Record Group 241, National Archives II.

21. *New York Times*, April 27, 1881, 12.

22. "Lawrence, Washington H." and "National Carbon," *Encyclopedia of Cleveland History*; Orth (1910).

23. See the following letters from Charles F. Brush: to Fred W. Wolf, February 15, 1907, Cecil Lightfoot, February 25, 1907, Carl Linde, February 25, 1907, and Carl Linde, August 5, 1907, Box 3, Folder 1, Charles F. Brush Collection.

24. See *Wall Street Journal*, February 15, 1917, 6, and October 3, 1917, 2; R. G. Dun & Company, *Mercantile Agency Reference Book: Ohio* (1910); Eisenman (1967).

25. This is another example of an ultimately unsuccessful venture that made a lot of money for its investors. The firm, which was organized in New York, was initially capitalized at $1 million, with much of this amount being issued in exchange for patents. Other early investors included officers of the Gramme Electric Company of New York. Brush Electric billed the new company for the equipment its employees built. See the testimony of Edward M. Bentley and Walter H. Knight in Rudolph M. Hunter v. Walter H. Knight, Interference Case 10,553. See also *New York Times*, September 9, 1884, 8, September 6, 1888, 8, September 1, 1889, 2, October 29, 1889, 1, and October 30, 1889, 9; Toman and Hays (1996); "Technology and Industrial Research," *Encyclopedia of Cleveland History*; Rose (1950).

26. Testimony of Eugene H. Cowles, Bradley and Crocker v. Cowles and Cowles, Case 12,615, Interference Case Files; *Biography of Alfred Hutchison Cowles* (1927); Cowles (1958); "Cowles Electric Smelting Furnace" (1886).

27. Matthiessen, who lived in Illinois, was invited by Alfred Cowles to invest in the firm, which he did after spending time in Cleveland observing the furnace in operation. Testimony of Frederick William Matthiessen, Lossier v. Willson v. Cowles & Cowles v. Rogers v. Darling v. Boguski v. Gratzel, Case 14,039, Interference Case Files. On Brush's reaction and subsequent use of the aluminum, see Cowles (1958), and *Scientific American*, May 15, 1886, 303. The Cowles brothers may also have gotten financial help from Eugene's father-in-law, the wealthy private banker E. B. Hale, before the marriage ended in a spectacularly messy divorce that made national news in 1890. See *New York Times*, June 10, 1890, 1, August 31, 1890, 1, December 24, 1890, 6.

28. The contracts licensing to Electro Gas patents owned by Alfred H. Cowles and the Electric Smelting and Aluminum Company are in the British Alcan collection, UGD 347/21/27/15, Glasgow University Archives. See also *Biography of Alfred Hutchison Cowles* (1927) and Cowles (1958).

29. For example, W. H. Bolton, founder of the Boulton Carbon Company (later National Carbon), was a foreman at Brush before organizing his spin-off enterprise in 1881. Funding for the venture was initially supplied by Willis Masters, son of Irvine U. Masters, owner of a prominent Cleveland shipbuilding firm, but the big infusion of capital owed to the efforts of Washington Lawrence, who managed the Brush enterprise during Boulton's period of employment. "Masters, Irvine U." and "National Carbon," *Encyclopedia of Cleveland History*; Rose (1950).

30. Lincoln started his firm with $200 or $250 (the accounts vary) that he earned from designing a motor for Herbert Dow. When he incorporated the firm in 1906, however, it was capitalized at $10,000. See Moley (1962); Dawson (1999); "Lincoln Electric Co." and "Reliance Electric Co.," *Encyclopedia of Cleveland History*.

31. Unfortunately, we have been unable to locate the employment contract between Short and the Walker Company. The assignment contract is dated July 1, 1896, but was not recorded at the Patent Office until January 5, 1898. See Liber

V-56, p. 322, Records of the Patent and Trademark Office, Record Group 241, National Archives II.

32. *New York Times*, November 24, 1897, 2, and September 18, 1898, 10; Moley (1962).

33. "The Baker Motor Vehicle Company," *Men of Ohio*; Wager (1986); "The Baker Materials Handling Co.," "Baker, Walter C.," and "Technology and Industrial Research," *Encyclopedia of Cleveland History*.

34. Hughes (1971); Cooper and Schmitz (1993); Rose (1950); "Electrical and Electronics Industries," *Encyclopedia of Cleveland History*.

35. *Cleveland Plain Dealer*, January 3, 1882, 8.

36. "Tracy, James Jared, Sr.," *Encyclopedia of Cleveland History*; Orth (1910).

37. "Herrick, Myron Timothy," *Encyclopedia of Cleveland History*; "Ohio Governors" (n.d.).

38. "Register"; and White Motor Company, "Important Milestones in White Motor History: Chronological Highlights of Present and Predecessor Organizations (1859–1949)," Container 4, Folder 39, Thomas H. White Family Papers Collected by Betty King, Ms. 4725, Western Reserve Historical Society Manuscript Collections. See also Hritsko (1988); Rose (1950); "White, Rollin Charles," and "White, Thomas H.," *Encyclopedia of Cleveland History*.

39. See the letters from Thomas H. White to Warner and Swasey dated January 12, 1881, January 28, 1881, May 25, 1881, and June 15, 1881, Box 36, Folder 3; Ambrose Swasey, "Address," May 19, 1920, Box 17, Folder 2; Reminiscences of Worcester R. Warner, January 11, 1927, Box 1, Folder 3, Warner and Swasey Collection, Kelvin Smith Library, Case Western Reserve University. Francis F. Prentiss, one of the partners with Jacob Cox in the firm that became the Cleveland Twist Drill Company, also helped to lure Warner and Swasey to Cleveland. See "Recollections of George D. Phelps," August 18, 1939, Box 20, Folder 7, and letter from Cox and Prentiss to Warner and Swasey, March 30, 1881, Box 36, Folder 3, Warner and Swasey Collection.

40. As his business grew, Kundtz expanded into new products from school desks to church pews to bicycle wheels, many of which were based on his own inventions. Later he also built automobile bodies for the Whites. By 1910, Kundtz headed a vertically integrated enterprise that employed 2,500 workers in five plants and was the largest consumer of hardwood in the state of Ohio. Eiben (1994); Hritsko (1988); "Kundtz, Theodor," *Encyclopedia of Cleveland History*; Rose (1950).

41. Report 3, Alice Lunn to Betty King, 29 December 1990, Container 4, Folder 34, Thomas White Family Papers. See also Hritsko (1988).

42. Orth (1910); "The Baker Motor Vehicle Company," *Men of Ohio*; Wager (1986); "The Baker Materials Handling Co." and "Baker, Walter C.," *Encyclopedia of Cleveland History*.

43. According to Hritsko (1988), Thomas H. White bought a steam car from Locomobile in 1899 and gave Rollin H. responsibility for maintaining it. Frustrated by the unreliability of the car's engine, Rollin developed an improved boiler and offered to sell his invention to Locomobile. When Locomobile refused to buy it, the Whites decided to develop their own car. See also White Motor Company, "Important Milestones in White Motor History"; "Twenty Years of Knowing How: Tracing the Development of The White Company and its Product Through Two Decades of Transportation Achievement," *Albatross* 9 (1921), pp. 4–5, in Container 4, Folder 39, Thomas H. White Family Papers; Wager (1986), 53–60; Rose (1950); "White, Rollin Henry," *Encyclopedia of Cleveland History*.

44. Letters from Henry Merkel to Betty King, January 4, 1991, and January 14, 1991; Report 3, Alice Lunn to Betty King, December 29, 1990; Report 10, Alice Lunn to Betty King, March 11, 1991; and photocopy, "28 Years of Constant Improvement behind Cletracs," Container 4, Folder 34, Thomas White Family Papers; Hritsko (1988); Wager (1986); Rose (1950).

45. See the notes to table 1.2 for the years included in each sample.

46. We broke down the data only for 1884–1886 and 1898–1902 because we did not collect patents for additional years for the 1910–1912 sample.

47. Assignments to national companies were disproportionately the work of productive inventors who were employees of those firms. For example, Clinton A. Tower, a foreman and pattern maker for the National Malleable Castings Company, assigned the bulk of the seventeen patents he obtained in the sampled years to his employer. National Malleable and Steel Castings Co. (1943).

48. For further discussion and analysis of these samples, see Lamoreaux and Sokoloff (1999a, 1999b, 2003, 2005, forthcoming).

49. Because the estimates in table 1.5 were computed over all of the patents received by the sample of *B* inventors, it is the behavior of the more productive inventors that is most clearly reflected in our results. They received the most patents and thus get more weight in these averages.

50. As our theory would predict, inventors whose patents were classified as being in sectors that would normally be considered as having more technical or capital-intensive technologies, such as electricity and telecommunications or heavy industry, were much more likely to assign their patents at issue and to make assignments to large firms or firms that shared their name, than those whose patents were in sectors such as light manufacturing or agriculture/food processing. The different patterns across sectors do not account for the regional differences, however. For example, inventors who received patents classified as electricity/telecommunications or as heavy industry were much more likely to assign them to large companies (as opposed to firms with the same name) if they resided in the Middle Atlantic than if they resided in the East North Central states.

51. Unfortunately, small cell sizes preclude a regional breakdown of this table.

52. These figures may somewhat understate the trend because we are likely to have undersampled patents for both the first career stage of the first subsample and for the last stage of the last subsample.

References

Adams, Stephen B., and Orville R. Butler. 1990. *Manufacturing the Future: A History of Western Electric.* Cambridge: Cambridge University Press.

Biography of Alfred Hutchinson Cowles, Sewaren, New Jersey: Scientist, Inventor, Economist. Pioneer in the Aluminum Process and in Electric Smelting. 1927. New York: Writers Press Association.

Brush, Charles F. 1905. "The Arc-Light." *Century Illustrated Monthly Magazine* 70 (May), 110–118.

Cleveland Stock Exchange Handbook. Cleveland, Ohio. Various issues.

Cooper, Hal D., and Thomas M. Schmitz. 1993. *A History of Inventions, Patents and Patent Lawyers in the Western Reserve.* Cleveland: Cleveland Intellectual Property Law Association.

Covington, Edward J. 1999. *Incandescent Lamp Manufacturers in Cleveland, 1884–1905.* Cleveland: Privately printed.

Cowles, Alfred. 1958. *The True Story of Aluminum.* Chicago: Henry Regnery Co.

"Cowles Electric Smelting Furnace." 1886. *Manufacturer and Builder* (February), 40–42 and (March), 64–65.

Cox, Jacob Dolson Sr. 1951. *Building an American Industry: the Story of the Cleveland Twist Drill Company and Its Founder.* Cleveland: Cleveland Twist Drill Co.

Croly, Herbert David. 1965. *Marcus Alonzo Hanna: His Life and Work.* Hamden, Conn.: Archon Books.

Cull, Robert J., and Lance E. Davis. 1994. *International Capital Markets and American Economic Growth, 1820–1914.* Cambridge: Cambridge University Press.

Davis, Lance E. 1965. "The Investment Market, 1870–1914: The Evolution of a National Market." *Journal of Economic History* 25, 355–399.

Dawson, Virginia P. 1999. *Lincoln Electric: A History.* Cleveland: Lincoln Electric Co.

Dictionary of American Biography. New York: Scribner's Sons, various years.

Dun, R. G. & Co. 1910. *Mercantile Agency Reference Book: Ohio.* New York: R. G. Dun & Co.

Eiben, Christopher J. 1994. *Tori in Amerika: The Story of Theodor Kundtz.* Cleveland: Ewald E. Kundtz, Jr.

Eisenman, Harry J. 1967. "Charles F. Brush: Pioneer Innovator in Electrical Technology." Ph.D. dissertation, Case Institute of Technology.

Encyclopedia of Cleveland History. Comp. David D. Van Tassel and John J. Grabowski. Bloomington: Indiana University Press, 1987. ⟨http://ech.cwru.edu/⟩.

Fogarty, Michael S., Gasper S. Garofalo, and David C. Hammack. N.d. *Cleveland from Startup to the Present: Innovation and Entrepreneurship in the 19th and Early 20th Centuries.* Cleveland: Center for Regional Economic Issues, Weatherhead School of Management, Case Western Reserve University.

Gorman, Mel. 1961. "Charles F. Brush and the First Public Electric Street Lighting System in America." *Ohio Historical Quarterly* 70, 128–144.

Hritsko, Rosemary Solovey. 1988. "The White Motor Story." Ph.D. dissertation, University of Akron.

Hughes, Thomas Parke. 1971. *Elmer Sperry: Inventor and Engineer.* Baltimore: Johns Hopkins University Press.

Israel, Paul. 1992. *From Machine Shop to Industrial Laboratory: Telegraphy and the Changing Context of American Invention, 1830–1920.* Baltimore: Johns Hopkins University Press.

Israel, Paul. 1998. *Edison: A Life of Invention.* New York: Wiley.

Kennedy, J. H. 1885. "The Brush Electric Light—The History of a Cleveland Enterprise." *Magazine of Western History* 3, 132–148.

Lamoreaux, Naomi R., and Kenneth L. Sokoloff. 1999a. "Inventors, Firms, and the Market for Technology in the Late Nineteenth and Early Twentieth Centuries." In Naomi R. Lamoreaux, Daniel M. G. Raff, and Peter Temin, eds., *Learning by Doing in Markets, Firms, and Countries* (19–57). University of Chicago Press.

Lamoreaux, Naomi R., and Kenneth L. Sokoloff. 1999b. "Inventive Activity and the Market for Technology in the United States, 1840–1920." Working paper W7107, NBER, Cambridge, Mass.

Lamoreaux, Naomi R., and Kenneth L. Sokoloff. 2003. "Intermediaries in the U.S. Market for Technology, 1870–1920." In Stanley L. Engerman et al., eds., *Finance, Intermediaries, and Economic Development* (209–246). Cambridge: Cambridge University Press.

Lamoreaux, Naomi R., and Kenneth L. Sokoloff. Forthcoming. "The Market for Technology and the Organization of Invention in U.S. History." In Eytan Sheshinski, Robert J. Strom, and William J. Baumol, eds., *Entrepreneurship, Innovation and the Growth Mechanism of the Free-Enterprise Economies.* Princeton, N.J.: Princeton University Press.

Lamoreaux, Naomi R., and Kenneth L. Sokoloff. 2005. "The Decline of the Independent Inventor: A Schumpeterian Story?" Working paper W11654, NBER, Cambridge, Mass.

Men of Ohio. 1914. Cleveland: Cleveland News and Cleveland Leader. ⟨http://www.cwru.edu/UL/DigiLib/CleveHist/MenOfOhio/Men.html⟩.

Miller, Carol Poh, and Robert Wheeler. 1990. *Cleveland: A Concise History, 1796–1990*. Bloomington: Indiana University Press.

Moley, Raymond. 1962. *The American Century of John C. Lincoln*. New York: Duell, Sloan and Pearce.

National Malleable and Steel Castings Co. 1943. *National Malleable and Steel Castings Company: 75th Anniversary, 1868–1943*. Cleveland: Privately printed.

"Ohio Governors." N.d. Ohio Historical Society. ⟨http://www.ohiohistory.org/onlinedoc/ohgoverment/governors/⟩.

Orth, Samuel P. 1910. *A History of Cleveland, Ohio*. Chicago: S. J. Clarke.

Rose, William Gansom. 1950. *Cleveland: The Making of a City*. Cleveland: World Publishing Co.

Short, S. H. 1989. "The First Electric Railroad." *Los Angeles Times*, May 14, p. 28.

Smith, S. Winifred. 1955. "Sidney Howe Short." *Museum Echoes* 28, 75–77.

Stockly, George W. 1901. "Some Early Arc Lighting Experiences." *Electrical Review*, January 12, p. 66.

Toman, James A., and Blaine S. Hays. 1996. *Horse Trails to Regional Rails: The Story of Public Transit in Greater Cleveland*. Kent, Ohio: Kent State University Press.

U.S. Census Office. 1870–1920. *Census of Manufactures*. Washington, D.C.: Government Printing Office.

Wager, Richard. 1986. *Golden Wheels: The Story of the Automobiles Made in Cleveland and Northeastern Ohio, 1892–1932*. 2nd ed. Cleveland: John T. Zubal.

2

The Organizing and Financing of Innovative Companies in the Evolution of the U.S. Automobile Industry

Steven Klepper, Carnegie Mellon University

Advances in automobile technology in the nineteenth century, before the commercial production of automobiles, were largely developed in Europe. The commercial production of automobiles was also initially more advanced in Europe than the United States. But within fifteen years of the start of the U.S. automobile industry, U.S. firms were in the vanguard of technological advances, and the U.S. industry grew at extraordinary rates, propelled by continual product and process innovations. Three great firms emerged to lead the industry, General Motors, Ford, and Chrysler, all based in Detroit, Michigan, the acknowledged capital of the U.S. automobile industry. The industry was so big, with General Motors at its helm, that its interests became entwined with the nation's, as reflected in the once popular saying, "What's good for General Motors is good for the country."

While the industry ended up being dominated by just a few firms, at its heyday the industry was composed of well over 200 producers, and advances came from numerous quarters. The main purpose of this chapter is to understand the origins of the firms that propelled the industry forward. Where did these firms come from, and how were they financed? Why were the leading firms so concentrated around one medium-sized city, Detroit, and what role did this play in technological advance? Why was the industry so successful for so long, yet in modern times the U.S. firms have fallen behind the technological frontier and have struggled to catch up with foreign competitors?

These questions have long occupied scholars of the U.S. automobile industry. An enormous literature has developed about the industry, reflecting both its historical importance and the fascination of hobbyists with vintage automobiles. As a result, unique materials are available to trace the origins of U.S. automobile producers, especially those in the technological vanguard. Like many other industries, the initial innovators were largely firms that toiled in related industries such as bicycles, carriages and wagons, and engines. Subsequently, these firms gave birth to a whole new generation of firms founded by employees of incumbent firms, called spin-offs. As a class, the spin-offs performed distinctively, accounting for most of the firms that made it into the ranks of the leading firms after the first ten years of the industry. The spin-offs were responsible for many of the important innovations in the industry, and they were key to its concentration around Detroit.

The spin-off process is described, with particular attention devoted to the origin and finance of the spin-offs and their contributions to technological advance. Experienced auto men not only founded the spin-offs but were also instrumental in their finance. Much like modern venture capitalists, they arranged financing for employee start-ups and helped direct them. Most of all, they provided an outlet for diverse ideas that were not well received within incumbent firms. Not all the spin-offs were successful, but the ones that were often made important technological and organizational contributions that moved the industry forward, sometimes compensating for deficiencies in the leading firms. The spin-off process exemplified some of the best attributes of capitalism that led Nelson (1990) to describe it as "an engine of progress." One can only speculate about the poor technological performance of the U.S. automobile industry in modern times, but it seems related to the virtual foreclosure of entry since the 1920s that has contributed to a long decline in the number of producers and an unhealthy technological inbreeding among the leaders (Halberstam 1986).

The chapter begins by reviewing the evolution of the market and geographic structure of the industry and discussing the origins of the entrants into the industry. It then reviews the history of the first four great firms that located in the Detroit area. These firms spawned most of the successful spin-offs in the industry, and the origin and financing of their spin-offs is discussed as well. It finally presents general themes regarding the finance and performance of the spin-offs.

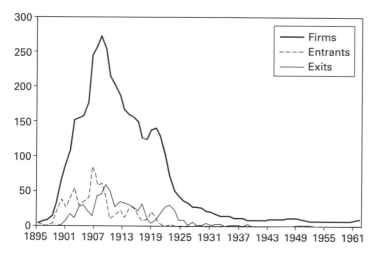

Figure 2.1
Entry, exit, and number of firms

History of the Industry

Smith (1968) compiled a list of every make of automobile produced commercially in the United States from the start of the industry in 1895 through 1966.[1] The annual number of automobile entrants, exits, and producers based on Smith is plotted in figure 2.1.[2] The first firm entered in 1895. Few firms entered in the next four years, but then entry increased sharply. By 1904 there were 155 automobile firms, but collectively they produced only 23,000 automobiles. Subsequently, the industry grew at extraordinary rates. In the decade 1909 to 1919, average annual output growth was 26 percent, with production reaching approximately 1.7 million automobiles in 1919. The number of firms peaked in 1909 at 272 and then declined sharply as entry decreased. By 1925 there were only 51 firms in the industry. Subsequently, entry was negligible, and the number of firms continued to decline, dropping to 9 by 1941. Not surprisingly, the industry evolved to be a tight oligopoly dominated by three firms: Ford, General Motors, and Chrysler, with Ford and General Motors the initial leaders. Their share of the industry's output rose from 38 percent in 1911 to over 60 percent in the 1920s, when Chrysler emerged out of two early entrants as the third leading firm in the industry. Together with Chrysler, General Motors and Ford accounted for over 80 percent of the output of the industry by the 1930s, and over time their joint market share increased further.

The growth of the industry was spurred by tremendous technological change. The original automobiles had low-power steam, electric, or gasoline engines. They were buggy-like contraptions with engines under the body, tiller steering, chain transmissions, open bodies, and hand-cranked starters. Numerous innovations led to cars with much more efficient and powerful gasoline engines mounted in the front, pressed steel frames, shaft-driven transmissions, automatic starters and electric lighting, closed bodies, steering wheels, and a host of other improvements, such as four-wheel brakes and safety windshields.

The production process was also improved greatly, which led to labor productivity growth of 12.5 percent per year from 1909 to 1921. Many new, specialized machine tools were developed, which led to great improvements in precision manufacturing and ultimately the interchangeability of parts. New methods were developed to manufacture major components, such as casting the cylinder block and crankcase as one unit. Firms also integrated backward into component production to ensure reliable supply. The layout of production was changed so that machinery and equipment were organized in sequence rather than by type. This eventually led to the moving assembly line, which enabled workers to specialize in repetitive, low-skilled operations, vastly improving labor productivity.

Demand was greatest for low-priced cars with rugged and reliable performance. This was epitomized by Ford's Model T, which was introduced in 1908. It was a four-cylinder car that originally sold for $850, but by 1916 Ford had reduced its price to $350. Few cars in this size range were competitive with the Model T. Most manufacturers, including ones that began with smaller cars, ended up gravitating toward the production of larger and higher-priced cars, whose market was limited.

Ford helped fuel the concentration of the industry around Detroit, but the process started much earlier with the entry of Olds Motor Works in 1901. The annual number of firms and the percentage of all firms located in the Detroit area from 1895 to 1941, when the number of firms reached a trough of nine, is presented in the bottom two panels of figure 2.2.[3] In the first six years of the industry, 1895–1900, there were sixty-nine entrants. Packard Motor Car Co. (originally Ohio Auto Co.) entered in 1900 and moved to Detroit in 1903, but otherwise no manufacturer was located in Detroit until Olds Motor Works in 1901. Subsequently, the number of firms in the Detroit area rose, reaching a peak of forty-one in 1913, four years after the peak in the number of firms in the industry. The number of Detroit-area firms subsequently declined along with the decline in the total number of automobile producers. After the entry of Olds Motor Works in 1901, the percentage of firms in the Detroit area rose to 15 percent by 1905, then fell back some in the next four years, after which it increased to 24 percent by 1916. It subsequently fell back again in the next eight years or so, after which it climbed to over 50 percent by 1935.

The concentration of activity around Detroit was actually considerably greater than the percentage of firms based in the Detroit area. The editors of the magazine *Automobile Quarterly* compiled a list of the leading makes of American automobiles beginning in 1896 based on production figures by make (Bailey 1971). The annual number of leading makes manufactured by Detroit-area firms for 1896 to 1928 is plotted in the top panel of figure 2.2. The one make listed for the Detroit area in 1896 and 1897 reflects one experimental car made by Ford and Olds, respectively, in these two years. The first listing of a Detroit-area firm that produced more than one car was Olds Motor Works in 1901, when it was credited with the manufacture of 425 cars. Olds was the only firm in the Detroit area listed as one of the (nine) industry leaders in 1901. Subsequently, the number of makes manufactured by Detroit-area firms increased through 1915, when it reached thirteen (out of fifteen makes listed), and then reached fifteen (out of eighteen) by the end of the period in 1928. With the leading makes accounting for well over 80 percent of the total industry output after 1910, firms in the Detroit area dominated the industry by the mid-1910s. Fourteen separate firms in the Detroit area populated the ranks of the industry leaders in the decade 1911–1920,[4] and Detroit-area firms continued to dominate the industry for the next forty-five years.

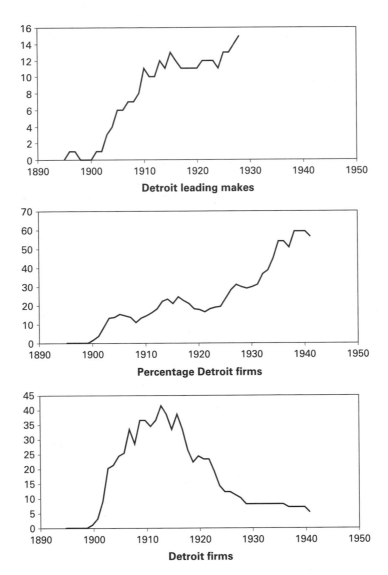

Figure 2.2
Concentration of the industry around Detroit

Table 2.1
Entrants by background and time of entry

Period	Total	Number of experienced firms	Number of experienced entrepreneurs	Number of spin-offs	Number of inexperienced firms
1895–1904	219	46	45	16	112
1905–1909	271	43	32	47	149
1910–1966	235	31	31	82	91
Total	725	120	108	145	352

Origins of Entrants

A unique resource is available to trace the background of every automobile entrant. *The Standard Catalog of American Cars, 1890–1942* and *1946–1973* (Kimes 1996, Gunnell 1992), provides a brief description of the heritage of every firm in Smith (1968). Four categories of entrants were distinguished. The first is firms diversifying from other industries, called experienced firms. The second is de novo entrants founded by individuals heading (and usually owning a substantial portion of) firms in related industries, called experienced entrepreneurs. The third is de novo entrants with one or more founders who had worked previously for another automobile firm, called spin-offs. The last category is a residual of de novo entrants with limited preentry experience related to automobiles, called inexperienced firms. Among the 725 entrants, 120 were classified as experienced firms, 108 as experienced entrants, 145 as spin-offs, and 352 as inexperienced firms.[5]

To convey how the origin of firms changed over time, the firms were divided into three entry cohorts of roughly equal number. The first cohort contains the 219 firms that entered between 1895 and 1904, the second cohort the 271 firms that entered between 1905 and 1909, and the third cohort the remaining 235 firms that entered between 1910 and 1966. Table 2.1 reports the number of entrants of each type in the three entry cohorts. Inexperienced firms were the largest group of entrants in each period, but over time, the percentage of entrants that were inexperienced fell from 51 percent in 1895–1904 to 39 percent in 1910–1966.

The next two biggest groups in the first period were experienced firms and experienced entrepreneurs. Each of these groups accounted for 21 percent of the entrants in the first period, but their significance also declined over time. By the last period, each accounted for only 13 percent of the entrants. The big change in the composition of the entrants came from the spin-offs. In the first period, spin-offs accounted for only 7 percent of the entrants, but by the last period, this increased to 35 percent.

The spin-offs, along with the experienced firms and experienced entrepreneurs, also performed distinctly well. Figure 2.3 presents survival curves for all the entrants and for each of the four types of entrants separately. For each group, separate survival curves for each of the three entry cohorts are reported. In each graph, the age of the entry cohort is plotted on the horizontal axis and the natural log of the percentage of firms in the cohort surviving to each age is plotted on the vertical axis. Among all the entrants, the survival curves are ordered by time of entry, with the earlier entrants having higher survival rates, particularly at older ages. This reflects that earlier entrants had a decided advantage, with most of the leaders of the industry entering early. The same ordering generally holds for each type of entrant. More important, the survival curves indicate that the inexperienced firms, which accounted for about half of all the entrants, survived much shorter than the other three groups of entrants. This reflects that preentry experience, in either a related industry or an incumbent firm, was especially valuable to compete in the industry.

While the survival rates of spin-offs were comparable to the experienced firms and experienced entrepreneurs, the spin-offs increasingly dominated the industry over time. This is reflected in figure 2.4, which presents the annual number of leading makes of automobiles in 1901–1919 that were originally produced by each of the four types of entrants. In the first few years, roughly half of the producers of the leading makes were experienced firms, with experienced entrepreneurs accounting for most of the other leading makes. Only one leading make was produced by a spin-off. Subsequently, the number of leading makes accounted for by spin-offs increased dramatically. It peaked in 1916, when spin-offs accounted for eleven of the fifteen leading makes, with spin-offs account-

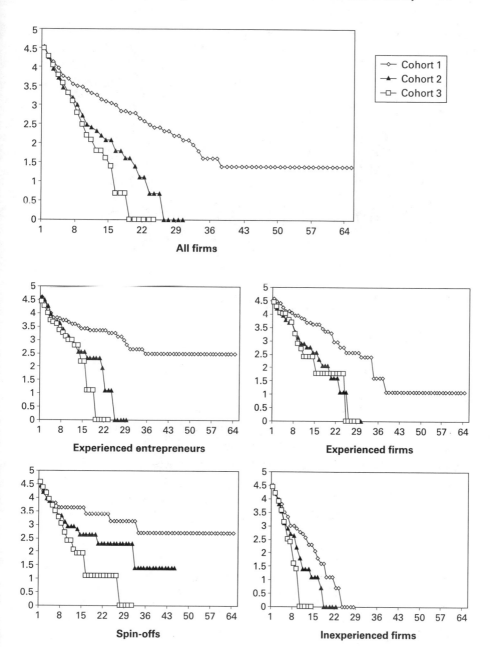

Figure 2.3
Survival curves by background and time of entry. x axis = age; y axis = natural log of the percentage of survivors

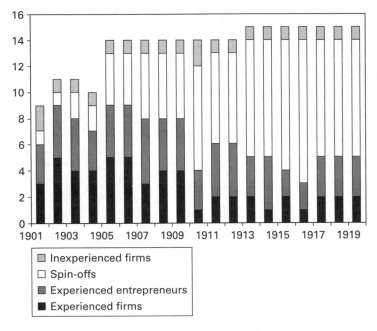

Figure 2.4
Backgrounds of leading firms

ing for nine of the fifteen leading makes by the end of the period in 1919. For most of the period, the number one make was produced by Ford, which was a spin-off, and many of the other top makes, such as Maxwell, Hudson, Huppmobile, Chevrolet, and Dodge, were introduced by spin-offs.

The Early Detroit-Area Leaders

Nearly all the spin-offs that produced a leading make were descended from four early entrants in the Detroit area: Olds Motor Works, Cadillac Motor Car Co., Ford Motor Co., and the Buick Motor Co. Olds Motor Works was the first to enter in 1901. Prior to Olds, only two firms sold more than 100 cars in a year: Pope Manufacturing Co. and Locomobile Co. of America (Bailey 1971). Pope manufactured primarily an electric-powered car, and the Locomobile was steam powered, but both electric and steam cars had serious limitations. The scene was set for a gasoline-powered car to take over leadership of the industry, and Ransom Olds obliged with the now famous Curved-Dash Runabout.

Ransom Olds ran Olds Motor Works, one of the leading engine producers in the United States, which was founded by his father as P.F. Olds & Son in 1880 in Lansing, Michigan.[6] Ransom Olds developed his first gasoline-powered vehicle in 1896. To finance his growing engine business and his experimentation with automobiles, Ransom Olds had to raise additional capital. This led to the company's being reorganized in 1899 as Olds Motor Works, with Olds losing majority control to one of his investors, Samuel Smith, who brought in his son, Fred, to help manage the company.

The company was moved from Lansing to Detroit, where the Smiths were based, and it experimented with different types of automobiles in 1899 and 1900 before settling on the Curved-Dash Runabout in 1901. It was a one-cylinder car accommodating two people and sold for $650. Olds envisioned a substantial market for a low-priced runabout (a small, inexpensive open car), and his car was instantly successful. There were other comparable cars on the market at similar prices to the Curved-Dash Runabout. Olds Motor Works, however, was a more experienced firm than its rivals, which enabled it to plan, produce, and promote the car in a manner its rivals could not (May 1977). Orders flooded in for the car after full-scale publicity for it began in February 1901, but production was stalled by a disastrous fire that burned down Olds's Detroit plant, ultimately leading the firm to return to Lansing. Nevertheless, 425 Runabouts were sold in 1901, which was second among all makes. By 1903 the Runabout was the number one seller in the industry at 4,000 units, with over 5,000 units sold in the next year (Bailey 1971).

The Runabout was largely an assembled car, which was true of most cars of its era. All of its major components, including engines and transmissions, were purchased from other firms. While many have implicated the fire as the impetus for Olds's contracting out for engines, it appears that Olds lacked the capacity to satisfy the growing demand for its regular engines, which was its main business, as well as build engines for its automobiles (May 1977). Olds's production was on a much greater scale than any prior firm, which required it to organize an elaborate system of subcontracting and assembly of parts. Abernathy, Clark, and Kantrow (1983) compiled a list of 631 innovations in automobiles over the period 1893–1981 that were ranked by their impact on the production process. Olds was credited with developing the first mass production system for automobiles, which was the first of only nine innovations that achieved the highest impact rank. Olds innovated in the assembly of cars by moving vehicles from station to station on platforms equipped with wooden casters. Workers were assigned specific tasks, and parts were made available close by in large bins (May 1977). These were some of the key ingredients of the famous moving assembly line that Ford would later develop.

Olds subcontracted for parts on an unprecedented scale. He contracted with Leland and Faulconer, a successful Detroit machining company with 160 employees, to solve a noise problem that Olds had with its transmissions. Leland and Faulconer was headed by Henry Leland, who was immortalized in a biography started by his son (Leland 1966) as the "Master of Precision." Leland was an experienced machinist, manager, and machine tool salesman who had worked for some of the machine tool companies that were in the forefront of precision manufacturing in the nineteenth century. The transmission problem was given to one of his talented employees, Alanson Brush (Kollins 2002d), and after it was solved, Olds contracted with Leland and Faulconer to produce its transmissions (May 1977).

Olds also contracted with Leland and Faulconer to produce engines. Ransom Olds did not put a high priority on precision manufacturing, and his cars were subject to problems on the road that Olds solved as they materialized. Leland was an expert in precision manufacturing. After tearing down Olds's engine and analyzing it, he was able to manufacture it much faster than Olds after machining its parts to such uniformity that they could be interchanged (May 1977). Alanson Brush also was able to improve the design of the engine to increase its horsepower at no extra production cost (Kollins 2002d). Olds did not adopt the improved design, though, because it would have necessitated costly retooling (Leland 1966).

Olds also contracted for his engines and transmissions with the Dodge Brothers, another Detroit machining company. John and Horace Dodge were expert mechanics who invented and patented an adjustable ball-bearing bicycle wheel hub and had co-organized a bicycle company that they sold in 1900. They set up a successful machine shop in Detroit in early 1901 (Kollins 2002a), and Olds subcontracted with them for both engines (May 1977) and transmissions (Kollins 2002a) on a large scale. Both Henry Leland and the Dodge Brothers would play important roles in the automobile industry and in the success of Detroit. So would a number of other Detroit subcontractors of Olds, including Barney Everitt, who produced bodies for Olds, and the Briscoe Brothers, Benjamin and Frank, who produced sheet metal parts for Olds, including radiators, gas tanks, and fenders.

Ransom Olds remained with Olds Motor Works until 1904, when he was pushed out of the company after a dispute with the Smiths. Olds did not see the need for standardized, precisely made parts that were later required to increase the speed and efficiency of the manufacturing process. The Smiths began to experiment with alternative methods of manufacture, which resulted in clashes with Olds and Olds's departure. Olds Motor Works had two more years of outstanding sales, but the Smiths were not successful in developing new products, and the company drifted off into larger and more expensive automobiles with limited markets. When they were approached to be part of the 1908 General Motors merger, the Smiths were anxious to sell out. They received $3 million, which reflected the value of existing plants and the Olds organization. Under General Motors's aegis, Olds Motor Works was again successful and regained the ranks of the fifteen leading makes by 1915.

The next great company in the industry was Cadillac Motor Car Company. Cadillac is rightly associated with Henry Leland, but it was started by Henry Ford under the name Henry Ford Company. Ford built experimental cars in the 1890s when he was employed at the Detroit Edison Illuminating Company, where he rose to the position of superintendent. He left Edison in 1900 to start his own automobile firm in Detroit, the Detroit Automobile Company, which was financed by a number of Detroit leaders, including its mayor. After $86,000 was expended in a year with twelve cars in various stages of completion and no sales, the company was dissolved (Szudarek 1996). Subsequently Ford developed a successful race car, and some of his original investors and others backed him in a second company, the Henry Ford Company. The investors put up $35,000 in cash on a capitalization of $60,000, but once again Ford had difficulty moving into production, and Henry Leland was brought in to advise the investors on the future of the company (Szudarek 1996).

Events are unclear, but Henry Ford soon left the company, and it was renamed Cadillac Motor Car Company after Leland advised the investors that the company was worth continuing.[7] He offered the company the improved version of Olds's one-cylinder engine that Olds declined to use, and he suggested reorganizing manufacturing operations to overcome production problems. The company's first car, introduced in 1902, incorporated Leland's engine and not surprisingly resembled the Curved-Dash Runabout. William Metzger, a successful bicycle and car dealer, was hired to head sales, and Cadillac was successful immediately. It became the third leading make in 1903 with a production of 1,698 units and the second leading make in 1904 with a production of 2,457 units (Bailey 1971). Subsequent production problems led Leland to become general manager and eventually to merge his machining company with Cadillac, after which Leland became a major stockholder as well as leader of the firm.

Cadillac prospered after Leland's increased involvement. It produced over 8,000 cars in the next year and a half. It was known for the reliability and quality of its cars, which ultimately led Cadillac to produce bigger and higher-priced cars. Although this restricted the market for the company's cars, causing it to drop from the number two position in the industry in 1904 to number seven by 1908, Cadillac remained successful and was sold to General Motors in 1909 for $4 million. Henry Leland remained with Cadillac and General Motors until 1917. During his tenure, Cadillac became the first company whose parts were fully interchangeable. The first reliable electric starter was developed by Charles Kettering for Cadillac in 1912, which was quickly adopted by other manufacturers. Cadillac developed the first large-scale production V-8 automobile engine, which was also one of the nine innovations to achieve the highest production rank in the list compiled by Abernathy, Clark, and Kantrow (1983). Leland left Cadillac to start his own firm to produce aircraft engines after the head of General Motors, William Durant, refused to get involved in the World War I effort.

It was Henry Ford who started the next great firm in the industry after he was ousted from the Henry Ford Co. Alexander Malcomson, a coal dealer who knew Ford from his days at the Edison Illuminating Company, financed Ford's efforts to develop another automobile, which was finished in December 1902. Malcomson could not make the payments to suppliers, which necessitated bringing in his uncle, John Gray, and other investors, who collectively paid in $28,000 in cash on a capitalization of $100,000 to start Ford Motor Company in Detroit in 1903. Ford and Malcomson each received 25.5 percent of the shares, and Gray was made the president with 10.5 percent of the shares. Perhaps the most significant investment in the company was made by the Dodge Brothers. Ford had failed twice because he was unable to get into production. This time he hired the Dodge Brothers to produce the major components of his car, including its engine, transmission, and axles. The Dodge Brothers took a considerable risk, committing to deliver 650 chassis at $250 each, for a total of $162,500. In return, they were to receive successive $5,000 payments as the order was delivered. They had to invest in machinery, tools, and materials to execute the Ford order, which required nearly their entire establishment of 150 people. In exchange for the capital they effectively invested in Ford Motor Company, they received 10 percent of the stock in Ford (Nevins 1954).

Ford's initial car resembled Cadillac's, and it was immediately successful. By the end of the model year in September 1904, Ford had sold 1,708 cars, vaulting it into second place in the industry with profits exceeding $250,000 (Kollins 2002a). Ford Motor Company continued to improve its cars, but a dispute arose between Henry Ford and Malcomson over new models. Ford wanted to produce the Model N, which was a simple, inexpensive car Ford had developed to sell for $500. Malcomson wanted to market a luxury car with a larger per unit profit. Ultimately Malcomson sold his stock to Ford and left to form his own company, which was unsuccessful (Sudzarek 1996).

Ford hired Walter Flanders to direct production of the Model N.[8] Flanders was a sales representative for three machine tool companies and an expert on the use of machine tools in manufacturing. Ford planned to center production around the Model N, and Flanders recognized that production had to be carefully planned to get the price down to the

$500 goal. He reorganized the factory layout according to sequential operations on parts rather than by type of machine, which was critical to the later evolution of the moving assembly line. Special-purpose machine tools were introduced to improve productivity. Flanders emphasized the importance of interchangeable parts for high-volume production and drilled the workforce to improve their efficiency. He developed an orderly production plan based on newly developed sales forecasts to reduce inventories held in the factory. Flanders succeeded in eliminating the confusion that had reigned on the factory floor, and Ford sold over 8,000 cars in Flanders's first year versus 1,600 the year before, vaulting Ford into the number one position in the industry.

The Model N was the precursor to the Model T, introduced in 1908. Like Olds's Curved-Dash Runabout, it was not a revolutionary car but was rugged, responsive, and priced competitively at $850. The chassis was made of vanadium alloy steel, which was much stronger than other steels. It made the Model T light and strong enough to withstand the poor country roads of the time. It had a planetary transmission operated with foot pedals that facilitated rocking out of mud holes. The motor was a four-cylinder, cast en bloc with a detachable head, making servicing simpler. It also had a magneto integrated into the flywheel, making the electrical ignition system part of the engine, thereby eliminating the need for storage batteries. The detachable cylinder heads and magneto integrated into the flywheel achieved the second highest rank in Abernathy, Clark, and Kantrow's (1983) list of innovations, and the vanadium steel components achieved the fourth highest rank, which collectively qualified Ford as the leading innovator in the industry up to that point. After 1909, Ford produced only the Model T chassis. Numerous innovations in the production of the Model T were developed, the most notable being the moving assembly line and branch assembly plants, two of the nine innovations on Abernathy, Clark, and Kantrow's list that achieved the highest impact rating. These innovations enabled Ford ultimately to lower the price of the Model T to $350, and in its peak year of 1923, Ford sold over 1.8 million Model Ts (Bailey 1971).

Ford was founded in 1903, the same year the fourth great firm in the Detroit area, the Buick Motor Company, was founded. It took some time, though, before Buick achieved the ranks of the industry leaders.

David Buick founded Buick Motor Company to capitalize on an innovative engine that had been developed in his engine firm, which was started in 1899 with funds Buick received from the sale of his successful plumbing supply business. Buick quickly encountered financial difficulty, and the Briscoe brothers, suppliers to Olds, lent Buick money and then sold their controlling stock to James Whiting and other directors of the Flint Wagon Works, a successful carriage company in Flint, Michigan.[9] While the production of carriages involved progressive assembly and a well-developed division of labor, Whiting and his associates had limited experience with machine operations and raising the large amounts of capital required for automobile production. Consequently, Buick floundered and was on the verge of bankruptcy when it turned to Flint's most successful businessman, William Crapo Durant.

Durant had distinguished himself at a young age as a superb salesman and astute judge of organizations. He became a millionaire through the formation of a company with a partner, Dallas Dort, to manufacture a cart with a novel suspension whose rights he had acquired. Dort oversaw production, while Durant raised capital and hawked the company's products. The Durant-Dort Carriage Company became one of the largest carriage producers in the United States, and it integrated backward through separate organizations into the production of many of its components.

Before acquiring control of Buick, Durant put Buick's car through rigorous testing for almost two months, just as he had done with the cart that launched his carriage firm. Convinced that he had a first-rate product, he expanded Buick's capitalization from $75,000 to $1.5 million using his contacts, reputation, and enormous energy to sell stock in Flint and elsewhere. He used the capital he raised to build the biggest factory in the industry in Flint and used Durant-Dort's carriage dealers to develop an extensive wholesale and retail distribution network. Durant was skillful in working with his top employees to overcome problems and achieve the ambitious objectives he formulated for Buick. By 1908, Buick was the number two producer, with an annual production exceeding 8,000 automobiles.

With Buick at its center, Durant organized General Motors in 1908. GM was a massive horizontal and vertical integration of twenty-seven different organizations. Dallas Dort did not join Buick, and without his steadying influence, GM was soon in financial trouble. Poor inventory control, shoddy practices resulting from little oversight of production, and a limited awareness of the capital needs of the new company led Durant to lose control of GM to bankers. The bankers quickly righted GM and restored it to great profitability, after which Durant was able to regain control of GM, only to lose it again for good. GM went on to introduce many of the innovations in Abernathy, Clark, and Kantrow's (1983) list, accounting for more product innovations than any other firm. It displaced Ford as the number one producer in the 1920s and remained number one thereafter.

A number of important themes emerge from the history of Olds, Cadillac, Ford, and Buick. Consistent with the survival graphs in figure 2.3, the preentry experience of the leaders of each firm had a substantial effect on the firms' performance. Ransom Olds, Henry Leland, and William Durant were leaders of very successful firms in industries related to automobiles, and they were able to capitalize on this success in automobiles. They were also able to use their prior experience to help structure their automobile firms. Olds was experienced at promoting his engines and in large-scale production, which he exploited in popularizing the Curved-Dash Runabout and quickly ramping up production. Leland was an expert in precision manufacturing, and he established organizational procedures to make Cadillac a leader in precision manufacturing and quality. Durant had successfully organized a large, vertically integrated organization in a related industry, and he quickly developed a large automobile firm with a state-of-the-art factory and impressive distribution network.

The absence of comparable experience seems to have been a major handicap for Henry Ford, David Buick, James Whiting, and Samuel Smith. Ford had never headed any business prior to the automobile industry. Although he was able to design two cars that performed well enough to attract backers, he was not able to develop an organization that could manufacture either car at a competitive cost until he teamed with the Dodge Brothers. Buick had succeeded at a small plumbing supply business but quickly dissipated his funds and was unable to raise further capital despite having a first-rate product. Whiting and his partners were not able to do much better than Buick. They had operated a modest-size carriage and wagon company, but they had little experience with the kind of machinery used in automobiles, and they too were unable to raise the kind of capital needed to develop Buick's car. Samuel Smith was a wealthy businessman, but he had no manufacturing experience to draw on to replace Olds's leadership.

The experience of Olds's two main subcontractors, Leland and Faulconer and the Dodge Brothers, and Ford's experience with Ford Motor Company suggests that experience with incumbent automobile firms was also valuable. Leland was able to capitalize on the superior engine it developed for Olds, while the Dodge Brothers' experience no doubt enabled them to satisfy Ford's needs. Ford's initial efforts were obviously unsuccessful, but this may well have provided a valuable lesson that he took to heart in his third start-up. The longevity of the spin-offs in figure 2.3 suggests that experience with incumbent firms was indeed valuable, which is a theme that will be developed further in the next section.

The experience and success of Olds's subcontractors and Durant in the carriage industry also played an important role in the finance of Cadillac, Ford, and Buick. The frustrated stockholders of the Henry Ford Company sought out Henry Leland's advice about their fledgling investment after ousting Henry Ford, much as modern investors rely on venture capitalists with a proven track record in a new industry. Similarly, William Durant was able to leverage his prior success to raise an enormous amount of capital from wealthy individuals, many of whom had helped finance his carriage company. Leland and the Dodge Brothers also merged their assets into Cadillac and Ford, becoming key investors. No doubt their experience as suppliers to Olds helped them judge the prospects of these investments. Presumably the Briscoe brothers also used their experience as suppliers to Olds in deciding to help finance Buick at an early critical juncture.

Olds's great success illustrates how one company can have a profound influence on the development of a region, much as Brush Electric had on the development of Cleveland around the same time (see chapter 1, this volume). Olds's subcontracting was instrumental in the formation and success of Cadillac, Ford, and Buick, which quickly became leaders of the automobile industry. None of them ventured far geographically, as was true as well of the firms associated with Brush Electric in Cleveland. The upshot was that four of the very best early automobile firms were located very close to each other in a region that would hardly be expected to become the automobile capital of the United States based on its size, natural advantages, or supply of companies in related industries (cf. Klepper 2002b). But combined with a powerful spin-off process, as described below, the early concentration of successful firms in the Detroit area fueled an extraordinary agglomeration of activity there.

Spin-Offs

The 725 entrants into automobiles were ranked according to the number of spin-offs they spawned, which equals the number of spin-offs for which they were the parent firm. The top five firms, with the number of spin-offs in parentheses, were Olds Motor Works (7), Buick/General Motors (7), Cadillac (4), Ford (4), and Maxwell-Briscoe (4). The first four firms were the pioneers in the Detroit area, and Maxwell-Briscoe, which is reviewed below, was a spin-off cofounded by Jonathan Maxwell, an employee of Olds Motor Works. Among the first four firms, not only did they have 22 spin-offs collectively, but in turn their spin-offs were responsible for 19 additional spin-offs. These 41 spin-offs constituted 28 percent of the 145 total spin-offs. Thus, the leading firms in the industry were especially fertile settings for spin-offs. This was confirmed by a statistical analysis of the factors influencing the annual rate at which firms spawned spin-offs (Klepper 2002b).

The spin-offs of the leading firms were also distinctly successful. The 41 spin-offs accounted for 11 of the 13 spin-offs after 1903 that introduced makes of cars that made it onto the list of leading makes after 1903. In a statistical analysis, it was found that the longevity of spin-offs was directly related to the performance of their parents, with better parents having longer-lived spin-offs (Klepper 2002b). Nearly all the spin-offs of Olds, Cadillac, Ford, and Buick/General Motors located in the Detroit area. Consequently, the four firms and their spin-offs were the primary force that led the industry to become so concentrated around Detroit (Klepper 2002b). The histories of the most successful spin-offs of the Detroit firms are traced to analyze the circumstances governing their founding, financing, and contributions to innovation.[10]

The first set of spin-offs were founded by employees of Olds Motor Works, including Maxwell-Briscoe, Reo, Thomas-Detroit, and Hudson.

Spin-Offs Founded by Olds Motor Works Employees

Maxwell-Briscoe

Jonathan Maxwell was the superintendent and chief tester for Olds Motor Works. He was born in Peru, Indiana, in 1864. Before working for Olds Motor Works, he had been a master railroad mechanic and worked for Haynes-Apperson, one of the first automobile entrants. William T. Barbour was the president of Detroit Stove Works, located next to Olds Motor Works in Detroit. He would accompany Maxwell on test drives of Olds Motor Work's cars. In 1902 Maxwell interested Barbour in starting a new company, which became Northern Manufacturing Company. Northern began with capital stock of $50,000, of which $38,000 was paid in. William Metzger, sales manager of Cadillac, also played a role in the formation of Northern and later left Cadillac to join Northern in 1906.

Maxwell, the chief engineer, designed a one-cylinder car similar to the Curved-Dash Runabout. Fifty were built and sold in 1903 according to one source (Kollins 2002d), and 200 in 1902 and 750 in 1903 were sold according to another source (Yanik 2001). Northern's car embodied a number of innovations, including three-point suspension of the engine, running boards, and a transmission integrated with the engine (Yanik 2001). All these innovations were listed in 1902 on Abernathy, Clark, and Kantrow's (1983) list of automobile innovations, with three having the third highest impact rating, making Northern an important early automobile innovator.

Apparently buyers wanted a larger car, and there was some dissatisfaction with Maxwell's design. In 1903 Maxwell left or was replaced by Charles King, who had worked for Olds Motor Works after Olds acquired King's marine engine firm. King was an early car developer in Detroit who was unable to secure financing to develop his car. Maxwell continued to be listed as director and stockholder in 1904, so it is not clear when he severed ties with Northern (Yanik 2001).

Maxwell subsequently developed a thermo-syphon cooling system for internal combustion engines, and he interested Benjamin and Frank Briscoe in manufacturing his system. Benjamin Briscoe asked Maxwell to assess his investment in the fledgling Buick Motor Company. Maxwell did not offer an opinion but indicated he had his own car ideas. In 1904 Benjamin Briscoe and Maxwell jointly organized the Maxwell-Briscoe Motor Company. Briscoe had limited success in finding Detroit capital, but secured $100,000 backing from J. P. Morgan in New York, who had previously helped finance the Briscoe Brothers sheet metal company that had supplied Olds Motor Works with radiators and other sheet metal parts. Initially Maxwell-Briscoe built its cars in a leased facility in Tarrytown, New York, close to Morgan.

Maxwell-Briscoe's first two cars were a two-cylinder runabout for $780 and a larger touring car for $1,450. The runabout had an uncommon integral engine crankcase and gear change housing of aluminum alloy cast in one piece. The touring car featured a sliding gear three-speed transmission with multiple disc clutch, for which Maxwell secured a patent. These innovations resembled ones on the 1902 Northern. The tooling of the factory occurred in 1904, and production largely got underway in 1905. Maxwell-Briscoe's cars were an immediate success, embodying fine engineering and good workmanship at a modest price. Maxwell had the eighth highest sales of all firms in 1905, fifth in 1906, fourth in 1907 and 1908, and third in 1909.

Maxwell-Briscoe was the key component of the 1910 United States Motor Company merger organized by Benjamin Briscoe to rival General Motors. This merger was unsuccessful, and Maxwell later emerged as the only intact component. (This is discussed further below in the context of the efforts of Walter Flanders.) Briscoe went on to form Briscoe Motor Co. in 1914, which experienced modest success before it exited in 1923.

Reo Car Company

After Ransom Olds was pushed out of Olds Motor Works in 1904 by Samuel Smith and his son, he joined Reuben Shettler to found a second automobile company, the Reo Car Company, in Lansing, Michigan. Shettler was an original stockholder in Olds Motor Works when it was organized in 1899. He moved to California and acquired the Olds distributorship for southern California. Shettler sold his distributorship in June 1904 and came to Lansing to organize Reo.

Reo was capitalized at $500,000, with Ransom Olds given 52 percent of the stock without paying in anything. Shettler was the number two stockholder and vice president. Shettler induced some prior Olds stockholders to subscribe, along with various men associated with Lansing banks and businesses. The attraction to invest was Ransom Olds.

Reo hired a number of Olds Motor Works employees, including Horace Thomas, an engineer at Olds who was made chief engineer, a position he would hold for thirty years; Richard Scott, a supervisor at the Olds Gasoline Engine Works since 1898, was brought in as plant superintendent; and Raymond Owens, former secretary at Olds Motor Works, was brought in as sales manager.

Reo's first model was markedly different from the Curved-Dash Runabout. It was jointly designed by Ransom Olds and Thomas and was the kind of car that the Smiths at Olds Motor Works had wanted to develop. It was a two-cylinder, 16 horsepower touring car with room for five people that sold for $1,250. Olds had always been linked to low-priced cars, but for the next thirty years, he would be linked to medium-priced cars. The Reo car was more rugged than the Curved-Dash Runabout, which was needed for increasing long-distance travel. It was not unlike Ford's efforts, but it lacked the technology that enabled Ford to sell his cars at much lower prices.

Reo was immediately successful. It was the number seven seller in 1905, number four in 1906, and number three in 1908, with sales of almost 4,000 units. This was the highest rank it reached, dropping subsequently but nonetheless remaining in the ranks of the leading makes through 1923 and appearing again for one year, 1927.

E. R. Thomas-Detroit

Roy D. Chapin was born in Lansing in 1880. He went to the University of Michigan for two years, where he became friendly with Howard Coffin and Roscoe Jackson, both engineering students. Chapin knew the

Olds and Smith families and went to work at Olds Motor Works in 1901. He engaged in a well-publicized trip from Detroit to New York in late 1901 to promote successfully the Curved-Dash Runabout. He rose to become sales manager at Olds Motor Works in 1904. Coffin and Jackson entered Olds Motor Works in the engineering department. By 1904 Coffin was chief engineer. Jackson switched to production and was promoted to factory manager in 1906. Chapin and Coffin became friendly with Frederick Bezner, who worked for NCR in Dayton, Ohio, before being employed in the purchasing department at Olds Motor Works. By 1905 Bezner had risen to the important position of purchasing agent.

In 1905 Coffin developed a new four-cylinder car that seemed to be a compromise between the Curved-Dash Runabout and larger cars favored by the Smiths. With some encouragement from the Smiths, the project advanced to Bezner, who negotiated contracts for all the parts for the car. Then the Smiths reneged. Chapin was the first to leave in 1906. Through a friend, he met and convinced a Buffalo automaker, E. R. Thomas, who sold a high-priced car through his E. R. Thomas Motor Company, to support a company to produce Coffin's car. The company, E. R. Thomas-Detroit, which was located in Detroit, was capitalized at $300,000, with $150,000 in stock initially subscribed, $100,000 by Thomas and $50,000 by Chapin, Coffin, Bezner, and a fourth Olds employee, Brady. Only 20 percent was put up initially by each stockholder, providing only $30,000 in paid-in cash. The four Olds employees were the active managers, with Thomas the unsalaried president.

Their car, the Thomas-Detroit, was an assembled car that utilized only $16,300 in plant equipment according to the 1907 balance sheet of the company (Renner 1973). Parts contracts were key to the success of an assembled car; parts manufacturers could fail to make the parts to specification or not deliver at all, which could doom an automobile producer. Bezner exploited his experience at Olds to set up contracts to minimize these possibilities, introducing an important organizational innovation (Renner 1973). They were able to finance operations by collecting advance payments from dealers, which were used as a form of working capital to pay off parts suppliers with a delay. By 1906, when the company was underway and successful, the five partners paid in their total subscription of $150,000.

The company marketed its cars through E. R. Thomas Motor Company's network of dealers. It sold its cars for $2,750 and was successful immediately, shipping 500 cars by the middle of 1907, qualifying E. R. Thomas-Detroit for twelfth place among the leading ranks in 1907. It produced 750 cars in 1908.

Hudson Motor Car Company

Chapin, Coffin, Bezner, and Brady did not like Thomas getting two-thirds of the profits of E. R. Thomas-Detroit and were also concerned about their dependence on Thomas for marketing. Furthermore, Coffin had an idea for a new car model that he believed could be manufactured cheaply enough to sell for $1,500. It was a four-cylinder car that had features associated with higher-priced cars but was smaller, offering the prospect of a larger market than Thomas-Detroit's car. E. R. Thomas had been influential in the credit Thomas-Detroit was able to secure, and Chapin, Coffin, Bezner, and Brady were not ready to strike out on their own. Instead, they recruited Hugh Chalmers, who had been a pioneer in marketing at NCR but had just left in a dispute, to buy out half of Thomas's stock. Thomas agreed, and the company also began marketing its own cars. Chalmers was made president at a large salary of $50,000 and was given control over marketing and sales. Chalmers increased the company's national promotion, and Coffin's car, known as the Chalmers-Detroit 30, was very successful, attaining the tenth rank among makes in 1910.

With only one-third of the company's stock, Chapin, Coffin, Bezner, and Brady felt they were not being amply rewarded for their efforts. There was also talk of building four-cylinder cars for less than $1,000, which Ford had just done. Coffin believed he could do something similar. Coffin, Chapin, and Bezner formed a partnership with Jackson, who by now had left Olds Motor Works, to develop such a car. Each invested some money in this partnership along with Jackson's uncle by marriage, J. L. Hudson, a Detroit department store magnate. They developed a car similar to Coffin's 1909 Chalmers-Detroit but with a less elaborate body and a simplified, cheaper design. Hudson was formed in 1909 to produce and sell the car. It was capitalized at $100,000, with the four partners getting 25 percent of the stock and Chalmers also getting about 15 percent of the stock. Only $15,000 was paid in initially, with $12,500 from Hudson. The balance due from each stockholder was apparently financed from subsequent dividends. Figures from April 1910 indicate a total investment of $35,000 in plant and office equipment.

Chapin, Coffin, and Bezner were still employees of Chalmers-Detroit. Chalmers had limited interest in Hudson's car even after it was immediately successful. He was used to dealing with a well-developed product at NCR and wanted to limit changes in his product. He could not grasp that the greatest challenges in the automobile industry at this time were in engineering and production and not sales (Renner 1973). He accepted a proposal to buy out Chapin, Coffin, and Bezner from Chalmers-Detroit, exchanging his shares in Hudson plus $788,000 for Chapin, Coffin, and Bezner's shares in what was now known as Chalmers Motor Co.

Hudson was very successful, earning $5.385 million in net profits in its first six years (Renner 1973). By 1910 it was the tenth leading producer and remained in the ranks of the leaders for many years. It introduced a number of innovations on Abernathy, Clark, and Kantrow's (1983) list, including a sedan-type body that attained the fourth highest impact ranking and an inexpensive closed car built of wood and steel that was one of the nine innovations to achieve the highest impact rating on Abernathy, Clark, and Kantrow's list.

Hugh Chalmers was also influential in the formation of the Saxon Motor Company, which was founded by a Chalmer's employee.

Saxon Motor Company

Harry Ford was born in Knobnoster, Missouri, in 1880. In 1905 he was employed in the advertising department of the National Cash Register Company, and he joined Chalmers in 1909 at Chalmers Motor Company as advertising manager. He worked his way up to assistant general manager in 1913.

Ford had been convinced for some time of the possibilities of a good two-passenger car selling at a price of $400. He convinced Chalmers of the merits of his proposal, and in 1913 Chalmers organized a syndicate of ten men, each of whom put up $10,000; Saxon was thus capitalized at $350,000, with $250,000 in common stock (Sudzarek 1996). Harry Ford was president and general manager, and while Chalmers provided financial support, he was not active in the company. Toward the end of 1915 Ford bought out Chalmer's stock (for $450,000), giving Ford full control of Saxon.

Saxon's initial car was an agile runabout priced at $395. It was sold through Chalmers Motor Co.'s dealers. Its low price and solid design attracted many buyers, and 7,100 cars were sold in 1914 and 19,000 in 1915, when Saxon attained eighth place among the leading makes. In 1916 the company reorganized and expanded considerably, but it built too large a plant and subsequently experienced cost problems. Ford had to retire due to illness and died in 1918, which started a decline that ended in 1922, when the company ceased operating.

Spin-Offs from Cadillac

The next three spin-offs came out of Cadillac. The first two were cofounded by Alanson Brush, who was one of Cadillac's top employees. The third was started by the head of Cadillac, Henry Leland.

Brush Runabout Company

Alanson Brush was born in Algonac, Michigan, in 1878. After graduating from high school, he became affiliated with Ransom Olds during the 1890s and helped Olds build his first car. Shortly after Olds moved to Detroit, Brush joined Leland and Faulconer and was assigned to design the transmission that Leland and Faulconer had contracted to build for Olds. Brush also designed a one-cylinder engine for Olds that was later used on the 1903 Cadillac car. When Henry Leland joined Cadillac at the end of 1904, Brush joined as well as chief engineer.

In 1905 Brush was unhappy about the way Henry Leland tried to control his patents at Cadillac. He left with a $40,000 settlement and pledged not to compete with Cadillac for two years. He next designed a smaller car to test out new ideas he had, such as a coil-spring suspension. Frank Briscoe felt Brush's car had promise, and he organized the Brush Runabout Company in Detroit with a capitalization of $200,000. The

Briscoe brothers' factory was used for assembly of the car. The car was simple but attractive at a low price of $500. It sold 500 units in 1907, qualifying it for thirteenth place, and then reached sales of 2,000 units in 1909, qualifying it for ninth place among the leading makes.

It was said that Brush did not remain involved in the company, and it was acquired as part of the U.S. Motors merger arranged by Benjamin Briscoe in 1910. It succeeded for one year after the dissolution of the merger and then ceased to exist.

Oakland Motor Car Company

Brush cofounded Oakland in 1907 in Pontiac, Michigan, with Edward Murphy, the head of the Pontiac Buggy Company. Murphy liked the Brush Runabout and asked Brush to design a car for a new company. Brush brought him a two-cylinder car that he had designed while at Cadillac (Szudarek 1996) but that Cadillac had rejected (Kimes 1996).

The Oakland car for 1908 was an enlarged Brush car with many patented Brush features. The company was bought by General Motors in 1908, and Brush stayed on as a consulting engineer for William Durant. The Oakland became a four-cylinder car in 1910 and regularly sold 3,000 to 5,000 units per year, placing it among the leading makes. Brush was issued many patents, including on the planetary gear set and disc clutch that were forerunners of today's automatic transmissions and overdrive units.

Lincoln Motor Company

Henry Leland wanted to be involved in the production of aircraft and other military materiel during World War I, and when William Durant was reluctant to make such a commitment, Leland left in 1917 to form Lincoln Motor Co. Leland and William Murphy, one of the initial financiers of Cadillac, arranged for $2 million in personal loans to start Lincoln. Lincoln began producing Liberty engines on a cost-plus basis for the military.

In 1919 Lincoln was reorganized to produce automobiles powered by a new V-8 engine, a considerable improvement over prior engines. Unfortunately, Lincoln bought equipment at peak prices, just before the recession of 1920. Leland and the board of Lincoln Motors endorsed $5 million in bank notes to address the ensuing financial crisis. The continuing depression in sales coupled with a government effort to recoup alleged excess payments made to Lincoln for the Liberty engines led to a bankruptcy and Leland's losing control of Cadillac and his fortune.

In 1922 Henry Ford purchased Lincoln, and Leland left the company after a disagreement with Ford. Ford maintained the Lincoln brand as a luxury automobile, and many years later Lincoln made it into the ranks of the leading makes in the industry.

Spin-Offs Founded by Ford Motor Company Employees

Ford Motor Company employees were instrumental in the next three firms, E-M-F, Hupp, and Dodge.

The E-M-F Company

E-M-F was founded by Barney Everitt, who was born in Ridgetown, Ontario, in 1872; William Metzger, who was born in Peru, Illinois, in 1868; and Walter Flanders, who was born in Rutland, Vermont, in 1871. Everitt had supplied bodies to Olds in Detroit in the Curved-Dash Runabout era and subsequently supplied bodies to Ford. He became general manager of Wayne Automobile Company soon after it was formed in Detroit in 1904. Metzger was Cadillac's first sales manager and was involved in the founding of Northern. Walter Flanders was Ford's production genius and the main force in the company.

The partners were able to secure control of Northern and Wayne, which they used to form E-M-F in Detroit in 1908. Additional financing was supplied by the stockholders of Northern and Wayne, which included some of the financial elite in Detroit. E-M-F was capitalized at $500,000, though far less was paid in. EMF contracted with Studebaker, a leading Indiana carriage producer that had been trying for many years to diversify into autos, to market half of its cars in the West and South through its large carriage dealer network.

Flanders developed a modern, high-quality car at a low price of $1,250, which he was able to achieve by exploiting the advantage of a large-output production process that he had helped develop at Ford. He introduced a number of innovations and innovative practices in the production process at E-M-F, including the widespread use of jigs to hold parts to improve the precision of manufacturing, machines to smooth multiple faces of a casting at the same time, and spindle drills that could drill more than one hole in a casting in each operation. He established a chemical and physical laboratory to evaluate parts from suppliers and had the first plant in the industry with dies and presses dedicated to parts production.

Everitt and Metzger left the company early in a dispute about Studebaker's role. They were involved in the founding of subsequent auto firms, none of which was as successful as E-M-F. Subsequently Studebaker tried to strong-arm Flanders to force him to sell out his shares cheaply. Flanders ended up selling out to Studebaker in 1910 and leaving two years later. Initially he reorganized the remnants of the U.S. Motors merger and later rejoined Everitt and Metzger in new firms they jointly founded. E-M-F's car was among the leading makes from its outset in 1908, and Studebaker was a leading producer for many years, earning large profits for the first ten years after it acquired E-M-F. In 1925, Studebaker moved its Detroit plant to Indiana, where it was based.

Hupp Motor Car Company

Robert Hupp was born in Grand Rapids, Michigan, in 1876. He joined Olds Motor Works as a laborer in 1902 and worked his way up to manager of the repair department before departing in 1905. He subsequently worked for Ford, initially in charge of the repair, claim, and accessory departments and then as an assistant in the purchasing department and to the production superintendent (Sudzarek 1996).

Hupp's experience at Ford gave him insight into factory and cost conditions. He left in 1908 and began designing a prototype for a medium-priced automobile. Hupp Motor Car Company was formed in 1908. Joining Hupp, who was president, were J. Walter Drake, who was appointed vice president and general manager; Charles Hastings, who was assistant general manager in charge of sales, service, and accounting; and Emil Nelson, who was the chief engineer. Drake, who was born in Sturgis, Michigan, in 1875, gave up a thirteen-year law practice to join Hupp. Hastings, who was born in Hillsdale, Michigan, in 1858, had worked for Olds Motor Works since 1901, rising to sales manager for

Olds's Detroit and Lansing offices and for Europe. In 1907 he joined the Thomas-Detroit Company as office manager and sales executive, but left when Hugh Chalmers took over and brought in his own people. Nelson had been an engineer at Olds and subsequently Packard and was most responsible for designing the first Hupp car (Kollins 2002d, Sudzarek 1996).

Hupp Motor Company was formed with a capital stock of $25,000, with only $3,500 initially paid in. Subsequently, Edwin Denby, who would become secretary of the navy, bought a 20 percent interest in the company for $7,500, increasing working capital to $11,000 (Sudzarek 1996). Charles Hastings was instrumental in the ability of the company to overcome its extremely limited initial funding. He took Hupp's initial prototype to the Detroit automobile show, where he secured orders for 500 cars and $25,000 in advance payments, which was enough to begin production (Davis 1988).

Hupp's first car, introduced in 1909, was a four-cylinder runabout priced at $750. It had a patented three-point suspension that made it comfortable for two large people. It was an instant success, with sales of 1,600 cars in 1909, 5,340 cars in 1910, 6,079 cars in 1911, and 7,640 in 1912, when it attained the sixth position in the industry. Years later Henry Ford noted, "I recall looking at Bobby Hupp's roadster at the first show where it was exhibited and wondering whether we could ever build as good a small car for as little money" (quoted in Sudzarek 1996, 165). In 1912, along with Oakland, Hupp introduced the first all-steel open car body based on a design by the body manufacturer Edward Budd, which attained the second highest rank in Abernathy, Clark, and Kantrow's (1983) list of innovations.

Hupp formed his own companies to improve the manufacture of his car, and these were consolidated in 1910 into Hupp Corporation, with Hupp as president and Charles Hastings as vice president. Hupp wanted to expand his companies along the lines of General Motors, which was resisted by his partners, Drake and Hastings. Hupp sold his stock and started subsequent companies that were never as successful as Hupp Motor Car Company, which continued to be among the industry leaders into the 1930s.

The Dodge Brothers Motor Car Company

John and Horace Dodge were born in Niles, Michigan, in 1864 and 1868, respectively. Their father was a machinist and iron worker, and they learned their trade in his shop. After their initial agreement with Ford Motor Company, they became Ford's principal supplier of engines, transmissions, and axles, and Ford was their only customer. Their dependence on Ford led them to suggest in 1912 that Ford buy them out, but after Ford dawdled, they started their own automobile firm in 1914 (Kollins 2002a).

They announced they would double the size of their own plant to enter automobiles (Sudzarek 1996). Dodge Brothers was capitalized at $5 million. In a famous lawsuit in 1917 in which Ford was enjoined from reinvesting most of his profits, it was revealed that the Dodge Brothers had received $5.5 million in dividends on their original 10 percent stock holdings in Ford Motor Company, valued at $36 million. They used these proceeds to fund nearly all of their initial capitalization.

They designed a car that was basically an upgraded Model T, with features conspicuously missing on the Model T. Their initial car was so good that only minor changes were needed in the next ten years. They learned from their Ford experience not to use Ford's planetary transmission and other Ford limitations (Kollins 2002a). The initial Dodge offering sold for $785. Approximately 45,000 Dodges were built in 1915, the best first year for any firm up to that point (Kimes 1996), vaulting Dodge into third place in the industry. It maintained its high position for a number of years, including throughout the lifetimes of the Dodge brothers.

Dodge teamed with the body manufacturer Edward Budd in the design of the steel body of their initial car, which was a considerable engineering challenge. Although Hupp had a head start, Dodges were popularly known as the first all-steel-bodied automobiles (Sudzarek 1996). On Abernathy, Clark, and Kantrow's list of innovations, Dodge was credited with the first mass production of an all-steel open car body, which achieved the fourth highest rank among innovations. In 1923, Abernathy, Clark, and Kantrow listed them as the producers of the first all-steel closed sedan car body, which achieved the third highest rank among innovations.

The Dodge brothers both died in 1920, and eventually their company was sold to Chrysler Corporation in 1928.

Spin-Offs Founded by William Durant
The next two companies were founded by William Durant after each of the times he was ousted from General Motors.

Chevrolet Motor Car Company
In 1909, after the formation of General Motors, the Model 10 was Buick's big seller, accounting for 8,100 of its total production of 14,606 cars (Dunham and Gustin 1992). It was William Durant's main foray into a less expensive car, priced below $1,000, with the goal of eventually competing with Ford's Model T (Weisberger 1979). Buick also introduced a high-wheeled buggyabout designed by Alanson Brush, who was working for General Motors after Oakland was acquired by GM. It sold for $450, but it was not successful and was abandoned after two years (Dunham and Gustin 1992). After Durant lost control of GM to bankers in 1910 and then left GM, GM oddly dropped the Model 10 and substituted less successful cars in its place. This may have been a reflection of the desire by the bankers that took over the management of GM to produce a bigger car (Dunham and Gustin 1992). It was the move that Durant disagreed with most, and when he left GM, he intended to develop a smaller car that could ultimately compete for the market largely captured by the Model T.

He organized three companies: Chevrolet Motor Car Company, Mason Motor Company, and the Little Motor Company in Detroit and Flint, Michigan. They were headed by Louis Chevrolet, Arthur Mason, and Big Bill Little, respectively, all of whom had been employees of Buick. Mason, which was organized to produce motors, and Chevrolet were capitalized initially at $100,000, but perhaps no more than $10,000 was paid into either. Little was capitalized at $1.2 million, with $823,200 in stock issued, but only $4,827 was initially made available to draw on, and in the long run only $36,500 of outside cash was put into Little (Weisberger 1979). Durant was able to get the three enterprises going by purchasing for $200,000 all the assets of the expiring

Flint Wagon Works, including a factory and inventories of wagon and auto parts, with Flint Wagon Works accepting a personal note from Durant (Weisberger 1979). Then in 1912, Durant-Dort Carriage Co., whose profits had been declining due to the decline of the carriage industry, sold its founder more work space, the Imperial Wheel Factory, for $200,000, which was almost certainly paid in Chevrolet stock. Later Chevrolet recapitalized at $2.5 million, with Durant-Dort Carriage Co., for which Durant was still the treasurer, taking half the stock (Weisberger 1979). Dort became a vice president of Chevrolet, although within a year he withdrew from all business arrangements with Durant.

Little initially designed an attractive but not durable small car, while Chevrolet designed an uneconomical large car. Durant combined these efforts under the Chevrolet banner, and he introduced the H series of cars, which included a touring car at $875 and a roadster at $750. These were well received, as was the Chevrolet Four-Ninety, which was introduced at the end of 1914 to sell at a price of $490. In 1915 Durant used Chevrolet to reacquire control of General Motors, and in 1916 Chevrolet sales were 62,522 cars, which increased to 125,004 in 1917 (Kimes and Ackerson 1986), vaulting Chevrolet into fourth place in the industry. Ultimately, in 1927 Chevrolet displaced Ford as the number one seller in the industry and propelled General Motors to become the leader of the industry.

Durant Motor Company

After another buying spree at General Motors, Durant lost control for good of General Motors in 1920. In 1921 he organized Durant Motors Company Inc. The company was capitalized at 1 million shares with no par value, and initially 500,000 shares were issued at $10 per share. These shares were fully subscribed to by sixty-seven personal friends of Durant, who in total pledged to buy $7 million in stock. Many of his prior associates at Chevrolet and General Motors joined Durant. Offices were set up in New York, and plants in Flint, Long Island City, New York, and later Muncie, Indiana, were established.

Durant's first product was a four-cylinder car that was well received. He was more successful with the Star car introduced in 1922 at prices as low as $319. In 1923 Durant Motors attained fifth place in the industry, after which it declined but remained among the industry leaders for the rest of the 1920s. Durant again wanted to sell a full range of cars, and he purchased Locomobile, which was producing an unsuccessful luxury car that proved unsuccessful as well at Durant Motors. Durant Motors started losing money in the latter part of the 1920s and ceased producing in 1932.

Other Top Spin-Offs
There were two other spin-offs, Paige-Detroit Motor Company and the Chandler Motor Company, that made it into the ranks of the leaders but were not descended from Olds, Cadillac, Ford, and Buick/GM. Two of Olds's descendants, Maxwell and Chalmers, also were combined into Chrysler Corporation, which eventually joined GM and Ford as one of three powerhouses of the industry. These three firms are reviewed.

Paige-Detroit Motor Company
Fred Paige was president and general manager of Reliance Motor Car Co. of Detroit, which built two-cylinder cars until 1907, after which it concentrated on trucks. It was sold to General Motors in 1909, and Paige left and enlisted an engineer, Andrew Bachle, to design an automobile with a three-cylinder, two-cycle motor generating 25 horsepower. Harry Jewett was born in Elmira, New York, in 1870. He managed coal mines and started his own retail coal distributorship in Detroit in 1895. In 1909 he teamed with Paige to found Paige-Detroit Motor Company, with Paige as president. The company was capitalized at $75,000, with Jewett and his friends the main investors (Kimes 1996).

The initial car was not successful, and Jewett replaced Paige as president. Through further experimentation and additional hiring, eventually a successful car was developed, and the company prospered through careful management. Paige-Detroit produced 4,631 cars in 1914, 7,749 in 1915, and 12,456 in 1916, attaining fifteenth place in the industry in 1916. It subsequently slipped below fifteenth place but returned to the ranks of the leaders in 1925 and was sold to the Graham Brothers, who had been leading truck producers, in 1928 (Sudzarek 1996).

Chandler Motor Company

Frederick Chandler was born in Cleveland in 1874. After two years of high school, he began working for H. A. Lozier and Company of Cleveland, a manufacturer of sewing machines and bicycles. Lozier sold its bicycle business and in 1905 introduced a high-quality car that was immediately successful. Chandler became the manager of Lozier's sales agencies in the United States and Europe. The principal stockholders of Paige-Detroit were at the heart of an investing group that raised $1 million to acquire control of Lozier in 1910 and move it to Detroit (Davis 1988) to compete in the luxury market with Packard of Detroit. Chandler became vice president and general manager in 1911.

Lozier sold some of the most expensive cars in the United States Chandler tried to get it to produce a lower-priced car but was unsuccessful. In 1913 he and four other Lozier executives, including the sales manager, experimental engineer, treasurer, and the New York branch manager, resigned and formed Chandler Motor Car Co. with an authorized capital of $425,000. They had temporary quarters in Detroit but Chandler located permanently in Cleveland. They introduced a light, six-cylinder car with many similarities to Lozier's more expensive car, and in 1915 they sold 7,000 cars. By 1926 Chandler was selling over 20,000 cars, which qualified it for seventeenth place in the industry. When it experienced hard times in 1928, it was sold to Hupp Motor Car Company.

Chrysler Corporation

Chrysler was not a spin-off but was an outgrowth of two earlier spinoffs, Mawell-Briscoe and Chalmers (originally E. R. Thomas-Detroit). After the failure of the U.S. Motors merger, Maxwell-Briscoe was effectively reorganized as Maxwell Motors. Walter Flanders traded stock in a failing company he had formed for cash and stock in Maxwell and assumed its leadership. Maxwell inherited numerous plants from U.S. Motors, but Flanders sold off many of them and concentrated manufacturing operations in the Midwest with headquarters in Detroit. He focused the company on a new Maxwell line that was successful, and Maxwell returned to the ranks of the industry leaders. In 1917 a deal was struck with Chalmers, which was experiencing hard times, to lease the plants and assets of Chalmers for five years for $3 million. Flanders retired in 1919 with the company in good shape, but it subsequently declined.

Walter Chrysler was born in Ellis, Kansas, in 1875. He came to Detroit to work for Buick and rose to become president of Buick, but quit in 1919 over a disagreement with William Durant about the management of Buick. He was brought in to reorganize Maxwell and Chalmers in 1921 when he was still reorganizing Willys-Overland, another major firm that was experiencing financial difficulties. He introduced a new car at Maxwell that resembled a car he had commissioned at Willys-Overland, and the company evolved into Chrysler. The car was very successful, and when Chrysler subsequently acquired the Dodge Brothers in 1928 it had the full slate of cars it needed to compete with GM and Ford.

General Themes

A number of patterns emerge from the histories of the spin-offs.

Strategic disagreements and control changes were common in the first twenty years or so of the automobile industry and played an important role in the formation of spin-offs. Indeed, the founding fathers of the industry in Detroit— Ransom Olds, Henry Ford, Henry Leland, the Dodge Brothers, and William Durant—all left their companies after a major strategic disagreement and started one or more prominent spin-offs. Other prominent spin-offs that resulted from disagreements include Maxwell-Briscoe, E. R. Thomas-Detroit, Hudson, Brush Runabout, Oakland, and Chandler. Strategic disagreements also led a number of the founders of prominent spin-offs, such as Maxwell at Northern, Coffin and Chapin at E. R. Thomas-Detroit, Everitt, Metzger, and Flanders at E-M-F, and Hupp at Hupp Motors, to leave their firms to found new spinoffs.

Many of the disagreements involved the types of automobiles that would be most profitable to develop. A number of the spin-offs were founded when the parent firm was not willing to depart much from its successful strategy even when the industry was evolving in new directions. This was particularly true when parents resisted suggestions by employees to produce a smaller variant of the parent's car. E. R. Thomas-Detroit, Hudson, Chevrolet, and Chandler are examples of such spin-offs. Brush Runabout and Oakland are also instances of an employee with expertise in smaller cars leaving when the parent, Cadillac, drifted over time into larger cars. Maxwell-Briscoe was started under similar circumstances. The reverse reason motivated the leading spin-offs that came out of Ford Motor Co. Ford channeled all its energy into the Model T and resisted incorporating many new developments into the Model T that might have necessitated the development of new, higher-quality models. Both E-M-F and the Dodge Brothers followed many of Ford's practices but produced higher-quality, more expensive cars, while Hupp began with a car similar to the Model T but soon migrated into larger cars.

The founders of the leading spin-offs were prominent men who had amassed some wealth, but generally not enough to finance their firms alone. Consequently, they had to seek finance elsewhere, mostly from wealthy individuals, as few firms were able to attract finance from banks or investment bankers such as J. P. Morgan. These wealthy individuals had little knowledge of either the technology or market for automobiles and thus had to rely on the advice of others to guide their investments. Not surprisingly, the men who played this advisory role came from the automobile industry itself, but unlike modern venture capitalists, they also generally participated in the management of the firms they helped to finance. Nearly all the spin-offs had such men, including Benjamin Briscoe (Maxwell-Briscoe), Reuben Shettler (Reo), E. R. Thomas (E. R. Thomas-Detroit), Roscoe Jackson (Hudson), Hugh Chalmers (Hudson, Saxon), Frank Briscoe (Brush Runabout), Henry Leland (Lincoln), Barney Everitt and William Metzger (E-M-F), John and Horace Dodge (Dodge Brothers), and William Durant (Chevrolet and Durant Motors). Like modern venture capitalists, a number of these individuals, as well as others such as Jonathan Maxwell, Roy Chapin, Howard Coffin, Alanson Brush, Robert Hupp, and Walter Flanders, were also involved in multiple spin-offs, many of which did not make it into the ranks of the industry leaders.

Thus, incumbent firms served as the breeding grounds for both employees who left to found their own firms and the men who helped finance these firms. This imparted a self-reinforcing character to the growth of the industry around Detroit beginning with Ransom Olds.

Indeed, Ransom Olds had a profound influence on the evolution of the industry in Detroit, much as Charles Brush had on the industrial development of Cleveland around the same time (see chapter 1, this volume). This began with Olds's subcontractors. It was presumably the experience that Henry Leland gained supplying Olds with engines and transmissions that led the frustrated stockholders of Henry Ford Company to turn to Leland for advice about what to do with their assets. Without Leland's counsel and later involvement in the company, almost surely the stockholders would have dissolved their company and Cadillac would never have emerged. Presumably it was also the experience the Dodge Brothers got supplying Olds with engines and transmissions that enabled them to evaluate the prospects of Henry Ford's third start-up and decide to become centrally involved in the production of its cars. Given Ford's failure to actually produce cars in his two prior ventures and the scanty finance he raised for Ford Motor Company, it seems likely that without the Dodge Brothers, there would never have been Ford Motor Company and later the Dodge Brothers Motor Car Company. Yet another of Olds's subcontractors, the Briscoe Brothers, played a prominent role in the early financing of Buick, Maxwell-Briscoe, and Brush Runabout, all of which became leading firms.

Ransom Olds's influence transcended his subcontractors. He himself later left Olds Motor Works and was involved in the founding of Reo, another major firm. The financing for Reo was organized by Ruben Shettler, one of the financiers of Olds Motor Works when it expanded into automobiles. Olds also employed Coffin, Chapin, Jackson, and Hupp, all of whom were involved in the founding of important spin-offs and played a role in attracting finance from wealthy individuals. Olds employees also started a number of other spin-offs and were prominent in the management of many top firms, leading one writer to dub Ransom Olds as the "schoolmaster of motordom" (Doolittle 1916).

Another important individual whose influence operated centrally through his financial and organizational skills was William Durant. Durant had made his original fortune in the carriage and wagon industry by capitalizing on a cart technology that he acquired the rights to and then raised a considerable amount of money to exploit in his newly formed firm, the Durant-Dort Carriage Company. He did much the same thing in Buick. He carefully tested the novel engine that the Buick Motor Company had developed and then, using his reputation and contacts, raised a great deal of money from wealthy individuals to finance an enormous expansion of Buick. Given how poorly Buick was performing up to Durant's involvement, it seems likely that without Durant, Buick would have died and General Motors would never have gotten started in Detroit. Moreover, Durant did much the same thing with both Chevrolet and Durant Motors, and it was Chevrolet that General Motors used to displace Ford and become the leading firm in the industry.

Other financiers also played key roles in a number of the leading spin-offs in Detroit. Hugh Chalmers, who was brought into the industry from NCR, helped finance Hudson and was instrumental in the finance of Saxon. Barney Everitt, William Metzger, and especially Walter Flanders were also key individuals who helped finance and manage E-M-F and other, lesser spin-offs. Like Chalmers, these men were drawn to Detroit from elsewhere, as were many of the important participants in the industry like Jonathan Maxwell and Walter Chrysler. A key part of the spin-off story is that although these men originated elsewhere, when they founded new firms, they invariably located them in Detroit or nearby. In part, this may have had to do with raising money from local men, who no doubt preferred to invest in companies they could more easily monitor. But it also seems likely that the broader economic and social roots of the founders of the spin-offs kept them from venturing very far. The net result was that Detroit became the capital of the U.S. automobile industry, driven in large part by the experienced automobile men who were central figures in the finance of the spin-offs emanating from Olds Motor Works, Cadillac, Ford, and Buick/GM.

The influence of these spin-offs transcended Detroit. Judging by the list of innovations compiled by Abernathy, Clark, and Kantrow (1983), the spin-offs were prominent innovators in the industry. Abernathy, Clark, and Kantrow identified fifty innovations between 1902 and 1925 that had a rank of 4 or greater (7 was the highest rank) in terms of their impact on the production process. These generally constituted the most important innovations in the industry. Twenty-six of the fifty were introduced by General Motors and Ford, with Ford itself a spin-off. Thirteen of the other twenty-four innovations were developed by seven of the spin-offs descended from Olds, Cadillac, Ford, and Buick; this increases to fifteen of the twenty-four innovations if E-M-F is included.[11] Automobile innovations tended to diffuse rapidly, suggesting that the social returns to innovation significantly exceeded the private returns. As such, the spin-offs generated substantial social returns through the innovations they introduced. At times, they also compensated for deficiencies in their parents that led their once-successful parents to decline when they were unable to respond to technological advances. As such, the spin-offs provided insurance against the obsolescence of their parents.

At the same time, the parents of the spin-offs were seemingly important influences on their offspring. While spin-offs produced different cars than their parents did, there typically was much overlap between their cars and those of their parents, suggesting they learned valuable lessons from their parents. Maxwell-Briscoe initially developed a car with similar innovative features to Northern's. E. R. Thomas-Detroit initially produced a car that was a cross between the Curved-Dash Runabout of its parent and higher-priced cars. Hudson's first car was a less elaborate and cheaper version of the car Howard Coffin had designed for F. R. Thomas-Detroit. Oakland initially produced a car that Alanson Brush had developed at Cadillac. E-M-F sold a high-quality car at a relatively low price by exploiting the advantage of a large-output production process that Flanders had helped develop at Ford. Robert Hupp designed a car that even Henry Ford envied. The Dodge Brothers produced an improved version of the Model T. Chandler produced a smaller and less expensive variant of Lozier's luxury car. More generally, the fact that nearly all the spin-offs that made it into the ranks of the leaders were descended from Olds, Cadillac, Ford, and Buick suggests that the leading firms were especially fertile places to learn about the organizational and technological challenges facing automobile firms.

If indeed spin-offs learned valuable lessons from their parents and then went on to develop innovations that contributed importantly to the technical advance of the industry, the process of economic growth involving the spin-offs would seem rife with externalities. Whatever lessons they learned from their parents were unintended by-products of their parents' efforts. To the extent the spin-offs developed important innovations that quickly diffused throughout the industry, they were not able to appropriate the full value of their innovations. Since 1925 there has been virtually no entry into automobiles, by spin-offs or any other firms. Consequently, the automobile industry has lost a force that compensated for the conservatism of many of its leaders and helped propel its technology forward. As the number of firms remaining in the industry has dwindled, authority for choosing which innovations to pursue has become concentrated in an ever smaller number of producers. This may help explain why the big three U.S. firms have been so vulnerable to international competition from later entrants and why Detroit and neighboring cities like Flint, once vibrant and growing, have stagnated. Without its spin-offs, the United States no longer seems to be the major source for technological advance in automobiles that it once was.

Notes

I thank Peter Thompson for many helpful discussions. Support is gratefully acknowledged from the Economics Program of the National Science Foundation, Grant No. SES-0111429, and from IBM through its faculty partnership awards.

1. This section draws from Epstein (1928), Federal Trade Commission (1939), Rae (1959), and Smith (1968).

2. Entry and exit dates are based on the first and last year of commercial production of all makes of a producer. Mergers and acquisitions were treated as continuations of the firm whose name was retained, or in the case of mergers the largest firm involved, with the other firms treated as exits.

3. In addition to Detroit, the Detroit area was defined to include the following locations in Michigan, all of which are within approximately 100 miles of Detroit: Adrian, Chelsea, Flint, Jackson, Marysville, Oxford, Plymouth, Pontiac, Port Huron, Sibley, Wayne, and Ypsilanti. The boundaries of this region were chosen to reflect multiple locations of some of the firms within the region.

4. Even these figures understate the dominance of Detroit, with two of the other three prominent non-Detroit firms having links to Detroit. One, Studebaker, entered initially by marketing the cars of a Detroit company, E-M-F, that it later acquired. The other, Nash (originally Jefferys), was a leading firm acquired by Charles Nash, the ex-president of General Motors, in 1916.

5. See Klepper (2002b) for a detailed description of how the firms were classified.

6. The history of Ransom Olds's enterprises is recounted in May (1977).

7. The history of Cadillac is recounted in Leland (1966).

8. See Nevins (1954) and Yanik (2001) for Flanders's role in Ford Motor Company.

9. See Dunham and Gustin (1992) and Weisberger (1979) for a history of Buick.

10. The histories were reconstructed from Kimes (1996), Kollins (2002a, 2002b, 2002c, 2002d), Sudzarek (1996), and various company accounts that are noted. All production figures are from Bailey (1971). One of the firms that is reviewed, E-M-F, was not classified as a spin-off because it built on two prior entrants, but it was sufficiently like a spin-off to warrant having its history recounted.

11. No other firm was especially innovative during the 1902–1925 period, with the other nine top-ranked innovations each introduced by a different firm.

References

Abernathy, William J., Kim B. Clark, and Alan M. Kantrow. 1983. *Industrial Renaissance*. New York: Basic Books.

Bailey, L. Scott. 1971. *The American Car since 1775*. New York: Automobile Quarterly.

Davis, Donald F. 1988. *Conspicuous Production: Automobiles and Elites in Detroit, 1899–1933*. Philadelphia: Temple University Press.

Doolittle, James R. 1916. *The Romance of the Automobile Industry*. New York: Klebold Press.

Dunham, Terry B., and Lawrence D. Gustin. 1992. *The Buick: A Complete History*. Kutztown, Pa.: Automobile Quarterly.

Epstein, Ralph. 1928. *The Automobile Industry*. Chicago: A. W. Shaw Company.

Federal Trade Commission. 1939. *Report on the Motor Vehicle Industry*. Washington, D.C.: U.S. Government Printing Office.

Gunnell, John, ed. 1992. *Standard Catalog of American Cars, 1946–1975*. 3rd ed. Iola, Wisc.: Krause.

Halberstam, David. 1986. *The Reckoning*. New York: Avon Books.

Kimes, Beverly R. 1996. *Standard Catalog of American Cars, 1890–1942*. 3rd ed. Iola, Wisc.: Krause Publications Inc.

Kimes, Beverly R., and Robert C. Ackerman. 1986. *Chevrolet, A History from 1911*. Princeton, N.J.: Automobile Quarterly Publications.

Klepper, Steven. 2002. "The Evolution of the U.S. Automobile Industry and Detroit as its Capital." Mimeo.

Kollins, Michael J. 2002a. *Pioneers of the U.S. Automobile Industry, Vol. 1, The Big Three*. Warrendale, Pa.: Society of Automotive Engineers.

Kollins, Michael J. 2002b. *Pioneers of the U.S. Automobile Industry, Vol. 2, The Small Independents*. Warrendale, Pa.: Society of Automotive Engineers.

Kollins, Michael J. 2002c. *Pioneers of the U.S. Automobile Industry, Vol. 3, The Financial Wizards*. Warrendale, Pa.: Society of Automotive Engineers.

Kollins, Michael J. 2002d. *Pioneers of the U.S. Automobile Industry, Vol. 4, The Design Innovators*. Warrendale, Pa.: Society of Automotive Engineers.

Leland, Wilfred C. 1966. *Master of Precision*. Detroit: Wayne State University Press.

May, George S. 1977. *R.E. Olds: Auto Industry Pioneer*. Grand Rapids, Mich.: William B. Eerdmans Publishing Co.

Nelson, Richard R. 1990. "Capitalism as an Engine of Progress," *Research Policy* 19, 193–214.

Nevins, Allan. 1954. *Ford The times, the Man the Company*. New York: Charles Scribner and Sons.

Rae, John B. 1959. *American Automobile Manufacturers*. Philadelphia: Chilton.

Renner, Gail K. 1973. *The Hudson Years: A History of An American Manufacturer*. Ph.D. dissertation, University of Missouri–Columbia.

Seltzer, Lawrence H. 1928. *A Financial History of the American Automobile Industry*. Boston: Houghton Mifflin.

Smith, Philip H. 1968. *Wheels within Wheels*. New York: Funk and Wagnalls.

Szudarek, Robert G. 1996. *How Detroit Became the Automotive Capital*. Detroit: Typocraft Company.

Weisberger, Bernard A. 1979. *The Dream Maker*. Boston: Little, Brown.

Yanik, Anthony J. 2001. *The E-M-F Company*. Warrendale, Pa.: Society of Automotive Engineers.

3

Why Did Finance Capitalism and the Second Industrial Revolution Arise in the 1890s?

Larry Neal and Lance E. Davis

In the last quarter of the nineteenth century and the first decade of the twentieth, the separate worlds of technology and finance were both transformed, much as they were in the last decade of the twentieth and continue to be in the first decade of the twenty-first. In technology, historians of science now regard the scientific breakthroughs that occurred in this period as defining the research agenda that spawned the scientific miracles of the twentieth century and continue to transform our daily lives. Concentrated in this short time span were inventions of electric generators, dynamite, photographic film, light bulbs, electric motors, internal combustion engines, steam turbines, aluminum, and prestressed concrete—and all this even before the turn of the century. The pre–World War I surge of invention culminated with airplanes, tractors, radio, plastics, neon lights, and synthetic fertilizers in the first decade of the twentieth century. So marvelous were these transformations of matter and energy that, arguably, scientists of the late eighteenth century would have had difficulty comprehending the devices that were in common use throughout the advanced industrial economies in 1914. But the conceptual grasp of an Edison or a Tesla, according to Vaclav Smil (2001), could probably take in discussions of modern scientific endeavors as related in, say, a recent issue of *Scientific American*. The scientific concepts that underlay the technological advances of the Second Industrial Revolution are still evolving today, but at the time they represented an unprecedented breakthrough in the advance of human knowledge.

This episode of explosive inventiveness remains a benchmark for technological progress to this day. Economic historians label this period as the Second Industrial Revolution, when the technological basis of modern industry began its shift from steam power to electricity, the fuel basis from coal to oil, transportation from rail to auto, and fabrics from cotton to synthetics of rayon and nylon. Financial historians label the same period either the classical gold standard or the spread of finance capitalism. The years 1880 to 1913 saw the spread of the gold standard throughout the industrialized world; the rise of joint stock banks in the United States and Germany to compete internationally with the great merchant banks of Britain; the explosion of new corporations whose shares were traded on the stock exchanges of London, New York, and Berlin; futures markets in organized commodity and currency exchanges; holding companies; and trust companies.

Financial historians have been struck for some time with the similarities in patterns of innovation, speculation, currency crisis, and financial panics that occurred in both the 1890s and the 1990s. The 1990s saw the full emergence of a global financial market; the 1890s witnessed the spread of finance capitalism. Historians of technology are already seeing analogies between the wave of scientific breakthroughs at the turn of the nineteenth century with those occurring at the turn of the twentieth century, even if the basis for scientific advance has moved from electro-chemical-mechanical interactions to biochemical and electronic interactions. The 1990s experienced the information technology revolution world wide; the 1890s underwent the Second Industrial Revolution based on electricity and petroleum. In short, there is ample reason to explore the interactions of economic growth, technological advance, and financial innovation of the 1890s in order to comprehend better the internal dynamics of the 1990s and the start of the twenty-first century. We see clearly that the rapid rise of information technology has interacted positively, and to excess, with the rapid rise of the global financial market. It seemed apparent to observers of events at the turn of the previous century as well that the rapid spread of new technology in electricity, telephony, wireless, and petroleum owed much to the concurrent rise of finance capitalism. Indeed, Joseph Schumpeter built his entire theory of business cycles and the internal dynamic of capitalism on the interaction of the worlds of finance and technology, based on his observations of developments in Europe and the United States in the late nineteenth and early twentieth centuries. More recently, Carlota Perez (2002) has expanded Schumpeter's ideas into a general theory of bubbles, both financial and technological, at the end of golden ages.

In both historical epochs, the leading country in developments in both finance and technology was the United States. Why this was so at the turn of the nineteenth century, and continues to be so at the turn of the twentieth century, we believe, lies in major part in the particular features of the American financial system with its complementary array of financial intermediaries and capital markets. In both epochs, the United States stood apart from the other leading industrial nations in the fragmentation and diversity of its banking system. The repeated banking crises under the national banking system of the late nineteenth century led eventually to the formation of the Federal Reserve System in 1914. The disintermediation that arose from the inflationary shocks of the 1970s led to a strengthening of the Fed's regulatory powers and independence from government control. While standard histories focus on the policy responses to the banking crises that occurred in each case, our attention focuses on the competitive responses that occurred in the capital markets of the United States as well. In the 1890s, new financial intermediaries arose—trust companies—that invested directly in securities traded on competing stock exchanges throughout the country. In the 1970s, new financial intermediaries also arose—money market mutual funds—that invested directly in high-yielding government securities trading on exchanges. The competitive quality of American capital markets contrasts as well as American banking fragmentation with the financial sectors in other leading industrial nations. While much of the theoretical literature of finance finds it useful to regard financial institutions and capital markets as substitute forms of financial intermediation, the historical experience highlights their complementarity in the long run.

Moreover, these distinctive features of the American financial sector meshed well with the peculiarities of the emerging technologies. Studies of new technology show that some of them require privileged information to become effectively operable, while others need open access to other innovators and potential users to become commercially viable. Privileged information technology required personal finance such as provided by relationship bankers; open access technology required impersonal finance available through open capital markets. When a new technology became commercially profitable, it attracted the attention of talented scientists and technical experts. In the American case, continued innovations and advances in productivity followed as a new technology diffused rapidly throughout the economy.

Liquid and deep capital markets are especially important for the finance of what Charles Perrow (1986) calls "loose coupling" or "nonlinear" technologies. Electricity was such a technology at the end of the nineteenth century, as was information technology at the end of the twentieth century. The earlier technologies derived from iron and steel and industrial chemicals were what Perrow calls "tight coupling" or "linear" in nature. Lending officers in banks could see clearly what a proposed innovation in the production processes of iron, steel, soda, and sulfur could accomplish at the level of an individual plant or firm. To realize the commercial prospects for electricity, telephony, or the Internet, by contrast, a firm needs complementary inputs from a social infrastructure. Financing the infrastructure of transmission lines or fiber-optic cables needed to make the new technology commercially viable requires access to large, liquid capital markets. We argue that a financial system concentrated excessively on either capital markets or banks cannot respond effectively to the financing needs of developing technologies. As technologies emerge from breakthrough scientific concepts to diffusion of generally useful applications, their financing needs to evolve as well—from personal or relationship finance of privileged technical knowledge to impersonal, capital market finance of networks of varied applications of the new technology.

In this chapter, we argue that the deflationary decade of the 1890s and the subsequent gold inflation decade of the 1900s provide a laboratory in which it is possible to study the distinct natural experiments that were being conducted with respect to institutional arrangements of finance. By institutions, we mean the rules, both formal and informal, that governed the operation of the global financial market of the time; by natural experiments, we mean the changes in regulations of the financial intermediaries and financial markets and the innovative responses by the financial intermediaries that operated on these markets.

These experiments in the financial sectors of the United States, Great Britain, and Germany were conducted at the same time that new technologies were exploding (sometimes literally) into each economy. Both the new technologies and the financial experiments undertaken in the 1890s created new opportunities for investors and their financial agents or intermediaries in all three countries. Our tentative conclusion is that the financial experiments that exploited the possibilities of expanding the scope of financial markets, as opposed to strengthening the security of the banks, succeeded in generating more rapid economic growth for their economies as the twentieth century began. But that conclusion is predicated on the assumption that banks and capital markets provided complementary financing for the emerging technologies in Germany and especially in the United States over the period 1880 to 1914. Banks developed relationships with inventive firms that were creating and exploiting proprietary information, while capital markets allowed large-scale systems of production and distribution to arise that propagated generic knowledge on a scale never before imagined.

The Financial Experiments of the 1890s

Comparing the different financial systems that had arisen in the core countries of the industrialized world in the earlier century and how they initiated or responded to the technological revolutions of that era will help us answer the question of private-public priorities and how they interact in the technological and financial spheres, at least for that past experience. For this purpose, we draw on our previous work dealing with the impact of the continued deflation that occurred in the first twenty-five years of the classical gold standard, 1873 through 1897, on financial markets in the advanced industrial countries, and the responses that occurred in the major stock markets of London, Berlin, and New York (Davis and Neal 2004).

The pressures of persistent and widespread deflation culminated in the 1890s, as every country adhering to the gold standard experienced directly or indirectly the effects of a series of financial shocks. After 1897, by contrast, a combination of gold discoveries, improved technology for exploiting known gold reserves, and the withdrawal of certain countries from the competition for gold reserves relieved deflationary pressures. Only the systemic crisis of 1907–1908 and the pressures of preparation for potential war, which did break out in 1914, marred the persistent expansion of the global financial system over the period 1897 to 1914. This period of gold inflation also spurred rapid economic growth in the core industrial countries and a continued outpouring of technical progress and rising productivity.

On all fronts—economic growth, financial expansion, and technical progress—the United States led the world then, as it has in the more recent period. Britain, which had earlier developed the institutions in finance and technology that made it the world's first industrial nation, seemed to falter when confronted with the challenges of financing the networks of infrastructure required to exploit the new technologies. Consequently, its rate of growth slackened, mainly because of lower rates of technical progress. Germany, by contrast, seemed to take up the American challenge in electricity, organic chemistry, and automobiles, but its reliance on financing led by the great banks eventually slowed its development of the new social infrastructure and new technology. German banks focused on rates of return, not rates of technical progress. The German capital markets reflected the political fragmentation of imperial Germany by marketing local and state government debt. Some excellent examples of electrical lighting and transport arose in some German cities as a result, but at the national level, finance was directed toward the military-industrial complex, not the scientific-industrial complex.[1]

To explain these differences among the financial sectors responsible for funding the development of new technologies at the time, we have focused in earlier work on the deflationary shock that each country experienced over the period 1873 to 1897, which culminated in the 1890s (Davis, Neal, and White 2004). We argued that persistent deflation in the Atlantic economies of the late nineteenth century created financial pressures on debtors, especially governments and railroads, that led to widespread financial crises in the 1890s. Faced with a common shock, the major stock exchanges in the first global financial market nevertheless responded in quite different ways—the result of differences in their political and legal environments and the persistence of their respective microstructures. Their distinctive responses to the challenges of deflation at the time provide a natural experiment for examining the results of different government policies and different financial innovations for the effective functioning of capital markets. We conclude that the New York and London stock exchanges were the most innovative, and their innovations the most beneficial for the long-run growth of their respective economies. New York's innovations, however, confronted directly the challenges of competition by other exchanges in New York and the rest of the United States, while London's self-interested responses constrained competition among its members. The regulatory reforms of the stock exchanges in Berlin and Frankfurt, by contrast, were the least helpful for the German economy up to the outbreak of World War I.[2]

The core industrial countries of the time—Great Britain, Germany, and the United States—each differed from the others in at least one important political or legal respect. Great Britain had a centralized political system with the financial power and the major capital market located in London, while Germany and the United States were more fragmented politically so that financial power and capital markets were more dispersed. There were also dramatic differences in their legal systems: Britain functioned on the basis of precedent-driven, judge-decided common law, overridden by statutory law of the central government when it was applicable. Throughout the nineteenth century, the central government restrained itself from any direct regulation of the stock exchanges, imposing temporary rules only at the outbreak of World War I. As a result of the dominance of common law with respect to stock exchanges in Great Britain, there were a number of securities exchanges. Although the numbers fluctuated, between nineteen and twenty-two provincial exchanges operated in each year between 1840 and 1914, and although the exchanges outside London tended to specialize in local issues, shunting of orders among the exchanges was common until 1912. Moreover, because of the internal rules of the London Stock Exchange (LSE), firms were limited in size, and as a result, there were no branches of London brokers or jobbers operating in Manchester, Bristol, or any of the other twenty provincial markets.[3]

Germany and the United States were fragmented, federal political systems with financial power dispersed and with regional capital markets that competed with each other. Moreover, Germany had its own version, or versions, of statutory civil law, and the United States had its own set of judges who interpreted common law in ways that increasingly diverged from the British cases that were originally taken as the binding legal precedents. While the United States was also based on common law, often extending back to British precedents, judicial decisions had to defer to statutory law as in Britain, but in the U.S., laws could and often did vary among the several states. In the United States over the years 1800 to 1970, there were some 200 local exchanges that operated at one time or another. That list includes such places as Spokane, Washington, hardly a major financial center. However, there were no constraints on the size on firms operating on the New York Stock Exchange (NYSE), and although there were local stockbroking firms operating in the regional markets, they faced direct competition from branches of firms with seats on the NYSE. The result of that competition can be seen in the distribution of business among the NYSE, other New York exchanges, and the major exchanges located outside New York. In 1910, the NYSE handled 68.5 percent of the total number of all stocks traded; other New York exchanges, 21.2 percent; and the regional exchanges in Boston, Philadelphia, and Chicago, 10.4 percent. In terms of the value of bonds traded, the NSYE handled 90.6 percent, other New York exchanges 1.5 percent, and the three regional exchanges, 7.9 percent (Michie 1987). In Germany, the unification of the Reich in 1871 under Prussian domination meant that the great universal banks concentrated their stock market activities in Berlin, at the expense of the Frankfurt exchange, which had been the leading stock exchange in Germany previously. The various regional exchanges in Germany soon lapsed into their respective niche markets, leaving the major market for government, railroad, mining, and industrial securities to the Berlin exchange (Gömmel 1992).

To sum up, one could argue that the fundamental institutional differences that created the distinctive American financial system came from the competitive environment of a federal political system combined with the permissive character of the American version of common law. Legal scholars have noted that American judges in the nineteenth century focused on allowing free access to a market, while British judges focused on allowing trade associations to set and enforce their internal rules— rules that often protected incumbent firms from competition by outsiders. It is ironic that in the case of the major stock exchanges, however, U.S. statutory law protected the NYSE from antitrust legislation on grounds that it was a club, while U.K. statutory law appeared to protect access of new entrants to the LSE. Competition from regional exchanges and alternative exchanges even in New York, however, kept the NYSE at the forefront of providing external finance for firms exploiting the new technologies. Gradual erosion of competitive forces within the LSE after 1890, as we explain below, limited its role in financing new technologies by British firms. Outsiders with a new technology to finance therefore stood a better chance of attracting the necessary funds from the variety of sources available in the American financial system rather than from a less competitive financial sector in Great Britain or Germany (Davis 1966).

The Effect of Deflation on the U.S. Financial Markets

According to Kenneth Snowden's analysis (1990), in the United States it was the continued effect of the deflation on the values of that nation's huge stock of railroad bonds that motivated the innovative responses of the NYSE in the 1890s. Snowden points out that the market response to persistent deflation in the United States—deflation that raised the real price of railroad bonds—first increased the wealth of existing bondholders, but then decreased the interest rate on bonds purchased by new investors. Because the U.S. railroad companies were private enterprises that lacked financial backing from the federal or state governments, they had originally offered very favorable terms to bondholders—terms that included not only high nominal interest rates but also a guarantee that the bonds would not be called or redeemed if their market price rose above par. As the price of more and more bonds did rise above par, railroads found themselves in the unpleasant position of having to continue to lay out high fixed nominal interest payments while at the same time they faced falling prices for their freight and passenger services. Moreover, they could not take advantage of the falling market yields to replace high-interest debt with new low-interest bonds, because they would have to buy the existing bonds at market prices; and they could not turn to the money markets to cover those costs because the value of the collateral available to back new bonds was declining due to the general deflation.[4]

The management of the American railroads responded in a variety of ways to this financial dilemma. Their strategies included attempts to maintain high prices through monopolistic cartel arrangements and financing further construction by selling stocks and bonds of newly incorporated railroad companies, rather than carrying out those operations through the established firms. Ultimately, however, their best recourse was to declare bankruptcy and throw themselves on the mercy of a judge's decision about the appropriate method for settling creditors' claims. At the time, there was no federal bankruptcy law, and therefore railroads declaring bankruptcy not only had the advantage of suspending interest payments while continuing normal operations during the time that they were in the hands of a receiver, but they also had some discretion in picking the judge, or at a minimum, the state that had jurisdiction over the legal proceedings and would decide the terms of reorganization (Campbell 1938). By the end of 1895, the series of competitive bankruptcies had put 25 percent of the total U.S. railroad mileage into the hands of receivers.

The suspension of interest payments to bondholders brought investment houses into the center of the reorganization schemes that were proposed in the series of attempts made to restore the long-run viability of American railroads. Three interrelated courses of action were developed and deployed: first, to replace the outstanding bonds with new bonds bearing a lower coupon rate; second, to write down the principal of outstanding bonds at the same coupon rate (essentially a partial default); and third, to substitute contingent income claims, usually in the form of preferred stock, for the existing bonds (Snowden 1990). The Union Pacific Railroad, which was not only the largest of the bankrupt roads but a railroad that was also the leading innovator in designing new financial assets, issued stock warrants—warrants that could be converted to bonds if the market price recovered. Across the board, the net result was to restore the profitability of American railroads, and profitability led to a new surge of investment in the period 1897 to 1907—investment that was focused on double-tracking, rail yards, and stations rather than on new routes (Neal 1969).

It has been argued that investors confronted with the uncertainty of the future yields on their holdings of railroad bonds turned to other possibilities for maintaining their rentier incomes. Snowden (1990), for example, concludes:

Had deflation and a reduction in yields not appeared in the late 19th century, as market participants expected, there would have been far less incentive for the stockholders of railroads to default.... In the absence of the delays created by the reorganizations, the rapid growth of rail capitalization that manifested itself between 1900 and 1913 would have continued to focus the attention of the investment houses and the bulk of investors primarily on the rails. The industrials, on the other hand would not have benefited from the change in investor attitudes that resulted from widespread rail bankruptcy. As a result, the market for industrial shares would have developed more slowly and been shaped to a larger extent by the individual promoters who began the process in the early 1890s. (405)

This scenario, however, while perhaps containing an element of truth, badly distorts the importance of the financial shenanigans of railroads in the evolution of the market for commercial and industrial securities. The sources for railroad finance were not severely limited after the 1890s, nor were the investment banks that had previously focused on railroad finance the leaders in developing finance for commercial and industrial corporations thereafter.[5] The effects of the railroad reorganizations, however, were reflected in some of the changes that occurred in the NYSE during the 1890s, as Snowden suggests. Although it would be another two decades before industrial and commercial securities became the center of activity on the NYSE, it is certainly true that the market for that sector's securities, especially preferred stock, became both formalized and important in the 1890s (Navin and Sears 1955). Moreover, Snowden is correct when he shows that the returns on these new securities were highly variable and provided investors with a high-risk but high-return alternative to their traditional railroad holdings. Finding out which alternative financial assets were most interesting to their respective customers was a challenge common to each of the stock exchanges in New York, London, and Berlin. Members of the NYSE and their competitors and collaborators in the regional exchanges responded in particular ways that proved especially beneficial for promoting the creation and spread of new technologies.

While the stock market panics of 1890 and 1893, which demonstrated to investors the risks now confronting them, produced government investigations of the stock market's operations in New York, only the state legislature was involved. The legislators in Albany were easily, and frequently, bribed into rescinding threatened regulations of the stock exchange. (It was also the case that regulators in New York City and Chicago were easily convinced to give electric power franchises to Edison and Insull, respectively.) The regulations of the NYSE were, however, revised, but the revisions were made by the operators of the exchange. The revisions came partly in response to the threat of competition from other exchanges—the Consolidated in New York and the regional exchanges elsewhere in the country. Over time, however, they were revised in large part because the competitive threat of other exchanges was reduced. As competition weakened, the threat of members deserting to other exchanges was reduced, and, as a result, the NYSE was able to impose more constraints on its members.

In the last decade of the nineteenth century, the exchange was able to institute two rule changes that strengthened its imprimatur of quality: a clearing mechanism for trades and listing requirements for new securities. In 1892, after three failed attempts, the governors finally established a clearing mechanism: a mechanism that was expanded until it included almost all listed securities by the end of the century (Sobel 1965, Wilson 1969). Clearing reduced the costs of completing trades, but was available only to members of the exchange and for listed securities. To list a security, the governing committee voted in 1895 to require that listed companies file annual reports, although it is clear that their word was still not law; they received no reports in either 1895 or 1896. By 1900, however, annual reports, including both audited balance sheets and profit and loss statements, became a prerequisite for both initial listing and retaining that listing (Sobel 1965). Both changes had been long considered by the members to be in their interests, but the competitive threats from other exchanges had previously prevented the governing committee from implementing them.

The NYSE's listing requirement had the desired effect of establishing the exchange as the blue chip market, creating an imprimatur of quality that has lasted to this day. The imprimatur greatly advanced the education of unsophisticated American investors of the late nineteenth century, and in so doing, it went a long way to solving the nation's capital accumulation and mobilization problems. The new requirements also greatly aided the exchange in its battle with its chief New York rival, the Consolidated Exchange. At the turn of the century, in terms of volume, about two-thirds as many shares were traded on the Consolidated as on the NYSE. Although competition continued through World War I, the NYSE's policies, which were designed to discourage members of exchanges located outside New York from dealing with the Consolidated and to deny the Consolidated easy access to the NYSE's prices, appear to have blunted, if not halted, the competitive threat. In renewal of the NYSE's contract with Western Union in 1900, a clause was inserted requiring Western Union to remove its telegraphs from the floor of the Consolidated Exchange. Although the clause was not upheld in the courts, the NYSE rigorously enforced rules against its members' dealing with members of the Consolidated and even opened an "Unlisted" department to take other business away from the Consolidated. After the stock market crash of 1907, only the curb market remained, and it was happy to serve as a complement to the NYSE, creating an initial market for new firms as a proving ground before they grew large and stable enough to be listed on the NYSE.[6]

The improvement in the NYSE's imprimatur of quality also made it possible to alter, although much less violently, its relationship with the New York Curb market, providing an alternative, less formal market for new securities issued by corporations including companies not listed on the NYSE for whatever reason. Previously, the Curb had existed somewhat uneasily alongside the NYSE. Between 80 and 90 percent of its business was carried out on behalf of members of the formal exchange. Gradually, as the Curb became a recognized part of the evolving securities market, its relations with the NYSE became better defined. In 1909, the representatives of the NYSE argued, "The curb market represents, first, securities that cannot be listed; second, securities in the process of evolution from reorganization certificates to a more solid status; and third, securities of corporations which have been unwilling to submit their figures and statistics to proper committees of the Stock Exchange" (New York State 1909, 44). By 1900, a listing on the NYSE provided a substantial guarantee of stability, and the Curb provided a market for riskier and more uncertain securities within the U.S. financial infrastructure. The two had become complementary rather than competitive organizations. Indeed, the Curb market provided the equivalent of what venture capitalists in the 1990s called "mezzanine finance," the intermediate stage of going public to a select groups of investors before listing on the Big Board.

While new technologies were emerging rapidly in the United States at the end of the nineteenth century, the accumulated effects of deflation on the secondary market for securities in the United States were leading to a series of innovative initiatives by businessmen engaged in stockbroking, especially those fortunate enough to be members of the NYSE. Driven primarily by the goal of restoring their incomes—incomes that had declined because of the loss of business as their wealthiest customers abandoned the stock market—the brokers took steps to retain their traditional customers and attract a wider customer base. To compensate for the disappointing returns now available in the dominant securities, railroad stocks and bonds, they widened the range of products available. Not only industrial and utility stocks in the new sectors, but also new forms of railroad securities—securities such as warrants, preferred stocks, and bond issues backed by specific forms of new capital—were promoted by the NYSE. To reassure their clients, they imposed listing requirements that steadily became more detailed and demanding. In order to reduce the costs of operating the exchange, they finally created a clearinghouse. To limit the threat of competition from competing exchanges, both within and outside New York, they tightened their control over access to the ticker tape, providing up-to-date price information.

Meanwhile, regional exchanges, especially in Chicago and Cleveland, brought shares to market of new companies in the emerging technologies (Deere, McCormick, Dow, Standard Oil) as well as in local electrical utilities and urban transport systems. The nation's unit banks vied in providing finance to local auto dealers so they could pay in advance for orders from Detroit's auto manufacturers.[7] In short, different financing techniques were available for different technologies at different stages of their respective developments. The beneficial effects (beneficial at least to the members of the NYSE) are seen in the turnaround in seat prices.

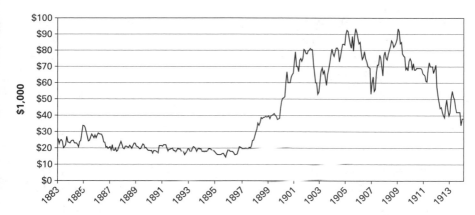

Figure 3.1
Seat prices: New York Stock Exchange, 1883–1914

From the lowest prices of $14,000 and $14,500 obtained on September 24, 1896, seat prices began a steady upward movement at the beginning of September 1898 and rose steadily until leveling off in 1906, before falling again after the crash of 1907 (see figure 3.1).

A similar sequence of events stemming from the general deflation in the gold standard countries of the world was traced out in the three overseas stock markets. In Germany and the United Kingdom, members of the leading stock exchanges sought ways as well to maintain their customer base, mainly by offering their clients new financial products. The characteristics of the new products, however, varied greatly across the exchanges, depending on differences in their political and economic institutions, and their internal microstructure.

The Effect of Deflation in Other Core Financial Markets

London

The effects of continued deflation on bond prices and market returns affected the LSE largely through the repercussions that the economic difficulties in the United States, South America, and Australia posed for British investors—investors who had increasingly diversified their holdings of securities in the global capital market that British merchant bankers had created in the middle of the nineteenth century. The 1890s began with a near escape from disaster brought about by the failure of the House of Baring. That failure, in turn, can be traced to Argentina's default on bonds underwritten by what had been Britain's leading merchant bank. The story is well known. The Bank of England under the leadership of its governor, William Lidderdale, organized a bailout financed by an ad hoc consortium of leading London bankers.

Lidderdale's response to the Baring crisis demonstrated the government's commitment to maintaining a regulatory and monetary environment within which the securities business could continue to flourish. There followed a continued expansion of both the business of the LSE and the size of its membership (see table 3.1 and figure 3.2). As in the United States, commercial and industrial shares were the largest gainers between 1893 and 1913. The impetus to their expansion came first in the form of so-called debenture shares. These shares permitted breweries to pledge the incomes they received from public houses tied to serve only their brews toward payment of the dividends on new capital. The new capital was issued precisely to purchase the exclusive vending rights to beverages that had been sold in previously independent, free pubs. Katherine Watson (1996) has documented the stock market boom in brewery shares that ensued.

By the late 1890s, British investors became enthused about small-denomination mining shares in the new claims on gold and diamond mines created in South Africa. The "kaffir" shares led to such an increase in trading business that special settlement days and procedures had to be created to cope with the ticket claims—claims that often had changed hands many times before the completion of the sale. Meanwhile, after-hours trading in American shares expanded to meet the competition that arose as New York brokerage houses established branches in London to serve their British client (Michie 1999).

These new activities clearly bore higher risks for the members undertaking them. Brutal evidence of the costs of risk taking for the small, numerous firms making up the memberships of the LSE comes from the accounts of the exchange's official assignees. These individuals were charged with administering the estates of defaulting members of the exchange until their creditors were paid off, at which time the defaulters could be readmitted to membership upon approval of the Committee

Table 3.1
Value of shares quoted in the London Stock Exchange official list, 1853–1913

Class of Security	1853	1873	1893	1913
Value of shares in millions of pounds				
British government and U.K. public bodies	853.6	858.9	901.6	1,290.1
Colonial and foreign governments and public bodies	69.7	486.5	1,031.5	2,034.4
Railways	225.0	727.7	2,419.0	4,147.1
Banks and financial institutions	13.1	113.2	199.5	609.1
Public utilities	24.5	32.9	140.3	435.8
Commercial and industrial	21.9	32.6	172.6	917.6
Mines, nitrate, oil, tea, and coffee	7.4	8.8	34.6	116.4
Total	1,215.2	2,260.6	4,899.1	9,550.5
Percentage of the value of shares				
British government and U.K. public bodies	70.2	37.8	18.4	13.5
Colonial and foreign governments and public bodies	5.7	21.4	21.1	21.3
Railways	18.5	32.0	49.4	43.4
Banks and financial institutions	1.1	5.0	4.0	6.4
Public utilities	2.0	1.4	2.9	4.6
Commercial and industrial	1.8	1.4	3.5	9.6
Mines, nitrate, oil, tea, and coffee	0.6	0.4	0.7	1.2
Total	99.9	99.4	100.0	100.0

Note: Foreign government bonds payable abroad but quoted in London are not included.
Source: Michie (1999, 89).

for General Purposes. By the end of the 1890s, the burdens of the official assignees had grown so onerous that their salaries were frequently raised and the size of their office staff enlarged. While the large number of members and the strict limits on firm size created failures on a regular basis in London, table 3.2 shows an exceptionally large number of defaults in the 1890s with an unusually high level of outstanding debts to be discharged by the defaulters.

To confront the problems raised by the increasing number and severity of failures among the members despite the absence of banking or financial crises, a series of protective measures were taken by the LSE. The

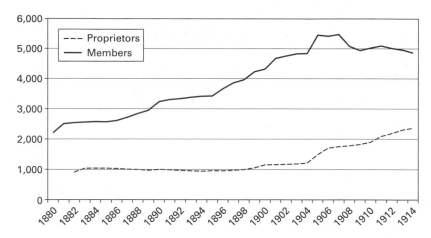

Figure 3.2
Members and proprietors of the London Stock Exchange, 1880–1914

body responsible for making the rules of the LSE was the Committee for General Purposes (CGP), elected annually by the members who had been admitted (by the same committee!) for the coming year. All of the CGP's rulings were designed to reduce risks for the majority of the members using the facilities of the exchange. Over time, the main increase in number of members came from members who were brokers, as opposed to members who were dealers (jobbers), or market makers on the exchange. Consequently, the new rules tended to favor brokers and to restrict the range of activities available to dealers, so they could not undercut brokers' commission by shunting trades to other exchanges. As dealers were the wealthier firms in the exchange, they had special influence with the owners of the exchange, who were holders of the initial shares issued to finance construction of the exchange. Often the dealers were shareholders as well, so they could use the power of the owners to set hours of operation and access to other exchanges, especially using the telegraph connections installed in the exchange. The proprietors, as the shareholders were termed, concerned themselves mainly with increasing the volume of business and the number of subscribing members on the exchange as their income derived from the dividends declared annually on their capital stock. The greater the number of members paying annual subscriptions, the higher were the dividends that could be paid out to the

Table 3.2
Size and number of failed members of the London Stock Exchange, 1879–1899

Year (ending March)	Total commissions	Number of failures
1879	£693:16:5	30
1880	£692:11:10	23
1881	£1,304:10:0	19
1882	£2,604:1:11	27
1883	£3,180:19:9	31
1884	£2,038:15:8	32
1885	£1,990:9:11	33
1886	£1,038:2:10	12
1887	£1,554:5:3	20
1888	£1,680:1:9	25
1889	£987:5:2	17
1890	£1,247:15:6	19
1891	£3,164:6:8	37
1892	£1,105:1:11	22
1893	£504:15:3	14
1894	£4,298:7:10	49
1895	£763:7:8	10
1896	£4,416:12:8	23
1897	£1,592:16:6	10
1898	£1,354:1:5	19
1899	£3,193:18:3	18

Note: Commissions are a fixed percentage of payouts to creditors.
Source: Guildhall Library, Ms. 14600/65, Minutes of the Committee for General Purposes, February 15, 1897.

proprietors. The subscribing members, however, were concerned mainly in maintaining their incomes from brokerage fees and occasionally from underwriting the flotation of new firms.

The turning point in the balance of power between the dealers and the brokers among the members came as early as 1875 and 1882 (see figure 3.2). The need to enlarge the facilities substantially led to a corresponding change in the balance of power between the members (the dues-paying traders operating on the exchange) and the proprietors (the dividend-receiving owners of the original exchange building). Changes in the deed of settlement (changes that were required to underwrite the finance needed to pay for the construction of larger facilities) increased the original 400 shares to 20,000 shares, and those changes stipulated that all new shareholders had to be members (Morgan and Thomas 1961). As the membership continued to increase over the following years, the interests of the proprietors and members tended to converge, but because most members were brokers, the convergence was toward the interests of the brokers. Eventually in 1904, the members voted to require that any new member purchase a nomination from a retiring member so that total membership would be capped at the existing level. The number of members then peaked at 5,481 in 1905 (Michie 1999).

Existing members, however, now had effectively a property right in their seat on the exchange, much as did the members of the NYSE. But the prices of the seats turned out to be surprisingly low, especially compared with those of the NYSE. Between 1905 and 1914, London member prices ranged between £15 ($73) and £150 ($731), compared to New York seat prices that ranged from $13,000 to $95,000 between 1879 and 1914. On the NYSE, a property right in seats had been established after the merger with the "open board" in 1869. The number of members admitted was set at 1,060, and only once in the years before World War I was this figure increased—by forty in 1879, to absorb the traders on the government bond market (Michie 1987). The monopoly arrangement with Western Union that limited access to the prices on the NYSE to member firms also produced excellent profits for the members holding seats on the exchange, which showed up in the sharp rise in their prices after 1898 (see figure 3.1).

By 1912, the members, now largely brokers, voted to enforce minimum commissions and outlaw the practice of the jobbers' shunting trades to outside brokers (Morgan and Thomas 1961). Even then, however, the vote was very close (1,670 to 1,551), and the new system worked much less well than its supporters had argued it would. The regional exchanges proved much more resilient than anyone had expected. As a result, they continued to compete effectively with the London brokers, undermining the revenues expected from the new fixed commissions. Jobbers who tried to maintain their revenues by engaging in arbitrage trades with foreign exchanges that listed securities traded on the LSE were strictly limited in number and forced to maintain elaborate records of their dealings. The fall in members after the 1907 panic in New York turned out to be the start of a continued decline in members and of revenues from trading. A system that for almost half a century had underwritten a very efficient national market was no longer effective; the national market was splintered into a number of only loosely connected regional markets. If World War I had not broken out, it is very likely that the minimum commission rule would have been repealed in 1914. As it was, between the problems engineered by the rule change and the Great War, the LSE never fully recovered the international position that it had previously held.

The relative values of access to the respective trading floors of London and New York show clearly that the New York market structure was superior in the face of even stronger competitors than faced the LSE. Over time, the trustees and members of the LSE passed other rules that also mimicked, but most often did not duplicate, key features of the regulations that governed the NYSE. These new rules included listing requirements that were intended to ensure customers of the quality of the securities available for purchase. But by the time new securities appeared on the Official List of London, they were already being traded at more competitive commissions in the markets where they had originally appeared. London, unlike New York, could not eliminate these alternatives for its customers.

The key differences that remained were the much larger membership, the greater number of securities listed, and the greater importance of foreign securities on the LSE compared to the NYSE. To provide incomes for the growing number of members of the stock exchange in London, the Committee for General Purposes created the Share and Loan Department in 1872, which was given the responsibility for confirming that the new companies asking to be listed were legitimate enterprises. The secretary of the Share and Loan Department, Henry Burdett, began publishing summaries of his efforts starting in 1882 with *The Stock Exchange Official Intelligence*. For the period 1882 through 1909, we can see the number and capitalization of all joint stock companies that were registered in the United Kingdom (figure 3.3).

It is evident that in the 1890s, and continuing especially though 1909, that the average capital of new companies declined greatly relative to

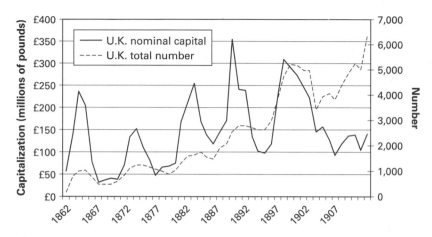

Figure 3.3
Registered companies in the United Kingdom, 1861–1909

their size in the 1880s, and this despite gentle inflation that began after 1897. The rise of big business was clearly more striking in both the United States and Germany. When we examine the new listings on the LSE itself, however, it is striking that the amount of new capital listed on the exchange always exceeds the amount of new capital registered in new British joint stock companies (figure 3.4).

When we break down the composition of the new listings in 1909, we find that the continued expansion of the LSE was due mostly to expanded issues of existing securities in traditional sectors: government bonds and railroads in Britain, the colonies, and the United States. While the number of new listings was greatest under the category "Commercial and Industrial," the average capital was quite small: £431,000. Electric power and light companies, not shown, were even smaller: £202,050 capitalization on average (figure 3.5). Compare this with the $1 million minimum required for a firm to be listed on the NYSE.

In Britain, unlike in the United States, the threat of financial crises caused by continued deflation and defaulting debtors was circumvented by the concerted efforts of the central bank and the central government. Nevertheless, as can be seen by the increasing number and severity of defaults, the difficulties of sustaining incomes for the members of the LSE increased throughout the 1890s. For different reasons than in the

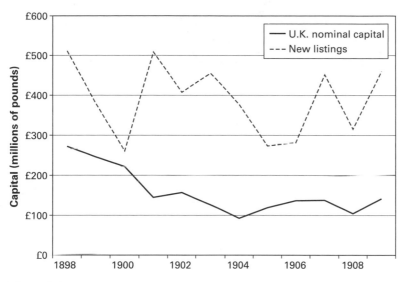

Figure 3.4
Nominal capital of companies registered in the United Kingdom and new listings of all kinds on the London Stock Exchange, 1899–1909

United States, then, the problem created for capital markets by continued deflation was left to the LSE itself to work out, as was the case for the NYSE. The solutions generated internally within each exchange were similar in that listing requirements were increased substantially for both exchanges and clearing arrangements were enlarged to reduce costs of trading among the members. Gradual limitations on the number of members and moves toward enforcing minimum commissions were initiated in London in the 1890s, in belated imitation of the success of New York, but could not take effect until after World War II, when the members and proprietors finally became one and the same, and the old company was replaced by one that combined the trustees and managers committee with the CGP into the new company's council. In the nineteenth and early twentieth centuries, however, the much larger membership of the LSE encouraged more attention to smaller capitalized firms than was true in the NYSE, and much attention was diverted to foreign listings, especially American companies and very small firms in the mining sector. Consequently, the LSE did not provide the continued life cycle financing for British firms embarking on a new technology, at least not

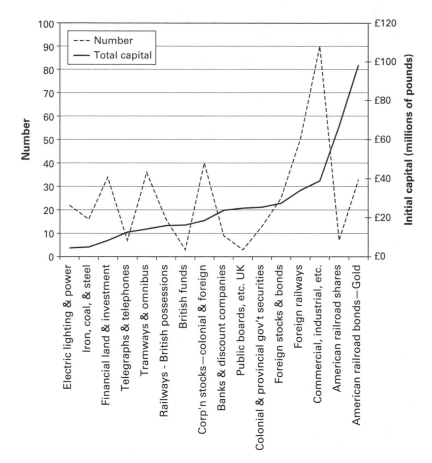

Figure 3.5
New listings in 1909, major categories

to the extent that the combination of regional exchanges and the Curb and NYSE in New York provided finance for American firms.

True, other factors must have accounted for some of the marked differences in technological advance between Britain and America at this time. A balanced assessment of the British situation with respect to financing of new technologies must conclude that the financial sector was not wholly responsible for Britain's lag in electricity, chemicals, and automobiles. For example, the relative lack of public support in Britain for basic science in research environments, whether within corporations or universities, must account for some of the lag, especially given the more rapid response of Germany with its competitive technical universities (Rosenberg and Landau 1992). The regulatory restrictions at the municipal level in Britain, restrictions condoned at the national level, limited the possibilities for creating even regional electric power grids (Hughes, 1983). Perhaps these factors bear the brunt of responsibility, as previous historians have argued. To the extent that the characteristics of the LSE fell short in providing the necessary finance for new network technologies, the small size of jobber partnerships on the stock exchange and the small numbers of jobber firms in determining the policies of listing by the LSE are the main culprits. But that the opportunities for British investors were much greater in American and German railroads, electric power systems, and municipal utilities than in the British counterparts was not the fault of the British financial sector, much less the LSE.

Berlin

In Germany, the explosion of incorporations that occurred both before and after the founding of the Reich and the receipt of 5 billion francs in reparations from the defeated French nation led to speculative manias that ended in the Gründungkrise of 1873.[8] That explosion was certainly aided and abetted by a law passed on June 11, 1870—a law that made it much easier to create a corporation. The passage of that piece of legislation was the high point of a series of attempts to liberalize the marketing of corporate shares, and it sealed the structural interdependence of that German nation's banks and industry, an interdependence that still exists to this day. The rise of new joint stock banks after the passage of the law is particularly noteworthy. In the first two years of the new German Reich, 107 joint stock banks were founded—banks with a total capital of 740 million marks (Gömmel 1992). By the end of 1873, 73 of the newly chartered banks were in liquidation (Gömmel 1992).

Faced with a crisis, and in an attempt to protect the earnings of the remaining corporations, the government's initial reaction was to raise customs barriers. In the longer term, however, as it tried to improve the economic robustness of the nation's business organizations, the government moved to restructure the internal organization of the nation's corporations. In 1884, a new law redefined the framework of governance of German corporations. Each corporation was required to adopt an institutional decision-making structure that consisted of three distinct committees, with each committee serving a different function. The managing board of directors (Vorstand) and a general assembly of stockholders (Generalversammlung) were features that were common to corporations in all four countries. The German law, however, added a third oversight board, the Aufsichtsrat—a board with a large majority of its members drawn from outside the firm. Those members represented not owners and managers but labor, the government, the general public, and the banks. The Aufsichtsrat was peculiar to Germany (Gömmel 1992).

The stock market crises of the early 1890s led to further major reforms in Germany. As occurred in other continental countries at this time, the German reforms outlawed the informal exchanges—the so-called Winkelbörsen—that had sprung up around the formal exchange, and they specified that only transfers validated on the formal exchange would have standing in legal disputes. The new law went further, however, by outlawing uncovered, or short selling, of securities. As a result, trading in corporate securities tended to move not merely out of Berlin, but out of all of Germany, to the friendlier purviews of the Amsterdam and London stock exchanges. In retrospect, it seems that the formation of the Kommission für den Börsenenquete, a commission that included only token representation from members of the stock exchange and a commission that was heavily weighted with representatives of agricultural interests eager to do anything to raise prices of farm products, was responsible for this outcome. But given that the concerns of a wide range of potential interest groups had been represented in the composition of the Aufsichtsraten, one of the three committees charged with overseeing the governance of a corporation, the broad composition of the commission reflected political reality, if not economic rationality.

As a result of the legal changes, trading on the German stock exchanges quickly became concentrated on public securities issued by German state and city authorities. At the same time, the great banks continued their efforts to develop new private sector business in adjacent, politically friendly countries. Both Austria and Italy were initial beneficiaries of the legal changes and the response of German investment banks to those changes. According to Cohen (1992) and Good and Ma (1999), the initial outcomes have been deemed beneficial for both countries, although more recent analyses of the financial sectors in each country suggest that there may have been few long-run benefits.[9] To continue their active trading on the securities held by their customers, the German banks appear to have had increasing recourse to the forward markets available in Amsterdam and the trading facilities of the LSE (Michie 1999).

In sum, the competitive responses of the regional exchanges in Germany to the growing dominance of the Berlin exchange over the period of continued deflation created an economic environment in which their clients felt comfortable in seeking alternative political jurisdictions for their activities whenever local legislation or regulation constrained them. The main clients, of course, were the great universal banks that arose after 1848 in Prussia. When the legislation of 1898 limited their possibilities for hedging risks, they found it natural to turn to external markets in search of diversified investments and higher returns for themselves and their clients. Extending their banking activities into Italy and the Austro-Hungarian Empire, and their stock trading activities into Amsterdam and London, with investment banking outposts in New York, they were able to offset in large part the effects of suppressive domestic legislation. By 1908, after the ripple effects of the 1907 panic in New York had passed through Berlin, much of the legislation was modified in favor of the great banks. The international diversity of German investors, however, proved a liability during World War I—a liability that was obviously not clearly foreseen by Germany's decision makers.[10]

Summary

By the end of the nineteenth century, the three major stock markets in the countries leading the Second Industrial Revolution (New York, London, and Berlin) had responded in different ways to the widespread deflationary pressures—pressures that affected both the traders and their customers in very similar ways. Hardly surprisingly, since the policies adopted were quite dissimilar, the economic productivity of the responses differed markedly among these countries. Both the German and British exchanges seem to have flourished in the aftermath of the microstructure reforms that took effect in the first decade of the twentieth century, but on closer examination, it appears that the observed resurgence in London and Berlin was more deeply rooted in the indirect complementarity that developed between the broker-dominated market in London and the jobber-, or bank-, dominated market in Berlin, rather than in the changes in the rules that governed the markets. Despite the trauma of the 1907 panic on the NYSE and the governmental investigations and reforms that followed in its aftermath, it would appear that the competitive responses of the NYSE to the challenges of deflation carried the most promise for the design of efficient capital markets in the decades that were to follow. The exchange's original draconian listing requirements—a listed firm not only had to show a history of profitability, but it had to be a representative of an industry with a history of profitability—was reinforced at the end of the century with the requirement that a listed firm must annually produce audited financial statements. Taken together, the rule changes hastened the process of investor education so much so that by 1914, American investors were at least as willing to hold the paper securities of private companies as were investors in the United Kingdom. Moreover, the innovation of continuous calls, coupled with the belated introduction of the clearinghouse, greatly simplified, and therefore reduced, the transaction costs involved in buying and selling securities. Again, daily settlement, as opposed to the LSE's practice of periodic settlements, made transactions on the NYSE much more transparent than their London counterpart, and that difference also helped relieve the uncertainty fears of investors. Finally, the development of the complementary relationship with the Curb market made it possible to maintain the blue chip character of securities traded on the NYSE, while at the same time providing a market for securities that could not qualify for the NYSE's imprimatur. Unfortunately, this rating of the relative efficiency of the microstructure of the rules circa 1914 still cannot be tested directly. The effects of massive government war finance required for World War I, to say nothing of the indirect impact of widespread military mobilization, overshadowed future developments in all three exchanges, particularly those in London and Berlin.

Conclusion

Thomas Parke Hughes (1983) has developed the military analogy where reverse salients, enemy outposts that remain within the lines of an advancing army, are the technically difficult problems that must be solved by inventors, whether scientists or technicians, before a new technology can be commercially profitable. Just as military commanders are obsessed in a campaign with straightening out the lines of battle while advancing against the enemy, so innovators must focus on overcoming the technical difficulties that arise in developing a system that incorporates a new technology effectively within an existing economy. And, of course, politicians must focus on overcoming regulatory difficulties that inhibit technical transformation while lawyers can make a comfortable living by defending the property rights of stakeholders in existing technological systems. The reverse salient metaphor developed by Hughes implies that unusually high rates of financial return can accrue to the inventor who conquers the technical difficulty that created the salient obstacle to the diffusion of a new technology. But the military metaphor leaves unanswered the operational question of whether conquest is best accomplished by clandestine operations or frontal assault—by privileged information or by public access.

Richard Nelson (1992) has argued persuasively that both proprietary information on processes and generic knowledge of scientists and technicians are necessary, and complementary, aspects of technological advance. Nelson focused on the complementarity of public-supported research in universities and research institutes that is available to all interested parties with private research that is kept under proprietary application by commercial firms. In this chapter, we have examined the financial implications of the complementarity of proprietary and generic technical knowledge. Reverting to the felicitous metaphor of reverse salients being successively captured as technological progress continues, we argue that if conquest is best conducted by privileged information, then access to bank finance is necessary. But if public access to the use of the technology is the best strategy for overcoming the reverse salient, as is the case in network technologies, then access to capital markets is necessary. Since the nature of each new technology is different and evolves differently over its life span, a financial sector that incorporates both intermediaries and markets will be the most effective in providing the necessary funding at each stage in the development of a new technology. The lessons for the new global financial markets at the beginning of the twenty-first century may be that self-interested innovations initiated by the leading participants in capital markets—innovations designed to improve the competitiveness of the formal exchanges (New York) or of the major investment houses trading in the exchanges (Berlin)—will be more beneficial in the long run to their respective economies than regulatory constraints imposed by governmental authorities (Berlin) or self-protective measures initiated by stockbrokers seeking secure commissions and joint-stock banks seeking guaranteed loans (London).

Notes

1. Broadberry (1997) shows that differences in manufacturing productivity among the United States, the United Kingdom, and Germany stayed roughly the same from 1850 to 1914. But he acknowledges that the United States and Germany outstripped the United Kingdom in overall economic growth, especially after 1890. In his accounting framework, the U.S. and German advances derived from greater shifts of labor from low-productivity sectors to higher-productivity sectors for those two countries compared to Britain.

2. In this regard, it is worth noting that the Parisian reforms imposed by the central government in 1898 were the most stifling for future financial innovations.

3. On the London Stock Exchange, the number of partners in a firm was limited; every partner in a London firm had to be a member of the LSE, and no member of the exchange was allowed to have any business other than broking and jobbing. For a list of the provincial exchanges whose records have survived, see Thomas (1973).

4. As an aside, it might be noted that the noncallable provisions in corporate bonds help to explain why the dramatic concurrent reduction in government debt did not elicit a "crowding-in" effect on private investment. For a discussion of this point, see James (1984).

5. See the discussion in Davis and Gallman (2001, 300–312).

6. Although on average the shares traded on the Consolidated were lower valued, between 1886 and 1913, the volume traded on the Consolidated averaged 64 percent of the volume of shares traded on the NYSE; and between 1888

and 1896 the figure was 95 percent, and it exceeded 100 percent in four of those years. Davis and Gallman (2001). For an analysis of the effect created by the NYSE's 1900 contract with Western Union, see White (2005).

7. The classic reference on the rise of the American automobile industry is Seltzer (1928).

8. The *Gründungkrise* of 1873, as Germans termed the crash that ended the speculative surge in stock market activity that began after the defeat of France in 1870.

9. The initial studies are reported in Cohen (1992) and Good and Ma (1999). The longer-term results are reported in Fohlin (1998) and Tilly (1998).

10. On this point, see Ferguson (1999), especially chapter 9.

References

Campbell, E. 1938. *The Reorganization of the American Railroad System, 1893–1900.* New York: Columbia University Press.

Cameron, Rondo, and V. I. Bovykin. 1991. *International Banking, 1870–1914.* New York: Oxford University Press.

Carosso, Vincent P. 1987. *The Morgans; Private International Bankers, 1854–1913.* Cambridge, Mass.: Harvard University Press.

Cassis, Youssef. 1994. *City Bankers, 1890–1914.* Cambridge: Cambridge University Press.

Cohen, Jon S. 1992. "Financing Industrialization in Italy, 1894–1914: The Partial Transformation of a Late Comer." In Rondo Cameron, ed., *Financing Industrialization.* Brookfield, Vt.: Edward Elgar.

Davis, Lance E. 1966. "The Capital Markets and Industrial Concentration: The U.S. and the U.K.: A Comparative Study." *Economic History Review*, 2nd ser. 19, 255–272.

Davis, Lance E., and Robert Gallman. 2001. *Evolving Financial Markets and International Capital Flows: Britain, the Americas, and Australia, 1865–1914.* Cambridge: Cambridge University Press.

Davis, Lance E., and Larry Neal. 2004. "Deflation, the Financial Crises of the 1890s, and Stock Exchange Responses in London, New York, Paris, and Berlin." In Richard Burdekin and Pierre Siklos, eds., *Deflation: An Historical Perspective.* Cambridge: Cambridge University Press.

Edwards, G. W. 1938. *The Evolution of Finance Capitalism.* New York: Longmans, Green and Co.

Ferguson, Niall. 1999. *The Pity of War.* New York: Basic Books.

Fohlin, Caroline. 1998. "Fiduciary and Firm Liquidity Constraints: The Italian Experience with German-Style Universal Banking." *Explorations in Economic History* 35, 83–107.

Fohlin, Caroline. 1998. "Relationship Banking, Liquidity, and Investment in the German Industrialization." *Journal of Finance* 53, 1737–1758.

Gömmel, Rainer. 1992. "Entstehung und Entwicklung der Effektenbörse im 19. Jahrhundert bis 1914." In Hans Pohl, ed., *Deutsche Börsengeschichte.* Frankfurt am Main: Fritz Knapp Verlag.

Good, David F., and Tongshu Ma. 1999. "The Economic Growth of Central and Eastern Europe in Comparative Perspective, 1870–1989." *European Review of Economic History* 3, 103–137.

Hughes, Thomas Parke. 1983. *Networks of Power: Electrification in Western Society, 1880–1930.* Baltimore: Johns Hopkins University Press.

Hughes, Thomas Parke. 1989. *American Genesis: A Century of Invention and Technological Enthusiasm, 1870–1970.* New York: Viking.

James, John. 1984. "Public Debt Management and Nineteenth Century American Economic Growth." *Explorations in Economic History* 21, 192–217.

Michie, Ranald C. 1987. *The London and New York Stock Exchanges, 1850–1914.* London: Allen & Unwin.

Michie, Ranald C. 1999. *The London Stock Exchange: A History.* New York: Oxford University Press.

Morgan, E. V., and W. A. Thomas. 1962. *The Stock Exchange: Its History and Functions.* London: Elek Books.

Navin, T., and M. Sears. 1955. "The Rise of the Market for Industrial Securities, 1887–1902." *Business History Review* 29, 105–138.

Neal, Larry. 1969. "Investment Behavior by American Railroads: 1897–1914." *Review of Economics and Statistics* 51, 126–135.

Nelson, Richard R. 1992. "What Is 'Commercial' and What Is 'Public' about Technology, and What Should Be?" In Nathan Rosenberg, Ralph Landau, and David C. Mowery, eds., *Technology and the Wealth of Nations.* Stanford: Stanford University Press.

New York State. 1909. *Report of Governor Hughes' Committee on Speculation and Commodities.* June 7.

Perez, Carlota. 2002. *Technological Revolutions and Financial Capital: The Dynamics of Bubbles and Golden Ages.* Cheltenham, UK: Edward Elgar.

Pohl, Hans, ed. 1992. *Deutsche Börsengeschichte.* Frankfurt am Main: Fritz Knapp Verlag.

Rosenberg, Nathan, and Ralph Landau. 1992. "Successful Commercialization in the Chemical Process Industries." In Nathan Rosenberg, Ralph Landau, and David C. Mowery, *Technology and the Wealth of Nations.* Stanford: Stanford University Press.

Rosenberg, Nathan, Ralph Landau, and David C. Mowery. 1992. *Technology and the Wealth of Nations.* Stanford: Stanford University Press.

Schumpeter, Joseph. 1939. *Business Cycles.* Philadelphia: Porcupine Press.

Seltzer, Lawrence H. 1928. *A Financial History of the American Automobile Industry: A Study of the Ways in Which the Leading American Producers of Automobiles Have Met Their Capital Requirements.* Boston: Houghton Mifflin.

Smil, Vaclav. 1994. *Energy in World History.* Boulder, Colo.: Westview Press.

Smil, Vaclav. 2001. *Enriching the Earth: Fritz Haber, Carl Bosch, and the Transformation of World Food Production.* Cambridge, Mass.: MIT Press.

Snowden, Kenneth. 1987. "American Stock Market Development and Performance, 1871–1929." *Explorations in Economic History* 24, 327–353.

Snowden, Kenneth. 1990. "Historical Returns and Security Market Developments, 1872–1925." *Explorations in Economic History* 27, 381–420.

Sobel, Robert. 1965. *The Big Board: A History of the New York Stock Exchange.* New York: Free Press.

Thomas, W. A. 1973. *The Provincial Stock Exchanges.* London: Frank Cass.

Tilly, Richard. 1998. "Universal Banking in Historical Perspective." *Journal of Institutional and Theoretical Economics* 54, 7–32.

Watson, Katherine. 1996. "Banks and Industrial Finance: The Experience of Brewers, 1880–1913." *Economic History Review* 49, 58–81.

White, Eugene N. 2005. "Competition among the Exchanges before the SEC: Was the NYSE a Natural Hegemon?" Working paper.

Wilson, John Grosvenor. 1969. "The Stock Exchange Clearing House." In Edmund Clarence Stedman, ed., *The New York Stock Exchange: Its History, Its Contribution to National Prosperity, and Its Relation to American Finance at the Outset of the Twentieth Century.* New York: Greenwood Press.

4

Funding New Industries: A Historical Perspective on the Financing Role of the U.S. Stock Market in the Twentieth Century

Mary A. O'Sullivan

In contemporary discussions of varieties of financial systems in the advanced industrial economies, what is typically regarded as important and distinctive about the United States is the capacity of its stock market to cater to the funding needs of new industries and, specifically, of the young firms that enter them. It is this characteristic that has convinced admirers of the stock market of its inextricable link with the processes of innovation and technological change that drive economic development in the United States. In this chapter, I present a historical perspective on the relationship between the financing activity of the stock market and the development of the broader economy in the United States, with a particular focus on its role in the emergence and development of new industries.

My analysis shows that a number of the sectors that have been prominent issuers of stock over the past century, notably the public utilities sector as well as the real estate and financial sector, are not ones that feature prominently in the literature on productivity growth and innovation. The industries that are emphasized in this literature are usually found within the manufacturing sector, but there is a dearth of systematic data available that would allow us to understand the importance of stock issues in specific manufacturing industries over the course of the century. To get around this problem, I rely on case studies of several new industries to understand the role that the stock market played in their development.

I focus on the period from the late nineteenth century until the late 1920s. Several important manufacturing industries emerged at this time and grew rapidly based on sustained processes of innovation and technological change. The period was also an important one for the development of the U.S. stock market, with a liquid market for industrial securities emerging from the late 1890s, and relatively high levels of stock issuance recorded, especially in the 1920s.

The development of some of the most rapidly growing and productive new industries of the time straddled the transition to a highly developed market for industrial securities. In these cases, the early development of the industry tended to be financed by local financial networks, as was the case for new industries earlier in the nineteenth century. The stock market became involved only later in their development to facilitate their consolidation. In contrast, a number of important industries experienced their initial takeoff after the establishment of a substantial market for industrial securities. I analyze the role of stock issues in the emergence and early growth of three such industries—the automobile, aviation, and radio industries—and find that it differed substantially in its importance and characteristics in each case.

The stock market was of limited significance to the early development of the automobile industry, with the overwhelming majority of its investment being funded from internal sources. When the stock market became involved in the development of particular automobile firms, it was often to facilitate consolidation through mergers and acquisitions. When it financed investment, it tended to do so only when the companies that it funded were already rather large.

In contrast, the stock market played an important role in the rapid development of the aviation and radio industries in the 1920s. In aviation, it funded large numbers of new companies that entered the industry at that time, and it also facilitated a wave of consolidation that consumed many of the independent players in the industry by the beginning of the 1930s. The stock market's role in the radio industry of the 1920s was predominantly one of facilitating entry, especially by new, young companies.

Based on these examples, it would seem that the 1920s was an important turning point in the involvement of the stock market in the development of new industries in the United States. Further studies of new industries are necessary to confirm this hypothesis, but to the extent that it is sustained, it begs the question of what caused this development in the role of the stock market. I argue that changes in the characteristics of the demand for corporate stocks played an important role in driving the growing enthusiasm for the stocks of new firms and new industries in the 1920s. In addition, the dynamics of the investment banking industry as well as the institutional structure of the U.S. stock market mediated the interaction between the demand for, and supply of, corporate stock in ways that encouraged the changes in the stock market's role.

Finally, there is the question of the effects on new industries of the stock market's active involvement in their early development. Although there is a strong tendency in the literature on finance and growth to assume that more external finance leads to more investment and, in turn, to more growth, I argue that the relationship between stock issues and economic performance is not nearly as mechanical as this virtuous cycle assumes. First, even to the extent that the proceeds from stock issues are invested in the formation of new capital, they may not foster better economic performance. Firms may overinvest or, more generally, make poor investments. As I show, these types of concerns are of considerable relevance for the aviation and radio industries of the 1920s.

Second, firms may use the proceeds of their stock issues for purposes other than capital formation, such as mergers and acquisitions, refinancing their existing obligations, and bolstering their treasuries for later use. The real effects of these various uses will not be reflected in higher contemporaneous levels of investment. For mergers and acquisitions, for example, their economic impact depends on the implications of changes in ownership on the productivity of the businesses and assets that change hands. In general, there is no a priori reason to assume that the consolidation that stock issues facilitate will be beneficial from the perspective of the overall economy. The case of the aviation industry confirms the need for skepticism in this regard.

These examples, in casting doubt on the existence of an automatic link between stock issues and improved economic performance, highlight the need for further research to understand the economic implications of the involvement of the stock market in the development of new industries. In this regard, what is important is not so much the impact of access to external funds on the companies that completed stock issues. Instead, the challenge is to analyze the impact on the industries in which these companies operate. To do so will require a much more extensive study than I have been able to provide here of the innovative dynamics of these industries and the role of stock issues in advancing or retarding the development of the markets that they served and the technologies that they employed.

These issues have a relevance that goes far beyond the 1920s. There was a resurgence of U.S. shareholders' enthusiasm for new industries and their entrants with a wave of hot issues in the late 1950s and early 1960s that was dominated by electronics companies. Many of the factors that seem to explain similar developments in the 1920s played a role, in somewhat different form, at that time. Questions can be asked, and indeed were asked by contemporary observers, about the economic implications for the companies and industries that were funded by the stock market at this time. Moreover, the wave of hot issues had a longer-term significance since it prompted regulatory reforms and, ultimately, the emergence of the institutional characteristics, notably the Nasdaq market, that are so closely associated with the funding of new industries in the United States today.

Historical Trends in the Financing Role of the U.S. Stock Market

Until 1890, the market for corporate securities in the United States was primarily a market for railroad securities, with coal and textile companies as the only important representatives of the industrial sector. However, in the 1880s and early 1890s, a number of developments took place that laid the foundation for the emergence of a substantial market for industrial securities.[1] Yet it was not until the close of 1897, with the end of the depression that began in 1893, that a broad-based market began to evolve (Navin and Sears 1955).

In Figure 4.1, I show the cash proceeds from stock issues in the United States in 2000 dollars for the period from 1897 to 2000.[2] They were already relatively high by the early part of the twentieth century, and the rapid growth that occurred in the 1920s brought them to an impressive peak by 1929.[3] However, stock issues collapsed in the early 1930s following the stock market crash and the onset of the Great Depression. Although there was a recovery in 1936 and 1937, it proved temporary, as

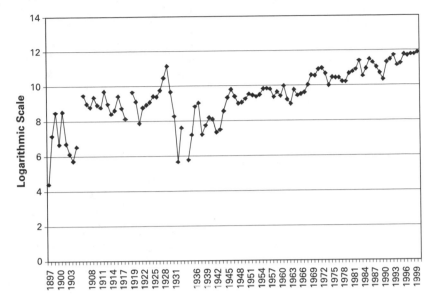

Figure 4.1
Real corporate stock issues in the United States, 1897–2000 (logarithmic)
Source: author's analysis based on data described in note 2

wartime led to another decline in the funds raised in the stock market. However, the postwar period, especially from the late 1940s, was marked by a fairly steady upward trend in the proceeds from stock issues. It culminated in the 1980s and 1990s when the real proceeds from stock issues reached their highest levels for the entire century.

However, as Figure 4.2 shows, if we take account of the growth in the U.S. economy during the century, it is the 1920s that stands out as the decade with the highest level of stock issuance. No other year before or after came close to 1928 and 1929 in the levels of stock issuance as a percentage of national economic output. More generally, over the course of the century, it is possible to identify at least three distinct historical periods of stock issuance.

The early decades of the century were characterized by high and rising levels of stock issuance. The crash and subsequent depression interrupted that trend, and although issues recovered, they remained relatively subdued for several decades. In the final decades of the century, however, stock issues recovered and rose to new heights, although still lower than those reached earlier in the century.

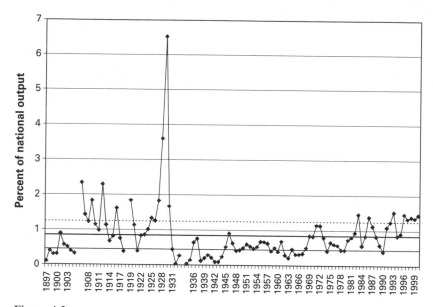

Figure 4.2
Corporate stock issues in the United States as a percentage of GDP, 1897–2000
Source: author's analysis based on data described in note 2

In addition to the major changes that occurred in the level of stock issues over time, there were also important developments over the century in the types of stocks that companies issued. In the early part of the century, issues of preferred stock tended to be favored over common stock. The prevalence of preferred stocks is usually attributed to the efforts of corporations and their financiers to make stocks look as much as possible like the bonds that U.S. investors were more used to holding (Baskin and Miranti 1997).

There was an important change in the composition of stock issues in the 1920s with a major increase in the importance of common stock. The trend accelerated in the late 1920s, with common stock accounting for 60 percent of stock issues in 1928 and as much as 75 percent in 1929. This development seems to be explained by investors' growing interest in participating in the large capital gains that were accruing to common stocks at the time. Tax changes also led to a decline in the relative attractiveness of preferreds.

Preferred stock returned to favor in the late 1930s and early 1940s, when they reached more than 60 percent of total stock issues. They declined in relative importance in the postwar period as the lure of capital gains during the stock market boom of the 1950s and 1960s increased the relative appeal of common stock; common stock reached a peak of 84 percent of total stock issues from 1966 to 1970. In the 1970s, the economic downturn once again induced increased interest in preferred stock, although this time, it reached only 29 percent of all stock issues from 1971 to 1975. The subsequent decade witnessed some waxing and waning in the popularity of preferred stock, and by the 1990s, it accounted for just over 20 percent of all stock issues.

The Stock Market and the U.S. Economy

Perhaps the most straightforward path to analyzing the relationship between the provision of finance through stock issues and the development of the U.S. economy is to study the relationship between total stock issues in the United States and aggregate economic indicators such as GDP or investment. In a provocatively titled paper, "The Stock Market and Investment: Is the Market a Sideshow?" Morck, Shleifer, and Vishny (1990) consider a number of different channels through which the stock market might affect economic activity, pointing out that "the most common view of the stock market's influence, says that the stock market affects investment through its influence on the cost of funds and external financing" (p. 158). While they find some empirical support for the view that equity financing influences aggregate investment, their evidence suggests that its role is very limited.

If we look at the data on the proceeds of stock issues in the previous section, it would seem that there is no simple relationship between stock issues and capital formation for the aggregate economy. For example, there was an enormous expansion in stock issues in the 1920s without any major increase in investment in the economy. The ratio of gross fixed nonresidential investment to GDP for the decade from 1921 to 1930 was 12.7 percent compared with the marginally lower 12.5 percent for the period from 1911 to 1920 and the much higher ratio of 15.7 percent for 1901 to 1910 (Maddison 1991). Yet during the two earlier periods, the stock market played a much less important financing role than in the 1920s. For the century as a whole, regression analysis reveals that there is no stable relationship between stock issues and national economic aggregates such as investment or output.

In contrast to the rather inconclusive results on the historical relationship between stock issues and investment at the national level, cross-country regressions have uncovered evidence of such correlations across country. Levine and Zervos (1998), building on the earlier work of Atje and Jovanovic (1993), developed a number of measures of the development of stock markets and analyzed their relationship to economic growth, productivity growth, and capital accumulation. They found that "stock market liquidity—as measured both by the value of stock trading relative to the size of the market and by the value of trading relative to the size of the economy—is positively and significantly correlated with current and future rates of economic growth, capital accumulation, and productivity growth" (Levine and Zervos 1998, p. 538).

The differences among historical and comparative studies in the size and significance of the correlations that they have generated suggest the need for further research. However, the basic methodology employed to generate these results has attracted a good deal of criticism. The primary concern is with its limitations for establishing a causal relationship between the development of the stock market and economic growth.

Scholars have proposed a variety of strategies for getting around this problem (Levine 2003, Zingales 2003). The one that I pursue here focuses on the microeconomic units, the industries and firms, that are supplied with funds by the stock market, and studies the implications of their access to funds for their investment behavior and, in turn, for their performance. Based on this approach, we should arrive at a better understanding of the way in which, and the extent to which, the stock market influences growth through its role in funding investment. As Levine (2003) put it, "More microeconomic-based studies that explore the possible channels through which finance influences growth will foster a keener understanding of the finance-growth nexus" (p. 31).

Studying industries and firms in analyses of the relationship between stock issues and economic activity has a further and perhaps more important methodological appeal. Macroeconomic research on finance and growth that explores the relationship between national aggregates of financial and economic activity implicitly assumes that economic development is an undifferentiated quantity generated by an aggregate production function. It is only from this perspective that the relationship between finance and growth can be understood in terms of the influence of the quantity of finance provided on the "amount" of economic growth.

In contrast, research by social scientists and historians on innovation and technological change highlights the central importance of the particular context in which these processes occur to their impact on productivity and economic development. The central implication of this observation is that the economic impact of financial systems may be reflected not only in aggregate rates of economic growth but also in the differential development at certain times of particular industries and types of firms. As a result, what may be important about the stock market is not so much the aggregate quantity of finance that it provides as the types of industries and firms that it funds, the stages of their development when it funds them, and the terms on which it makes funds available to them (for a more detailed discussion, see O'Sullivan 2004).

Stock issues can play a role in the development of an industry in one of two basic ways (Rajan and Zingales 1998). First, they may be important in facilitating aggregate investment in a particular industry, thus allowing it to grow faster than would otherwise be possible. Depending on the characteristics of the industry in question—if it is a new industry, heavily engaged in research and development, or otherwise highly innovative—this process may prove to be particularly beneficial to the process of economic development. Alternatively, stock issues may fund an industry whose expansion, as a low-productivity industry, is detrimental to the development of the economy since it diverts resources that could be more fruitfully employed elsewhere.

Stock issues may also influence the development of an industry through their influence on its competitive structure. It may have an impact on the relative positions of incumbents in the industry if, by making funds available to some, it allows them to enhance or consolidate their positions. It may also facilitate an important change in the population of firms that compete in an industry by funding new entrants. If the firms that the stock market favors, be they incumbents or entrants, are particularly innovative firms, then it will contribute to economic improvement. Alternatively, if the stock market displays poor judgment, by allocating funds to laggards and lemons, then it will undermine the performance of the economy.

These issues are particularly relevant ones to explore for the U.S. case. For many observers, the role of the U.S. stock market in funding new industries and its capacity to cater to the financing needs of young, high-growth enterprises that enter them are among its most important and distinctive contemporary characteristics. In the next section, I begin with a general analysis of the industries that were important users of the stock market for raising capital over the course of the century. It reveals that a large proportion of stock issues conducted in the United States at that time were for the benefit of industries that are not normally associated with high rates of innovation or technological change.

I then focus on the historical role of the U.S. stock market in providing funds to support the emergence of new, innovative industries. My primary concern is with the period from the end of the nineteenth century to 1929, when technological and innovative fecundity coincided with high levels of activity in the U.S. stock market. In particular, I analyze the role of the stock market in funding the automobile, aviation, and radio industries. I show that the importance of the stock market, in terms of the scale of funds that it provided, differed markedly across these industries, as did the impact it had on the competitive structure of these industries.

The Role of Stock Issues in Industry Development

To determine which industries were important for stock issuance in the United States, I begin by looking at sectoral breakdowns of the proceeds of gross stock issues based on data that that are available from the *Commercial and Financial Chronicle* for the periods from 1919 to 1956 and from 1961 to 1995. One important message from these data is that sectors that are not usually associated with rapid productivity growth and innovation, notably the public utilities and financial sectors, were very important in determining overall levels of stock issuance in the United States. In the late 1920s, for example, utilities and investment trusts accounted for almost 50 percent of the record stock issues that took place at that time.

It was not just in the 1920s that these sectors were significant issuers of stock. The public utilities sector was important throughout the period from 1919 to the early 1980s, accounting for at least 20 percent of all stock issues during this time and, in several periods, for as much as 45 percent of all stock issues. Besides its prominence in the 1920s, the real estate and financial sector was also an important issuer in the second half of the twentieth century, especially in the late 1980s and 1990s, when it accounted for about half of all stock issuance in the United States.

For both of these sectors, their stock issuance was large relative to their importance in the overall economy. For the period from 1961 to 1995, for example, the public utilities and the real estate and financial sectors displayed the highest dependence on the stock market of all sectors in the economy. The proceeds from their stock issues amounted to 15.8 percent and 16.5 percent, respectively, of their fixed investment compared with an average of 10.3 percent for all sectors.

It may be that these sectors deserve greater prominence in the literature on productivity growth and innovation than they are currently accorded, but it is not the task of this chapter to pursue this issue. Most of the industries that tend to be associated with rapid productivity growth and technological change are found in the manufacturing sector. Manufacturing companies were certainly important issuers of stock in the first part of the century, accounting for nearly 32 percent of stock issuance from 1919 to 1956. Manufacturing's share declined thereafter, in line with its receding share of the national economy, to an average of 20 percent from 1961 to 1995.

Unfortunately, the overall quantity of money channeled into the manufacturing sector tells us little about whether the stock market was funding productivity growth and innovation. The manufacturing sector comprises a heterogeneous collection of industries, ranging from industries that are fast growing and highly innovative to those that are slow growing with low levels of productivity. What we really want to know is to what extent the stock market made funds available to the most innovative and productive industries within the broader manufacturing sector. The evidence available on the financing of new and innovative industries in the nineteenth century suggests that they were financed through networks of local capital (see, for example, chapter 1, this volume). To what extent, then, did this pattern change with the emergence of a national market for industrial securities from the 1890s?

The question is particularly relevant for the period from the late nineteenth century to 1929. As I have already noted, the period was a very active one for the U.S. stock market, with particularly high levels of financing taking place in the 1920s. It was also a period of very rapid growth in the manufacturing sector. Industries like electrical equipment, transportation equipment, rubber and allied products, chemicals and allied substances, and petroleum refining experienced extremely high rates of capital formation (U.S. Bureau of the Census 1976). In many cases, their growth was achieved on the strength of sustained processes of innovation and technological change.

Systematic data on stock issuance are not readily available at a sufficient level of detail to address the general question of whether the stock market was important in funding the development of the fast-growing and innovative industries of the period. Nevertheless, by focusing on a small number of prominent new industries that emerged during this period, it is possible to understand the financing role that the stock market played in their early stages.

The development of some of the most rapidly growing and productive new industries of the period from the late nineteenth century to the late 1920s straddled the transition to a liquid national market for industrial securities. In these cases, the financing of the industry in its early stages tended to be based on local financial networks. By the time a national market appeared, these industries were ripe for consolidation, and it was this process, and to some extent the subsequent financing of the dominant players that emerged from it, that the U.S. stock market facilitated when it became involved in their development (Doyle 1991, O'Sullivan 2005).

In contrast, a number of important industries experienced their initial takeoff following the establishment of a national market for industrial securities. In the discussion that follows, I focus on three such examples: the automobile industry, the aviation industry, and the radio industry. If there were industries that would inaugurate the U.S. stock market's role in funding the emergence of new, innovative industries, these would seem to be good candidates for the job. In fact, as I show, the significance and role of the stock market was markedly different in all three cases.

The Automobile Industry

The automobile industry was established in the United States in the mid-1890s when the first motor vehicle was built here, but it experienced its first major growth spurt in the early twentieth century. Total capital invested in the motor vehicle industry grew from just over $50 million in 1929 dollars in 1904 to $267 million in 1909, then to $1,936 million by 1919, and eventually to $2,742 million by 1929. As a result, the motor vehicle industry's share of total manufacturing capital in the United States rose from only 0.2 percent in 1904 to 4.4 percent by 1929 (U.S. Bureau of the Census 1976).

It was an industry characterized by rapid growth as well as high levels of innovation and productivity. In its early decades, the industry experienced high turbulence, with large numbers of entries and exits. As Klepper (2002) notes, the number of entrants to the industry increased from the establishment of the industry to reach more than 80 firms in 1907 alone, and then flattened out at that level until 1910. The rate of entry then declined sharply to an average of 15 firms per year from 1911 to 1922, becoming negligible thereafter. By the early 1920s, considerable consolidation had occurred and the number of firms had fallen from a peak of 272 to just over 50 firms (Klepper 2002; see also Smith 1968). In 1923 the top ten producers accounted for 90 percent of motor vehicle production in the United States compared to 52 percent in 1912 (Seltzer 1928).

In his book, *A Financial History of the American Automobile Industry*, Lawrence Seltzer (1928) focused on these dominant players to analyze the financing of the U.S. automobile manufacturing industry from its origins until the late 1920s. His research clearly shows that local financial networks continued to play a crucial role in the early development of the automobile industry as they did in the earlier genesis of other industries. With respect to the role of financial markets, he concluded that "individual financiers made sporadic investments in automobile enterprises from time to time, but the organized fixed-capital markets, until very recent years, played a relatively small role in financing the expansion" (52).

The Ford Motor Company represents an extreme case in its autonomy of the financial markets. The company was incorporated in June 1903 with a share capital of $100,000 in common stock. A majority of this stock, $51,000 in total, was issued in equal amounts of $25,500 to Henry Ford, the company founder, and Alexander Malcolmson, a Detroit businessman who had supported Ford's work in the past, in compensation for the assets and services that these men brought to the company. The remaining $49,000 was issued for cash to a group of local investors, which included Malcolmson's lawyers, clerk, and bookkeeper, although only $28,000 was actually paid in (Seltzer 1928).

The company was successful from the beginning, and its early and sustained profitability allowed it to fund its rapid expansion entirely out of retentions. Commenting on the company's financial situation after only three or four years in operation, Henry Ford made it clear that access to financial resources was not an important concern: "We had plenty of money. Since the first year we have practically always had plenty of money. We sold for cash, we did not borrow money, and we sold directly to the purchaser" (quoted in Seltzer 1928, 93).

The extent to which Ford was able to rely on internal funds convinced some of the company's minority shareholders that he was doing so at their expense. When Ford announced the enormous River Rouge expansion in 1916 and declared that the company would fund it entirely from retentions, John F. and Horace E. Dodge, who owned 10 percent of Ford Motor Company stock, took the company to court to compel it to restrain its expansion and distribute the accumulated earnings to them and other stockholders. The case dragged on for several years until, in 1919, the Supreme Court mandated the payment of a special dividend by Ford (Seltzer 1928).

This development provoked Ford to buy out his minority shareholders and take the company private. On this occasion, Ford turned to outside investors to raise the funds he needed for the buyout, but he repaid the loan within the year. The company, then wholly owned by the Ford family, continued with its financial policy of funding its investments from retentions.

While Ford is an extreme case, Seltzer shows that it was not unique in its heavy reliance on internal funds to finance investment. The Hudson Motor Car Company, the Reo Motor Car Company, the Dodge Brothers Inc., and the Nash Motors Company all fall into the same category. Among the leading automobile companies, only the Packard Motor Car Company and the Willys-Overland Company raised substantial funds from the financial markets to finance their development. However, in both cases, most of their fundraising took place when they were already of a substantial size, with annual sales of $15 million or more.

Besides these cases, the financial markets played a limited role in boosting capital formation by the leading motor companies in the early years of the automobile industry's development. However, they were actively involved in the industry's early development through several attempts to consolidate its competitive structure. In particular, following the panic of 1907, several projects to merge the leading U.S. automobile companies were conceived. The establishment of the General Motors Company represented the first successful consolidation plan to get off the ground.

GM was established in 1908 by William Crapo Durant to acquire the capital of the Buick Motor Company. Durant's success as a businessman in the wagon and carriage business in Flint, Michigan, led to a request for him to undertake the reorganization of the Buick Motor Company whose financial difficulties posed a threat to several Flint businesses. Durant's reorganization proved highly successful, and by 1908, when GM took it over, the Buick company was the largest manufacturer in the U.S. automobile industry (Seltzer 1928).

GM acquired Buick from Durant in exchange for its own newly issued common and preferred shares.[4] Over the next two years, GM acquired more than twenty other companies, producers of automobiles or their parts, for a total value of $13.2 million. The most important of these acquisitions were those of the Cadillac Motor Car Company for $4.7 million and the Olds Motor Works for $3 million (Seltzer 1928).

Although most of Durant's acquisitions were paid for in stock, they quickly placed heavy demands on GM's cash reserves since many of them were undercapitalized and required funds for fixed and working capital. To secure these funds, GM concluded an agreement in

November 1910 with a syndicate of three banks—Lee, Higginson and Co. of Boston, J. & W. Seligman, and the Central Trust Company of New York—for the underwriting of a public issue of $15 million in 6 percent notes. The bankers negotiated very favorable terms for themselves, and control of the company was placed in the hands of a voting trust that they dominated for the period during which the notes were outstanding. As Seltzer points out, "The stringent terms exacted by the banking syndicate for the underwriting of this note-issue offer abundant testimony of the uncertain, speculative character of the automobile industry" (Seltzer 1928, 163, 164).

For the next five years, GM's affairs were dominated by the bankers who exerted strict restraint over GM's investments and focused primarily on increasing profits. During their tenure, the company's financial position was greatly improved, but GM's market share also shrank substantially, and the bankers were criticized by some for their myopic approach to business. Eventually Durant won back control of the company using the Chevrolet Motor Company to take it over with the help of prominent backers, notably Pierre du Pont of E. I. du Pont de Nemours and Louis Kaufman, the president of the Chatham and Phenix National Bank of New York. In 1916 he resumed his position as president of GM in the company and embarked once again on a strategy of external growth.

In contrast to GM, the example of the United States Motor Company represented a failed attempt to develop a successful business through the consolidation of existing players. It brought together a number of companies in the motor industry that were having difficulty raising funds and it was run by Benjamin Briscoe. The company was backed by powerful financiers, some of whom were also involved with General Motors, but it fell apart in 1913 as a result of disagreement among its backers. The assets of the company were subsequently reorganized to form the Maxwell Motor Company.

Prominent financiers were also involved in the establishment and development of the Studebaker Corporation. It was formed in February 1911 through a merger between the Studebaker Brothers Manufacturing Company, a manufacturer of carriages and wagons that was founded in 1868, and the Everitt-Metzger-Flanders Company, an automobile company established in 1908.⁵ As a result, it started out with a substantial business; it had net tangible assets of more than $15 million at its foundation and as much as $35 million in sales by 1912.

Studebaker assumed responsibility for the debts as well as the assets of its constituent companies. To recapitalize these obligations, it sold 7 percent cumulative preferred stock to raise cash of $8.3 million through a banking syndicate comprising Goldman Sachs and Company and Lehman Brothers, both of New York, and Kleinwort Sons & Company of London. Henry Goldman, Arthur Lehman, and H. H. Lehman were all members of its board of directors. The *New York Times* noted that "the new corporation will remain in the hands of the Studebaker people through the ownership of the common stock, though it is expected that the bankers will have a strong voice in the management."⁶

These examples illustrate that investment bankers and the financial markets were willing to be actively involved early on in the automobile industry when significant assets were involved. Their role in facilitating the consolidation of the industry was familiar since it was the one they had played earlier, in other new industries, such as electrical equipment. There was some diversification of that role during a motor stock boom that began in 1915, with several companies coming to Wall Street to seek new funds.

By early 1915, the stocks of General Motors, Maxwell, Studebaker, and Willys-Overland were listed on the New York Stock Exchange; these companies, with the Ford Motor Company, were the leading firms in the automobile industry at the time. Once the Willys-Overland stock graduated to the NYSE in February 1915, the New York Curb found itself bereft of motor stocks. The leading regional exchanges boasted of only three motor stocks besides those of the industry leaders: Packard Motor and Chalmers Motor were traded on the Detroit Stock Exchange, and Peerless Motor's stock was traded on the Cleveland Stock Exchange.

There were signs of a broadening of financial interest in the industry with a motor stock boom that got underway in March 1915. More automobile companies listed on the U.S. exchanges, and, in addition, some of them were able to sell common stock to the public, whereas in the past the only public issues that had taken place were of debt and preferred stock. Given the automobile industry's displayed success, it was hardly surprising that it would eventually attract greater attention from investment bankers and the organized financial markets. By 1915, the automobile industry was already very large; it produced almost one million cars and trucks, and the wholesale value of motor vehicle production was $702 million (Seltzer 1928, 75). The industry leaders, Ford and General Motors, had sales of $121 million and $94 million, and total assets of $89 million and $59 million, respectively, placing them in the ranks of the largest fifty industrial companies in the United States at that time (Moody's 1920; Seltzer 1928, 128, 230).

During the boom there was a marked expansion in the number of motor companies whose securities were traded in the public markets. In 1915, the stocks of four automobile companies—Chalmers, Chandler, Chevrolet, and International Motors—started trading on the NY Curb Exchange. In 1916, Chandler graduated to the NYSE and was joined there by White, Stutz, and Saxon, which listed directly on the Big Board. In the meantime, the stocks of fourteen additional automobile companies were admitted to trading on the Curb. In 1917, two of these—Fisher Body and Pierce-Arrow—moved up to the NYSE, but four more automobile stocks were added to the Curb in that year.

Some of these companies took advantage of the boom to raise substantial amounts of funds. In total, nearly fifty security issues were undertaken by automobile companies from 1915 to 1917 to raise cash of more than $100 million in new financing. Some of these issues were undertaken by the larger, established companies in the industry, including Willys-Overland, Studebaker, and Maxwell. There was also a second tier of issuers, including Chalmers, Fisher Body, Paige-Detroit, Peerless, and Pierce-Arrow, which had been around for a while and had grown into significant companies with sales of $10 million or more. In addition, a group of recent entrants to the automobile business, twenty of them in total, raised funds from the financial markets at this time. (Author's analysis based on data from the Commercial and Financial Chronicle).

Some of these new players, like Chevrolet, made their entry through the acquisition of existing companies in the industry, but others were entirely new companies. In both cases, they took advantage of the public appetite for motor stocks to drive their expansion. In total, these new players raised about $40 million in new financing during the three years from 1915 to 1917. These offerings were taken up with great enthusiasm by investors, with most issues oversubscribed several times. These companies were late entrants to the automobile industry, and the majority of them came to a sorry end. By 1924, thirteen out of twenty of them had exited the automobile industry; nine of them were bankrupt, one of them had retired from business, and three of them had discontinued the production of automobiles (Poor's 1939). However, there were six survivors, including Chevrolet, which had been founded by Durant and was taken over by GM after he took back control of that company from the bankers.

Although the motor stock boom petered out by 1917, there was a revival of investors' enthusiasm for the automobile industry after World War I, especially as the postwar conversion went more smoothly than expected. There was a temporary surge in entry into the industry at the time, and the boom in motor stocks, which began in mid-1919 and lasted until mid-1920, induced more young companies to come to the market to raise funds. The value of motor stocks then declined relative to other leading industrials until the end of 1924, when another boom took hold.[8] By then, entry into the industry had reached low rates, and a major shakeout had been underway for some time. As a result, there were richer pickings for financiers and stockholders in participating in consolidation as well as in encouraging established, successful companies with closely held ownership, such as the Ford Motor Company and the Dodge Brothers, to float their stocks.

On three occasions in 1923 and 1924, John Prentiss, the head of Hornblower and Weeks, one of the biggest investment houses in New England, offered to buy Ford's holdings for $1 billion with the intention of reselling securities in the company in the public markets. Although Ford turned him down on all three occasions, Prentiss was involved with the recapitalization of other automobile companies in the 1920s, notably General Motors, Hudson Motors, Dodge, Chevrolet Motor, and Chandler-Cleveland, and was quoted as saying that "one of the easiest tasks of the financiers of this generation has been the sale of securities of automobile companies to investors. This is primarily due to the fact that they have been very big money makers and to the fact that they are such big advertisers that their trade-marks and names become well known."[9]

His observation was proved beyond doubt by the successful flotation of the securities of Dodge Brothers in 1925. The company had been closely held from the time of its incorporation in 1914. However, following the death of both of the Dodge brothers, the trustees of their estate sought to liquidate the family's holdings in the company. The business was sold to the investment bank Dillon, Read & Company for $146 million. The bank then recapitalized the company in 1925 through the flotation of $85 million in preferred stock and $75 million in convertible debentures on the public markets. To facilitate a transaction of this magnitude, Dillon Read organized national selling syndicates that relied on the services of more than 400 financial houses across the United States. The deal was an overwhelming success, with both issues being heavily oversubscribed.[10]

The Aircraft Industry

The aviation industry was established only a few years after the automobile industry, with its origins usually traced to the first successful "powered" flight by the Wright brothers in December 1903. However, in contrast to the automotive industry, the development of aviation as a commercially viable industry took a long time to achieve. In its early years, it was largely concerned with the manufacture of small numbers of planes for airplane enthusiasts. Investment in the industry was limited, output was low, and the number of competitors was small during the first decade of the industry's existence.

As far as financing is concerned, as Rae (1965) noted, "The pioneering period of aircraft manufacturing in the United States shows a pattern characteristic of the growth of a new industry: that is, the emergence of a number of small companies financed from individual or local resources" (99). There were a couple of exceptions to this pattern with renowned inventors like the Wright brothers and Glenn Curtiss receiving the backing of nationally prominent financiers and industrialists (Rae 1965).

The outbreak of World War I stimulated the first major expansion in the industry's output with a dramatic increase in demand for airplanes from foreign governments and the U.S. government. The anticipation of U.S. involvement in the war led to some important changes in the competitive structure of the industry. In particular, in 1916 the Wright-Martin Aircraft Corporation was formed through the merger of the Wright Aeronautical Corporation, the Martin Company, and the Simplex Automobile Company. In the same year, a merger took place between Curtiss and the Burgess Aeroplane Company to form the Curtiss Aeroplane and Motor Corporation (Rae 1965).

The expectation and reality of higher military demand also brought new players into the aviation industry (Dodd 1933). For example, with a view to securing government contracts, the Dayton-Wright Company was formed in 1916 by Edward A. Deeds of National Cash Register and Charles Kettering of GM, among others, with Orville Wright as a consultant (Freudenthal 1940). William Boeing also entered the aviation industry in 1916 after a highly successful career in the lumber business. He changed the name of his company, Pacific Aero Products, to Boeing Airplane Company in 1917 and made training planes for the navy during the war. At this stage, most new entrants to the industry continued to rely on their own funds or those invested by local financiers (Rae 1965).

Estimates differ of the amount spent by the U.S. government on military aeronautics during World War I, but Rae puts the figure at somewhere over $400 million for the period from April 1917 to November 1918. This amount represented a huge influx of funds given the industry's small scale, and the results, in terms of the delivery of suitable aircraft for the war effort, fell far short of expectations (Freudenthal 1940, Rae 1965). Nevertheless, when hostilities terminated, enough output had been generated to contribute to a major industry bust as a large number of surplus planes was dumped on the market (Dodd 1933). Since commercial markets for airplanes had not yet been established to compensate for the decline in military demand, the result was a collapse in industry output and investment.

In 1921, fewer than sixty companies were recorded in the *Thomas Register* as being active either in the production of airplanes or of airplane motors (Thomas Register 1921). Among the companies specializing in the aviation industry, only Wright Aeronautical, Curtiss Aeroplane and Motor Corporation, and Sturtevant Aeroplane were covered by Moody's, the best available source for information on listed or otherwise actively traded stocks in the United States. Of these three, only Wright and Curtiss were traded on an exchange: the New York Stock Exchange (NYSE) for Wright and the New York Curb Exchange for Curtiss. As a measure of the stock market's limited involvement in the aviation industry, Dodd (1933) estimated that until 1921, the total amount of publicly offered securities amounted to no more than $15 million.

Producers in the industry struggled to stay afloat in the early 1920s. Some companies went out of business (Dodd 1933). Others sought to stay afloat by merging with their competitors; for example, the Consolidated Aircraft Corporation was formed in 1923 through the merger of the struggling Gallaudet Aircraft Company with what remained of the Dayton-Wright Company. Other aviation companies diversified; Boeing, for example, produced furniture in the early 1920s to remain in business. By 1924, the total amount of capital invested in the airplane manufacturing industry had actually declined from where it had been in 1920 (U.S. Bureau of the Census 1976, Dodd 1933).

However, in the second half of the 1920s, there was a dramatic turnaround in the industry's fortunes. By 1929 the total capital invested in aviation manufacturing reached $118 million in 1929 dollars compared with $19 million in 1919, and the number of wage earners employed had reached 9,856 compared to only 1,395 in 1921 (U.S. Bureau of the Census 1976, Dodd 1933). In addition, important investments had been made in the development of the nascent airline business as well as in airports and the delivery of other services to the aviation industry.

By the end of the boom in the late 1920s, aviation was still a relatively small industry in the United States. However, the decade from 1925 to 1935 was a crucial one for the industry's development. It was a period of significant technological change, which proved to be a boon to military production and laid the foundations for the development of major commercial markets.

Several factors contributed to the expansion of the aviation industry in the second half of the 1920s. By then, postwar dumping had run its course, and demand began to pick up again. There was some increase in airplane orders from the U.S. military, but much more important, a new source of business for the private sector emerged in the form of airmail service. Until 1925 the airmail service was operated by the government and, specifically, the Post Office Department. However, with the passage of the Kelly Bill or Air Mail Act of 1925, the postmaster general contracted with private companies to carry the U.S. airmail, thus furnishing private companies with a flourishing and lucrative new business.

Appropriations by the United States Post Office to domestic airmail service increased from a mere $100,000 in 1918 to $2.75 million in 1925 and then to $12.4 million by 1929 (Dodd 1933). Initially, the post office paid for contracted services on the basis of the postage revenues that it received. However, it changed this arrangement in 1926 to pay its contractors based on the weight of what they carried. As a result, the post office ended up providing a substantial direct subsidy to private contractors for the provision of airmail service (Freudenthal 1940)[11] It also boosted demand for airplanes to service these airmail routes; U.S. aircraft production peaked in 1929 at 6,193 planes of which 5,516 planes were for civil use (U.S. Bureau of the Census 1976).

The improvement in industry conditions led to an increase in the profitability of incumbent producers and also induced a new wave of entry to the industry. Although there had been some entry into the industry in the early 1920s, the number of producers of aircraft and parts was still relatively low in 1925 (Dodd 1933), and there were few operators since commercial aviation was still in its infancy. By 1929, however, the number of producers had more than doubled (Dodd 1933), and there had been significant entry by operators competing for the emergent commercial airline business, companies building and running airports, and other suppliers to the aviation industry. By 1931, there were an estimated 459 companies operating in all branches of the aviation industry (Dodd 1933).

Some financiers understood the improved prospects for the aviation industry from an early stage. Clement M. Keys was a leading example.

Keys was a Canadian who left Toronto for New York in the early 1900s to take a job as a journalist, and later the railroad editor, for the *Wall Street Journal*. Ten years later he set up his own investment bank, C. M. Keys and Company (van der Linden 2002).

In 1919, on a trip to Europe as a member of the American Aviation Commission, Keys became convinced of the potential of commercial aviation. When he returned to the United States, he bought Curtiss Aeroplane and Motor Company, which was in trouble, and borrowed money to keep it going until he turned it around. In response to the passage of the Kelly Act, Keys established National Air Transport, the first company that was founded as an airline enterprise, in May 1925. He loaded the company's board with prominent industrialists and financiers and capitalized the company at $10 million, an enormous sum for the aviation industry at the time. The company issued no stock to the public at the time, but it raised so much money that it had surplus far beyond its investment needs (van der Linden 2002).

Pratt and Whitney was also formed in 1925. Its founder was Frederick B. Rentschler who left Wright Aeronautical in 1924 and established a rival to that company with the approval of the navy. Like Keys, he enjoyed the backing of powerful financial interests, not least because of the help he received from his brother, Gordon, who was a director of the National City Bank. Later in the 1920s, Clement Keys and Frederick Rentschler would become rivals in the race to consolidation that dominated aviation.

However, it took longer for most investment banks and the country's financial markets to get directly involved in the dramatic developments in the aviation industry's competitive structure; practically no new securities were publicly issued from 1921 until the middle of 1927 (Dodd 1933).[12] Initially, the public did not seem particularly interested in investing in the aviation industry. As Freudenthal put it, "Flying was still generally considered a stunt, largely because of the postwar period of unrestricted flying which had resulted in many accidents. Serious flying seemed to be limited to the Army and Navy, which had captured a majority of the international air trophies" (1940, 89).

Charles Lindbergh's transatlantic flight in May 1927 changed all of that, contributing to great public excitement about the industry's prospects and massive speculation in aviation stocks. At the time, however, there were few such stocks from which investors could choose. Wright Aeronautical was the only aviation company with a listing on the NYSE. It made the engine that powered Lindbergh's plane, and its stock soared from $25 per share in April 1927 to 94\frac{3}{4}$ by December 1927 (Freudenthal 1940). In October 1927, Curtiss Aeronautical and Motor Company moved from the Curb to join Wright Aeronautical on the NYSE, presumably to participate more centrally in the aviation boom, and its stock price also rose dramatically.[13]

In September 1927 Boeing Airplane Company completed the first public issue by an aviation company for more than two years, but of debt rather than equity, selling 6 percent notes to raise $600,000. The *Wall Street Journal* described the issue as "another indication that investment demand has been stimulated" by Lindbergh's flight, noting that the entire issue was sold out in an hour or two, with more demand left unfulfilled.[14]

However, no further transactions occurred until March 1928, when Curtiss-Robertson Airplane Manufacturing Company sold shares to raise $500,000. Then the floodgates opened. From March 1928 to June 1930, 124 public offerings of stock were conducted by aviation companies to raise more than $300 million.[15] Almost all of these issues were of common stock, a fact that, as Freudenthal (1940) noted, reflected "a tacit acknowledgement of the uncertainty the companies felt about their future. No bonds were issued because the underwriters and issuing houses were unwilling to obligate the companies to pay interest charges and retire their bonds on definite fixed dates. Common stocks, mostly of no par value, committed the management to no specific return on investment and, besides, made it possible to draw in capital far beyond what was warranted by the physical assets and current income" (92).

The total amount of money raised by aviation companies through the sale of stock was enormous relative to the amount of invested capital in the industry at the time. It was also significant relative to overall levels of stock issuance, which themselves were at all-time highs. In the first eight months of 1929, the aviation industry was reportedly the fourth most important industry in terms of stock issuance, generating proceeds of $158.7 million compared with a total of $4.2 billion for U.S. industry

(Dodd 1933). This ranking is remarkable given the small size of this emerging industry.

The issuers were involved in the whole range of activities in the aviation industry, including the manufacture of frames and engines, the operation of airlines, the production of aircraft parts, the provision of aircraft services including runways, as well as the business of investing in aviation companies. The proceeds from these issues ranged in value from a low of $50,000 to a high of $40 million. The average issue was $2.6 million, with a standard deviation of $5.5 million, and the median issue was $1 million.

A couple of issues of stock were undertaken by established companies, like the Wright Aeronautical Company and the Glenn Martin Company, which had been around for more than ten years. However, in general, the aviation firms undertaking these issues were very young. Their average age was only 1.6 years, with a standard deviation of 2.5 years, and their median age was only 0.4 years.[16] Many of these companies were small companies that were new entrants to the industry; a total of forty-six issues raised proceeds of less than $1 million. Some of the aviation companies sold their stock on organized exchanges, especially the New York Curb Exchange, as well as regional exchanges such as Los Angeles, Chicago, Detroit, and Boston. However, many of these stocks were initially sold in the over-the-counter market and seasoned there before they graduated to one or more of the exchanges.

The wave of entry was a testament to how low the barriers to entry were in the aviation business at the time. However, some of the industry's most influential participants believed that this would soon change as more sophisticated technologies led to rising development costs. From their point of view, vertical integration was the appropriate strategy with which to confront this eventuality (van der Linden 2002). The U.S. Post Office created an added incentive for such integration by making the award of airmail contracts contingent on the reliability of the operator, which in those days was closely linked to its manufacturing experience.

A dramatic process of consolidation got underway in the aviation industry in the late 1920s, which revolved around the creation of large holding companies that controlled both manufacturing and transportation companies. The stock market was heavily implicated in this process. Some of the very youngest aviation companies undertaking stock issues were holding companies established to take active interests in existing companies. Moreover, the capacity of the most prominent holding groups to get access to resources from the stock market allowed them to extend their tentacles beyond all expectations.

The process of consolidation got underway in late 1928. One of the first major holding groups to emerge was North American Aviation, which was founded by Clement M. Keys, who was still president of the Curtiss Aeroplane and Motor Company. He had the backing of "about 40 prominent bankers and executives who have played important parts in advancing the aviation industry in his venture."[17] In particular, the company enjoyed the support of the Bancamerica-Blair Corporation and Hayden, Stone and Company of Boston. Blair & Company was the lead underwriter for a $25 million issue of the company's stock in the month of its formation.

At the same time, in his capacity as president of Curtiss Aeroplane, Keys was also involved in the merger of that company with another aviation pioneer and its long-time rival, Wright-Martin. The product of that merger, the Curtiss-Wright Corporation, took its place among the leading holding companies in the industry. In addition to its own direct holdings, Curtiss-Wright had important links to several important transportation companies like National Air Transport.

The most important competitor to the Keys' projects was the United Aircraft and Transport Corporation, which was incorporated in December 1928. It was formed from the merger of Boeing Airplane and Transport Corporation, Pratt and Whitney Aircraft Corporation, and Chance Vought Corporation and headed by Frederick Rentschler. The holding company was backed by the National City Company, a subsidiary of the National City Bank of New York, which was run by Rentschler's brother, Gordon (Freudenthal 1940). In fact, Gordon Rentschler was made president of that bank in part because National City Company made huge profits on the launching of this holding company. It underwrote a $15 million issue of units of stock comprising ten shares of preferred stock and four shares of common stock in January 1929 (Schwarzchild 1925–1928).

Other important holdings companies included the Aviation Corporation, more generally known as AVCO, which specialized in transportation, and the Detroit Aircraft Corporation, which absorbed Lockheed in 1928. Most of the merger and acquisition activity that these companies undertook was conducted through an exchange of stocks, but on several occasions in the late 1920s, they also sold stock to raise large amounts of capital for investment. Their stock issues represented by far the largest issues in the industry in terms of the proceeds that they raised.

These holding groups exercised an enormous influence on the competitive structure of the aviation industry. As Freudenthal put it, "The flood of mergers engulfed all the old pioneers and aviation engineers who had managed to carry their own companies through the previous years of erratic development. They could not compete, they realized, with the large combinations backed by powerful interests that were now dominating the industry. So they either sold out to the financiers, as Loening and Boeing did, or, like Douglas and Consolidated, admitted some financial interests" (Freudenthal 1940, 99).

By the end of the 1920s, these holding groups dominated the aviation industry, soaking up the large majority of its revenues. United Aircraft and Transport accounted for 42 percent of all plane and engine sales over the period from 1927 to 1933 and Curtiss-Wright for a further 39 percent, with independents such as Douglas, Glenn Martin, Great Lakes, Consolidated Aircraft, and Grumman together accounting for the remaining 19 percent (Freudenthal 1940). Holding groups exercised an even more overwhelming dominance over the operating side of the aviation business: United received 57 percent of domestic airmail payments in 1929, the North American Group accounted for 23 percent, and the Aviation Corporation for a further 12 percent, bringing the total share of these three holding groups to an extraordinary 92 percent (Freudenthal 1940).

Notwithstanding their dominance, the holding companies were not able to resist the fall in profitability that affected the aviation industry at the end of the 1920s. Industry profitability peaked in 1927 but then declined and turned to losses by 1930 as the depression hit (Dodd 1933). Aircraft production collapsed from 6,193 planes in 1929 to 1,396 at its lowest point in 1932, with the decline being driven by an implosion in the production of civil aircraft, which fell from their peak of 5,516 in 1929 to only 803 in 1932. Aircraft production for the U.S. military remained much more stable at 593 planes in 1932 compared with 677 planes in 1929 (U.S. Bureau of the Census 1976).

Revenues for the operating side of the business proved more robust at the height of the depression largely because the U.S. Post Office continued to increase its expenditures on airmail service. In 1932 it spent $19.9 million on domestic airmail service compared with $11.2 million in 1929. This was true notwithstanding the fact that the revenues it received from its airmail service increased from only $4.3 million in 1929 to $6 million in 1932.

Passenger revenues also grew from negligible amounts to $4.9 million by 1932, a development that was also attributable to the efforts of the U.S. Post Office. In 1929, when Walter Folger Brown was appointed to the position of the U.S. postmaster general, he lamented that "almost all of [the airmail carriers] were refusing to carry passengers and were depending wholly upon the Post Office department and we were getting nowhere in the development of airplanes" (quoted in Lazonick and Ferleger 1994). The collapse in aviation stocks increased his concern. As van der Linden (2002) notes, "The stock market crash revealed glaring problems in the airline industry even with the well-financed holding companies such as North American and AVCO. Following the October economic debacle, Postmaster General Brown became increasingly aware of the need for active governmental intervention to save not only the air mail carriers but also the entire air transportation industry" (p. 112). See also Vander Meulen 1991.

His response was the McNary-Watres Act of 1930, which amended the Kelly Act to pay airlines on the basis of the space available for transporting mail rather than the weight of the mail that they actually carried. The intention behind the 1930 Act was to subsidize the airlines to develop their capacity to carry passengers so that they would become less dependent on airmail revenues (Lazonick and Ferleger 1994). In 1930 Brown allocated contracts to create integrated transcontinental airmail routes among the companies that he believed would and could use their airmail subsidy to invest in the development of passenger transport. He privileged the well-financed airlines, which meant that the holding group companies were the main beneficiaries of the "spoils conference," as it was subsequently labeled. Although there were changes in their individual shares of the business, United, Aviation Corporation, and North American still controlled more than 91 percent of domestic airmail payments by 1933 (Freudenthal 1940).

Most of the holding companies survived the depression even if they had to retrench dramatically to do so. The main exception was Detroit Aircraft, which went into bankruptcy in 1931, dragging Lockheed Aircraft with it. Lockheed was bought in receivership by a group of investors who included its former general manager. It raised about $450,000 in two stock issues in 1933 and 1934. However, its development costs continued to mount as it sought to develop a new plane, the Electra, but with the help of a loan of $200,000 from the Reconstruction Finance Corporation it managed to survive.

Besides the Lockheed issues and a small stock issue in 1935 by Menasco Manufacturing, stock issues by aviation companies dried up in the first half of the 1930s. It was not until 1936 that aviation companies returned to the market in significant numbers to raise money. By then, the competitive structure of the industry had changed in response to a government decree mandating the breakup of the holding groups.

The widening gap between U.S. Post Office revenues for, and expenditures on, airmail service had become the subject of growing controversy, as had the "division of spoils" from the business among the holding companies. At the beginning of 1934, all of the airmail contracts from 1930 were cancelled by Roosevelt's government. In June 1934, Congress passed the Black-McKellar Bill or the Air Mail Act of 1934, which repealed all previous airmail legislation, including the McNary-Watres Act.

The act also mandated the separation of the manufacturing and transportation segments of the aviation industry. As a result, United Aircraft & Transport split itself into three parts: United Airlines, United Aircraft (which included Pratt & Whitney, Chance Vought, and Sikorsky), and Boeing. North American sold off its operator businesses to Transcontinental and Western Air and focused on manufacturing aircraft. AVCO reconstituted itself as a diversified company with interests within and beyond aviation and sold its operator businesses to American Airlines.

The delivery of the U.S. mail was temporarily turned over to the U.S. Army. Expenditures by the U.S. Post Office, which had already leveled off in 1933, plummeted in 1934 and 1935. However, by then, passenger revenues had risen, in large degree as a result of the McNary-Watres Act, and they surpassed airmail revenues for the first time in 1935. A stable foundation for the commercial aviation industry, as well as a strong contributor to the US military, had thereby been created (Vander Meulen 1991).

The Radio Industry

The origins of the radio industry are typically traced to the formation of the Marconi company in Britain in 1897. The U.S. industry got off the ground when an American subsidiary was launched two years later. Besides Marconi, the other key players in the early development of the U.S. radio industry were companies established by two other prominent inventors, Reginald Fessenden and Lee de Forest. Financing did not prove to be a problem for any of these ventures. As Maclaurin (1971) put it, "The most difficult ingredient to supply proved to be effective management rather than capital. Marconi, De Forest and Fessenden— all succeeded in obtaining venture capital in substantial quantities to back their wireless enterprises."

However, the commercial success of the early radio companies was rather modest, and as late as 1920, the radio industry was of limited economic importance. That changed dramatically in the following decade as the radio industry entered a period of rapid growth. The boon to the commercial potential of radio during this period was the development of public broadcasting in 1920 (Maclaurin 1971).

From its inauguration, radio broadcasting showed spectacular growth, and sales of broadcasting equipment and radio sets soared. By 1929, 69 million broadcast tubes and 4.4 million broadcast receiving sets were sold compared with 1 million and 100,000, respectively, in 1922. The total sales of equipment for broadcast reception rose from $60 million in 1922 to $843 million in 1929 (Maclaurin 1971).

By far the most important player in the radio industry at this time was the Radio Corporation of America. It was established in 1919 at the initiative of General Electric, with the approval of the U.S. government, to bring all of the important patents in the U.S. radio industry, primarily those of American Marconi, General Electric, American Telephone and Telegraph, and Westinghouse, under one roof. The dispersion of radio patents in the United States was increasingly seen as an obstacle to the country's ability to compete with other nations, especially Britain, in the development of international communications (Maclaurin 1971).

RCA was initially not intended to be a manufacturer of radio equipment. Instead GE and Westinghouse would supply radio transmitters and receivers. Both companies started producing radio sets after World War I, but they were not able to keep up with the massive increase in demand that followed the launch of commercial broadcasting. As a result, the boom in the radio industry brought hundreds of entrants into the industry. As Maclaurin (1971) notes, "Little capital investment was required to assemble radio sets. And there were many companies which possessed the 'know-how' to manufacture the principal component parts" (260). He estimated that more than 700 firms entered the industry in the period of four years from 1923 to 1926.[18]

Many of the early entrants were engaged in the production of crystal radio sets where patents could be worked around. However, it soon became clear that vacuum tube sets were the wave of the future in radio, and RCA exercised a dominant position in patents for all aspects of these sets from the circuit design through to the loud speakers (Maclaurin 1971). Some companies entered to do business on the basis of alternative patents for these technologies. Others competed by disregarding patent restrictions.

The wave of entry into the radio industry was accompanied by a boom in stock issues of radio companies. An expression commonly heard at the time was "a new radio stock a day," with pundits speculating that at least one share of stock was sold for every receiving set sold.[19] The most prominent new radio companies traded on the New York Curb, and by 1925 there were eighteen of them represented there. A couple of them, such as De Forest Radio and Duplex Condenser and Radio, had been around for more than ten years, but most of them were new companies established in the early 1920s. Sometimes they boasted the involvement of men with long experience in radio technology and a few of them came into existence to acquire a predecessor company with established operations.

The stock market's enthusiasm for the radio industry dissipated in early 1925 largely because of the pressure on profitability that high entry had caused. The leading radio stocks lost 60 percent of their value from December 1924 to May 1926.[20] If we exclude RCA's stocks from the calculation, the decline in the radio companies' market value was even more dramatic: 92 percent. The crash was a response to the overcrowding of the industry when, as one contemporary study put it, "all hands found the market tired and inventories at the peak. There was a rush to unload, and receiving sets were sold by the hundreds of thousands often at less than cost; but those who had bought the stocks of many of these companies found it impossible to get out without devastating losses. Four companies went into receivers' hands in the last year and application for receivership was announced for two others."[21]

The Curb stocks may have been the most prominent of the radio stocks, but they represented only a portion of the total number of these stocks sold in the United States in the early 1920s. As *Barron's* pointed out, "This loss in market value in 18 different companies' stocks ... is not the only loss suffered in radio stocks for other issues were floated in various sections of the country some of which were never actually listed on any market."[22] Some commentators attributed real effects to the crash; the *Wall Street Journal* claimed that "with a rush by all hands to scoop up some of the cream, the result was disastrous, inasmuch as it confused the public and actually stopped the boom in the sale of sets.... Overproduction, followed by cautious buying on the part of the public, astonished the promoters of the dozens of new companies."[23]

Entry declined dramatically from its peak of 258 companies in 1925 to only 26 firms in 1927, 16 in 1928, and 26 in 1929 (Maclaurin 1971). In part, the decline was a response to disappointed expectations, but the more important reason for the change was the success of RCA's attempts to enforce its patent rights. Concerned about the implications of hyper-competition for the profitability of the radio industry, RCA had been working through the courts to secure its future based on the control of key patents. By 1927, as Maclaurin (1971) put it, "The RCA group had … established a strong patent position in all the major branches of the radio industry, and an RCA license was considered essential for the manufacture of any up-to-date set or modern vacuum tube" (131). RCA was willing to license its technologies but only to a limited number of companies, which contributed to the reduction in the rate of entry to the business.

Following the crash of radio stocks in 1925, there was a lull in the public stock offerings of radio companies that lasted for almost three years, from July 1925 through February 1928. The only exception was an issue by Gold Seal Electrical in November 1928. Then, in March 1928, Grigsby-Grunow launched another wave of radio stock issues when it raised $500,000 through the sale of common stock. From March 1928 to September 1929, twenty-five public stock offerings were undertaken by radio companies to raise a total of $38.4 million.[24]

All of these issues were common stock offerings, with the only exception being an issue of cumulative preferred stock by Stromberg-Carlson that raised $1 million in July 1928. On average, these issues raised $1.6 million with a standard deviation of $1.6 million, and the median issue was $1.1 million. The issuing companies were on average 4.4 years old with a standard deviation of 6.3 years and a median age of 2.8 years.[25] Moreover, on closer inspection, some of the very youngest issuers came into existence to acquire the assets of existing corporations. As a result, the radio companies completing public stock issues at this time tended to be older than their aviation counterparts and older than earlier issuers of radio stock.

There was another bust in radio stock prices from 1929. The stock market crash and subsequent depression played crucial roles in precipitating the decline, but industry observers also blamed another overexpansion of the industry. As *Barron's* pointed out, "Increased competition, narrower profit margins, and reduced sales forced some of the weaker concerns to the wall. Among them were Kolster Radio Corp, which went into receivership in January, 1930; Earl Radio Corp., which suffered a similar fate in November, 1929; and Freed-Eiseman Radio Corp., which experienced receivership in December, 1929."[26]

Explaining the Role of the Stock Market in New Industries in the 1920s

The roles that the stock market played in the three industries that I have discussed differed substantially in their importance and characteristics. The stock market was of limited significance to the early development of the automobile industry, with the overwhelming majority of its investment being funded from internal sources. With only a few exceptions, when the stock market became involved in the development of particular automobile firms, it tended to be to facilitate consolidation through mergers and acquisitions or to fund investment when they were already rather large.

The stock market played a much more important role in the early commerical development of the aviation and radio industries. Although the aviation industry initially had limited contact with the stock market, this changed in the second half of the 1920s when the industry experienced its first major peacetime growth spurt. The stock market played an important role in channeling money into the industry, especially in the last three years of the decade. In doing so, it allocated funds to large numbers of new companies that were entering the industry at the time, but it also facilitated a wave of consolidation that consumed many of the independent players in the industry by the beginning of the 1930s. The stock market also had a high profile in the development of the radio industry in the 1920s, where its role was predominantly one of facilitating entry, especially by young companies.

Based on these three industries, it would appear that the 1920s inaugurated the substantial involvement of the U.S. stock market in the development of new industries. However, the stock market's involvement in the aviation and radio industries was not necessarily representative of its general role in the U.S. economy at the time. From his analysis of all of the security issues conducted in 1929, George Eddy (1937) concluded that "new productive enterprises were financed directly via public security issues in 1929 to the extent of no more than a few hundred million dollars at most" (86). Yet even if the stock market's participation in the development of new industries in the 1920s was of limited quantitative significance, it represented an important qualitative development in its role in the economy.

Of course, further research on the emergence of other new industries around this time is required to confirm the importance of the 1920s as a turning point for the importance of organized financial markets in the development of new industries in the United States. To the extent that this is the case, then the question to be addressed is why this happened. Specifically, what changes in the market for corporate stocks in the 1920s facilitated this development?

There is no question that changes in the demand for stocks in the 1920s facilitated the funding of new industries. In the early twentieth century, the ownership of industrial stocks in the United States was largely confined to the wealthiest households in the country. However, from the late 1910s through the 1920s, the dispersion of stockholding increased rapidly. In 1917 and 1918, the U.S. Treasury raised an unprecedented amount of $17 billion through the sale of Liberty bonds in the public markets, and in 1919 it raised a further $4.5 billion with its Victory loan (Carosso 1970). In doing so, it brought the savings of a whole new tier of American households into the securities markets, and many of them were persuaded of the merits of holding industrial securities as the 1920s unfolded (Carosso 1970). By the end of the bull market in 1929, the available estimates suggest that as many as 6.25 million Americans, that is, 5 percent of the U.S. population, owned stock (Goldsmith 1969).[27]

Perhaps as important as the growth in the number of investors participating in the stock market was their increased willingness to take risks for the prospect of capital gains. Until the 1920s, preferred stocks were favored over common stocks for their promise of a steady income stream. However, common stocks gained ground in the 1920s, especially in the second half of the 1920s, as investors displayed increasing interest in the prospect of speculative gains, even if they came at the expense of steady dividends. This attitude was crucial in allowing the aviation and radio companies to raise capital in the public markets in the 1920s. Almost all of them issued common stock to raise funds, and most of them took advantage of the discretion that this instrument accorded them to pay no dividends.

The willingness of stockholders to bet their money on the future prospects of aviation and radio reflected their growing realization of the gains to be made from investment in industrial stocks. By the 1920s, the success of companies that dominated industries like electrical equipment and automobiles was very clear, as were the benefits that their stock-holders had derived from that success. It generated great enthusiasm for getting in on the ground floor of new industries, even before their prospects were proven, and there was widespread discussion of the heuristics that might be used to select the stocks that would someday become the General Electric or General Motors of these industries.

As far as the supply of corporate stock, there seems little doubt that developments in the supply of stocks available to investors were an important precondition for their involvement in the funding of new industries. In the cases of the aviation and radio industries, the stock market became actively involved in their development by riding on a wave of entry that had been set off by changes in the technological and market structures of these industries. In both industries, the stock market's willingness to invest was set off by a public event—the start of commercial broadcasting in radio and Lindbergh's flight in aviation—that drew attention to their technological and market dynamics and stimulated huge run-ups in the stock prices of aviation and radio companies while the speculative momentum lasted.

Nevertheless, it is hard to argue that these types of supply-side factors were behind the stock market's greater propensity to fund new industries in the 1920s. After all, it was not the first time they had manifested themselves. In particular, the automobile industry had experienced high levels of entry in its early development and, as a consumer-oriented industry, had attracted considerable attention from the public without drawing the kind of funding from the financial markets that the aviation and radio industries did.

Besides changes in the demand for stocks, there were important developments in the way in which the demand for, and supply of, corporate stock were brought together during the 1920s that played an important role. The dynamics of the investment banking industry merit attention in this regard. Investment bankers had played an important role in facilitating the governments' efforts to sell its Liberty and Victory loans. In the process, they had developed selling practices to reach as many potential investors as possible. For example, there was a marked expansion in the size of selling syndicates that they used to distribute securities, and the nationwide syndicate made its appearance at this time. In general, the investment banking industry experimented with more aggressive sales techniques and transferred these techniques to the sale of industrial securities in the 1920s (Carosso 1970).

Some of the well-established investment banks participated in these developments in the 1920s, but as Carosso (1970) noted, "Most long-established houses, while continuing to grow, maintained their conservative policies in the face of growing competition. They did not adopt aggressive sales tactics or openly solicit business. They sold few common stocks even though these were becoming the more popular securities" (255). For example, common stock accounted for just over 3 percent of the securities that the House of Morgan, which remained the leading investment bank in the United States in the 1920s, distributed to the public from 1919 to 1933 (Carosso 1970).

However, with the major expansion of the investment banking business that occurred during the 1920s, there was a growing population of newcomer investment banks that were willing to be much more aggressive. An analysis of the lead underwriters of aviation and radio stock issues in the 1920s reveals few of the investment banking firms that had featured prominently in the Pujo hearings. Instead, many of these underwriters were relatively small, and often young, players operating out of New York. In addition, there were a considerable number of regional players that had initially made their mark by underwriting local issues. The investment banking affiliates of commercial banks also made an appearance. The aggressive originating and selling tactics of these types of players were largely what facilitated and fueled the sale of stocks that previously had been considered too speculative for public consumption and remained so for the most prestigious U.S. investment banks.

Besides the role of investment bankers in bringing the demand for, and supply of, corporate stock together, the institutional structure of the trading markets in which stocks were bought and sold in the United States also facilitated the funding of new industries in the 1920s. By then, the NYSE dominated the U.S. stock market in terms of trading activity and capitalization. It had secured its position by developing a reputation for listing the highest-quality corporate securities in the

United States and by aggressively defending it in the face of any direct competition. With the demise of the Consolidated Exchange in the 1920s, the Big Board faced no direct challengers to its position (Garvy 1944). However, other trading markets survived in the United States and prospered during the 1920s by carving out niches that allowed them to avoid direct challenges to the NYSE's position. It was largely through their efforts in the 1920s that so many new companies were able to get access to public financing.

The New York Curb Exchange, which started out some time before the Civil War, was the second most important exchange in the United States at the time. Some of the issues traded on the Curb Exchange were formally listed on it; a listing department was established in 1911 with responsibility for vetting the applications of issuers that sought to have their securities traded there. However, most of the trading on the Curb was in unlisted issues, that is, securities that were admitted to trading at the request of exchange members without the involvement of and, in some cases, over the objections of, the issuer. In some cases, unlisted issues were listed on other exchanges but the Curb Exchange never traded stocks that were listed on the NYSE to avoid a direct challenge to the Big Board's position (Sobel 1970, Twentieth Century Fund 1935).

Having witnessed the aggressive tactics that the NYSE used to eliminate its direct competitors, the Curb was careful to avoid looking as if it was competing for the Big Board's business. It tended to list smaller, more speculative issues than those admitted to trading on the NYSE and served as a testing ground for some of these stocks, which then graduated to the Big Board once they had proved themselves; well-known companies like General Motors, Du Pont, Montgomery Ward, and Goodyear started out on the Curb Exchange (Leffler 1957). Once this transition had occurred, the Curb stopped trading these stocks.

There was an important change in the operation of the Curb in 1921. Whereas previously it had operated outdoors, whence it derived its name, it moved indoors to a building in Lower Manhattan close to where it had always operated. That move seemed to impart it with an air of greater respectability, and its level of activity soared as the decade unfolded (Bruchey 1991). As I noted, many of the stocks of radio and aviation companies initially traded on the Curb in the 1920s, and in general, the decade proved to be a flourishing one for the exchange. By the early 1930s, more than 2,000 stock issues traded on the Curb Exchange, of which 400 were listed issues and about 1,700 were unlisted issues (Twentieth Century Fund 1935).

In addition to the NYSE and the Curb, there were also more than thirty organized exchanges outside New York City. The largest of them in terms of the volume and value of securities traded were in Chicago in the Midwest, Philadelphia and Boston on the East Coast, and San Francisco and Los Angeles on the West Coast (Twentieth Century Fund 1935). These exchanges had developed to provide markets for the securities of local corporations, but as the more successful of these companies graduated to the Big Board, they were increasingly concerned about generating new sources of business.

Their problems were largely solved in the 1920s when they experienced a flood of new listings; for example, from 1926 to 1929, Chicago's listings more than doubled from 237 to 535, and Boston's listings increased from 300 in July 1925 to 437 in July 1930 (Securities and Exchange Commission 1963). As the SEC Special Study noted, "Perhaps the most important single reason for many listings on the regional exchanges...was the desire of issuers to secure the special privileges afforded to listed securities by many State blue sky laws. Such laws generally required registration of securities offered for sale in the State, but accorded exemption to securities listed on an approved securities exchange (Securities and Exchange Commission 1963, 915–916).

In addition to the organized exchanges, the United States also had a substantial over-the-counter (OTC) market, which also served as a seasoning ground for new securities. The OTC market was really a collection of dispersed markets in which broker-dealers conducted business with each other and with members of the public. When someone wanted to sell or buy a stock, a broker would seek to find a matching offer to buy or sell that stock and negotiate a price for the trade. Some dealers were described as market makers in particular securities to the extent that they carried inventories of these securities and stood ready to buy and sell them to other broker-dealers. The market was particularly useful for trading stocks, like unseasoned stocks, which had thin markets. One informed observer estimated that by the early 1930s, about 30 percent of all outstanding stocks were traded in the OTC market (Twentieth Century Fund 1935).

The Effects of the Stock Market on the Development of New Industries

Having considered a range of explanations for the growing involvement of the stock market in the development of new industries in the 1920s, there remains the question of the economic implications of that involvement. As I shall explain, the task of thoroughly addressing this question requires research that goes beyond the scope of this chapter. However, the evidence that I have compiled already suggests some questions that need to be considered in further work.

In their research on finance and growth, macroeconomists are prone to assume a virtuous cycle from greater access to external finance through more investment to higher productivity and growth. Although it is possible that such a cycle may occur, we should not assume, pending evidence, that it will necessarily occur. Corporations have choices about how they allocate the monies they generate from stock issues, and different corporate uses of the proceeds of stock issues have distinct implications for the real economy.

It may be that companies use the proceeds of stock issues to fund investment in the accumulation of fixed or working capital, in which case they contribute to the process of capital formation that generates new assets in the economy. Firms may also allocate the proceeds to refinance existing financial obligations, acquire existing assets through acquisitions, and even bolster their treasuries for later use. Stock issues that fund these uses facilitate changes in the ownership of, or other financial claims on, existing assets in the economy. They may well have important implications for the real economy, but they will not be the same as those associated with an increase in capital formation.

Even to the extent that the proceeds from stock issues are invested in the formation of new capital, we ought not to assume that they will necessarily foster better economic performance. To the extent that speculation is an important determinant of stock issuance, it may encourage either a wave of overinvestment, if more funds are supplied to an industry than can be productively used within it, or poor investment if funds are allocated to inefficient firms. Concerns about the negative impacts of stock market hubris were often voiced in the late 1920s. As Eddy (1937) noted, "The reputed ease with which funds were obtained by security financing in 1929 is often cited as the cause of the formation of a great many new enterprises that ought never to have been undertaken" (86).

These concerns are particularly relevant for industries like aviation and radio in which new firms raised large amounts of capital in the 1920s. So what was the impact of the capital raised on the fortunes of these industries? As far as the radio industry is concerned, Maclaurin (1971) argues that many of the young companies that entered the industry in the 1920s were imitators rather than innovators: "Most of these concerns were largely imitative and were therefore not important 'transmitters of newness'" (262). Even before the full onslaught of the depression was felt, a number of companies had gone bankrupt. By the middle of the 1930s, very few of the companies that had entered the radio industry in the 1920s were still in existence (Maclaurin 1971).

These observations raise questions about the economic value of entry to the radio industry and, by association, with the role of the stock market in facilitating that process. Of course, it is possible that the stock market, in funding entrants, was effective in choosing the winners among the entrants or in increasing their chances of survival. In fact, few of the companies that emerged to dominate the radio industry by 1940 were represented among the young companies that completed stock issues in the 1920s (Maclaurin 1971). In the case of the radio industry, therefore, it is difficult to make the case that the significant involvement of the stock market in the 1920s in funding entrants to the industry was crucial for facilitating innovation. In fact, some observers believed that it may even have undermined that process.

As far as the aviation industry is concerned, the aggregate amount of funds generated by the aviation industry through stock issues, relative to its extant capital at the time, also suggests cause for skepticism about the possibility of these monies' productive use. In contrast to the experience of the radio industry, however, young entrants to the industry seem to have been innovators as well as imitators. Unfortunately, given the extent to which the wave of consolidation in the industry absorbed these companies, there is no straightforward way to analyze their subsequent performance in the 1930s and beyond.

Turning now to other uses of the proceeds of stock issues, we should not expect their real effects to be reflected in higher contemporaneous levels of investment. As far as mergers and acquisitions are concerned, for example, their economic impact depends on the implications of the change in ownership on the productivity of the businesses and assets that change hands. Again, there is no a priori reason to assume that the consolidation that stock issues facilitate will be beneficial from the perspective of the overall economy until we have evidence to show that this is the case.

Most research on the economic impact of mergers and acquisitions measures the abnormal returns that accrue to the stockholders of target and acquiring firms. Leeth and Borg (2000) have undertaken a study along these lines for the merger wave of the 1920s, and their findings echo those from similar studies that focus on recent periods of merger activity: target firm shareholders generate large abnormal returns, in excess of 15 percent, while acquiring firm shareholders break even. These types of findings are usually interpreted as positive overall evidence in favor of mergers. However, the appropriateness of the methodology depends on the assumption that stock prices are accurate predictors of the real economic impact of mergers. That assumption seems particularly questionable for the U.S. stock market in the late 1920s.

An alternative approach to using stock valuations as proxies is to focus directly on the real effects of merger and acquisitions. In her research on the Great Merger Movement at the turn of the twentieth century, Lamoreaux (1985) studied the long-term economic effects of mergers by analyzing their capacity to maintain their dominant positions in their respective industries. She found that it was limited, showing that in many cases, the consolidators were outcompeted by stronger rivals, and concluded that the movement cannot be understood as being driven by an inexorable economic imperative.

A similar study for the 1920s would undoubtedly shed light on the economic impact of the mergers and acquisitions undertaken in that decade. However, the analysis is complicated by the stock market crash and the depression. It is plausible, even likely, that these shocks to the U.S. economy could have contributed to the demise of some consolidators that would have functioned effectively in normal economic times. More generally, as the example of the aviation industry shows, it is difficult to disentangle the effects of consolidation from other factors.

In the aviation industry, the consolidation that took place late in the 1920s was heavily criticized not only for fostering excessive concentration in the industry but also for placing financial over productive interests in the development of the industry (Freudenthal 1940). However, as I have already noted, all of the holding companies survived the hardships of the early 1930s with the exception of Detroit Aircraft. Nevertheless, that outcome was driven as much by politics as economics. So too, as I have explained, politics were the undoing of the holding companies in the mid-1930s.

Therefore, it is difficult to know what to conclude with respect to the economic impact of consolidation from the history of the aviation industry. In a similar vein, the impact of changes in ownership on target companies is typically difficult to identify. In these cases, it is hard not only to separate the effects of other factors but also to disentangle the performance of targets from the performance of the larger entities into which they were absorbed. Even if the targets reappear later as independent entities, it is difficult to control for the positive or negative effects of having had a corporate parent, even if only for a short time.

Clearly there is a need for new research to evaluate the implications of the active involvement by the stock market on the development of new industries like radio and aviation. While I have highlighted some of the challenges of assessing the impact of stock issues on the fortunes of particular companies, the real task is more challenging still. If our concern is with innovation and productivity growth, then what really matters are the effects of stock issues on the development of technologies and markets at the level of the industry as a whole rather than the fortunes of individual companies.

The possibilities and problems of such work are evident in two related but conflicting studies of the contemporary hard disk drive industry. In "Capital Market Myopia," William Sahlman and Howard Stevenson (1985) argued that venture capitalists and the stock market massively overinvested in the industry during the period from 1977 to 1984, with negative consequences for innovation and financial returns. However, another study of the disk drive industry claims that that the diagnosis of capital market myopia is suspect in long-term perspective.

Bygrave, Lange, Roedel, and Wu (2000) accepted that many players in the disk drive industry failed. However, they argued that along the way, they contributed to an extremely dynamic process of innovation that brought quality up and costs down to unprecedented degrees. Moreover, they claimed that the survivors ultimately enjoyed sufficient commercial success to justify the financial bets that were made on the industry. It is these types of debates that we need to stimulate, with respect to the historical evidence, to advance from here to better understand the role of the stock market in the development of new industries.

Beyond the 1920s

The stock market crash in 1929 and the economic crisis of the 1930s interrupted the stock market's relationship with new industries. There was an upsurge in stock issues in 1936, but it came to an end with the recession of 1937, and then World War II took its toll. From what we currently know about the composition of stock issues at this time, few of the stock issues that took place from the mid-1930s until the late 1950s were initial public offerings.[28] Companies seeking funds from the stock market during this period were, for the most part, already publicly traded.

New markets and new technologies emerged in the postwar period. However, it took some time before investors were willing to participate in their development, especially when it relied on young entrants. The bull market that got underway from the early 1950s was primarily focused on the stocks of established companies that had already built up a track record in developing new technologies and markets. Investors were reportedly skeptical about the capacity of young firms to survive in competition with incumbents, especially in high-technology industries (Sobel 1977).

It was not until the late 1950s that substantial numbers of small, high-tech companies once again sold their stock to the public. An important catalyst for the change occurred in October 1957 when the Soviets put the first *Sputnik* into orbit, an event that was greeted with shock and dismay in the United States (Sobel 1977). The U.S. federal government was already allocating substantial resources to the development of technologies, especially electronic technologies, that could be employed in building its military capacity. *Sputnik* galvanized the U.S. political elite to make even greater financial commitments to the development of technology. Liberal government support for emergent high-technology companies was thus ensured for a long time to come.

A boom in initial public offerings got underway in 1959 and continued until the decline of the stock market in early 1962. During this period, as the SEC (1963) put it, "The distribution of securities by companies that had not made a previous public offering reached the highest level in history" (487). For example, in 1959, 63 percent of common stock issues were unseasoned; in 1960, 72 percent; and in 1961, 76 percent.[29] By comparison, less than 30 percent of the stock offerings during the late 1940s were unseasoned (SEC 1963).

The largest number and volume of unseasoned issues were in the manufacturing sector. Many of these IPOs were popularly described as "hot issues," with the hottest of them in new industries, especially the emergent electronics industry. Other glamour stocks were found in the fields of scientific instruments and research, photography, printing and publishing, sporting goods and amusements. As the SEC (1963) put it, "In many cases there was little to support the public enthusiasm for a particular issue except magic words indicating membership of glamorous industry in name or description of business" (486). More than 85 percent of registered issues in these glamorous industries were companies with less than $5 million in assets. No breakdown of registration A issues is available but these companies were undoubtedly even smaller.[30]

The wave of hot issues in the late 1950s and early 1960s was important for what it signaled about the renewal and generalization of U.S. investors' interest in financing new industries and new entrants to them. Similar demand-side factors to those that were at work in the 1920s, including an increase in stock ownership and a growing enthusiasm for common stocks, played an important role in driving this development.

In addition, high rates of entry into the underwriting and brokerage industries were important, with many of the unseasoned stock issues being originated and sold by new firms. Finally, the structure of the trading markets played a role.

In contrast to the 1920s, the regional exchanges no longer played an important role in the distribution of new issues. In response to the speculation of the 1920s, most states enacted laws prohibiting unseasoned issues from trading on these exchanges (SEC 1963). Most of the new issues of the late 1950s and early 1960s were seasoned in the nation's OTC market, which experienced rapid growth in the postwar decades. The structure of federal regulation of the securities markets contributed to the postwar growth of the OTC market, although this was neither intended nor desired by the regulators. In particular, many of the companies whose securities traded in the OTC market did not have to comply with the periodic disclosure, proxy, and insider trading provisions of the 1934 Securities Exchange Act, in contrast to companies listed on a registered exchange (O'Sullivan 2006).

The speculative fervor that surrounded the hot issues of the late 1950s and early 1960s focused attention on the operation of the OTC market and prompted the first major study of the U.S. securities markets since the early 1930s. The SEC's Special Study of the Securities Markets was a massive undertaking in terms of the resources devoted to it and the comprehensive nature of the analysis that it undertook. The study highlighted a wide range of problems with the operation of the OTC market and provided a long list of recommendations for its improvement.

Its greatest concerns were with the flow of information about trading in, and issuers on, the OTC market. These concerns led the SEC to recommend the automation of OTC operations "to assemble all interdealer quotations and instantaneously determine and communicate best quotations for particular securities at any time" (Hazard and Christie 1964, 263). As far as issuers were concerned, the study recommended that all issues held by public shareholders, whether they were traded on an exchange or in the OTC market, should comply with the Exchange Act's ongoing reporting, proxy, and insider trading regulations.[31] As I have described elsewhere, these recommendations were the foundations for the establishment of Nasdaq in 1971. It is, of course, this market that is today so closely associated with the development of young companies and new industries in the United States (O'Sullivan 2006).

Conclusion

In recent times, the U.S. stock market has gained considerable prominence for its capacity to fund new industries and the young firms that often enter them. For many commentators, this capacity is one of the great strengths of the U.S. system of capitalism contributing, in particular, to its dynamism in innovation and productivity. Yet there are also skeptics who point to the damaging influence of the speculative excesses of financial markets. In this chapter, I have looked backward, over the past century of stock market activity in the United States, to shed light on this discussion.

Based on case studies of three new industries in the early part of the twentieth century—automobiles, aviation, and radio—I show that the role of the stock market in their early development differed substantially in terms of its importance and characteristics. It was only in the latter two cases that the stock market was heavily implicated in their early financing. In aviation, it funded large numbers of new companies that entered the industry at that time and also facilitated a wave of consolidation that consumed many of the independent players in the industry by the beginning of the 1930s. The stock market's role in the radio industry of the 1920s was predominantly one of facilitating entry, especially by new, young companies.

These examples suggest that the 1920s was an important turning point in the involvement of the stock market in the development of new industries in the United States, and I suggest several reasons—the demand for stocks, the dynamics of investment banking, and the institutional structure of the U.S. stock market—that account for the enthusiasm for the stocks of new firms and new industries at this time. I also address the question of the effects on new industries of the stock market's active involvement in their early development. The relationship between higher stock issues and improved economic performance is not the automatic one that optimists tend to assume. In fact, the examples of the aviation and radio industries provide much more grist for the mills of the skeptics, although more research is required to work out the effects of stock issues on the innovative dynamics of these and other industries.

These issues have a relevance not only to the 1920s. Although there was a hiatus in the relationship between the stock market and new industries in the 1930s and 1940s, it came to an end with a wave of hot issues from the late 1950s. This wave was important for what it signaled about a renewal and expansion of U.S. investors' interest in funding new industries and their entrants and many of the factors that seem to explain that interest in the 1920s played a role, in somewhat different form, in the late 1950s and early 1960s. Questions can be asked, and indeed were asked by the SEC, about the implications for the companies and industries that were funded by the stock market at this time. But the wave of hot issues also had a longer-term significance in terms of the regulatory reforms that were wrought in response to it and the institutional changes that they begat. In short, the aspects of the current institutional structure of the U.S. stock market that are closely associated with the funding of new industries, especially the Nasdaq market, largely owe their existence to these reforms.

Acknowledgment

The research for this chapter was conducted, in part, with the help of funding from the ESEMK project, funded by the European Commission under the 6th Framework Programme (contract CIT2-CT-2004-506077), as well as from INSEAD, Fontainebleau, France.

Notes

1. Navin and Sears (1955) draw particular attention to the emergence of the trust movement and the related issuance of trust certificates in the 1880s, the subsequent conversion in the early 1890s of many of these trusts into corporations through the issuance of stock, a broader merger movement that involved many existing corporations, as well as an increased willingness on the part of company owners to take advantage of the emergent market for stocks to liquidate their investment.

2. In constructing my time series, I relied on several different sources of data. For the period from 1934 to 2000, I used data on gross stock issues compiled by the

Securities and Exchange Commission (SEC) and reproduced in the *Federal Reserve Bulletin*. The SEC series measures the cash proceeds for all new issues of stock that raised more than $100,000, whether or not they are registered with the SEC, and it includes the proceeds from stock issues sold through private placements as well as public offerings. For the period from 1919 to 1934, I used data compiled from an analysis of "New Capital Flotations" published in the *Commercial and Financial Chronicle*. These data are also reproduced in the *Federal Reserve Bulletin* and the U.S. Bureau of the Census (1976, Series X 510–515, p. 1006). They measure the cash proceeds of stock issues sold through public offerings by domestic corporations in the United States and some stock issues sold through private placements, although coverage of the latter is not comprehensive. Both of these series exclude cash issues of stock to employees, pension funds, and customers, under option schemes and those that are privately placed with investors other than financial institutions. They also exclude all noncash issues of stock, intercorporate transactions, and stock issues sold continuously such as stock sold by open-end investment companies (Goldsmith 1955). Besides the coverage of private placements, the most important difference between the SEC and Chronicle series is that the former includes stock issues by foreign corporations in the United States, whereas the latter excludes them. Finally, for the period from 1906 to 1918, I used data compiled by the *Journal of Commerce* and reproduced in Goldsmith (1955). Although the *Journal of Commerce* series is regarded as the best available source of data on stock issues for the period, information on exactly what is included and excluded from the series is limited. The series apparently includes most cash issues of stock, but it does not cover private placements (Goldsmith 1955). The treatment of noncash stock issues by the *Journal of Commerce* is uncertain (Goldsmith 1955).

3. That peak was not surpassed until 1983.

4. Durant also invested $500,000 in the new company in return for $500,000 in preferred stock and a bonus of $250,000 in common stock (Seltzer 1928).

5. The Studebaker Manufacturing Company had acquired one-third of the stock of the EMF Company in 1909 and the remainder in 1910 (*New York Times*, March 1, 1911, 13).

6. "Bankers in Studebaker Company," *New York Times*, February 2, 1911, 13.

7. "Willy-Overland's Convertible Preferred," *Wall Street Journal*, January 11, 1916, 6.

8. "The Swings in Motors and Industrials," *Wall Street Journal*, November 1, 1926, 9.

9. "Why Prentiss Bids for Ford Plant," *New York Times*, February 13, 1927, xxii; "Prentiss' Rise Spectacular," *New York Times*, February 3, 1927, 2.

10. "To Offer $75,000,000 Dodge 6s Saturday," *Wall Street Journal*, April 10, 1925; "The Dodge Transaction," *Wall Street Journal*, May 21, 1925, 1.

11. Freudenthal (1940) estimates that the U.S. Post Office provided a subsidy of 163 percent of the airmail revenues it received in 1929.

12. The only exception was the Aero Supply Manufacturing Company, which raised $375,000 in July 1925. Aero Supply was incorporated in Delaware in July 1925 as a successor to a company that had started business in 1920. It was listed on the New York Curb.

13. Dodd (1933) reports that an index based on ten aviation stocks moved from 104.9 in April 1927 to 1147 in May 1929 compared to an increase from 110 to 192.6 over the same period for an index of 337 industrial companies.

14. "Broad Street Gossip," *Wall Street Journal*, September 2, 1927, 2.

15. These estimates are based on data I compiled from volumes of stock issues publicly offered by banking and investment houses in the United States (Schwarzchild, various issues). They are the most comprehensive source of data on public stock issues in the United States for the period from 1925 to 1936. In addition, twenty-seven stock issues by aviation companies, which were sold to the public without using the services of an underwriter, which were recorded in Dodd (1933), were also included.

16. Data on the date of incorporation for issuing companies were obtained from Moody's. However, precise data were available for only ninety-one of the companies that issued stock.

17. "To Buy Aviation Stocks," *Wall Street Journal*, December 8, 1928, 5.

18. One hundred eighty-five firms entered the industry in 1923, 144 in 1924, 258 in 1925, and 161 in 1926 (Maclaurin 1971).

19. "The Smash in Radio Shares," *Barron's*, May 3, 1926, 9.

20. Ibid., 9.

21. "Radio Stocks down $96,281,650 in Year," *New York Times*, May 2, 1926, E11.

22. "Smash in Radio Shares," 9.

23. "Radio Expansion Hurts Radio Trade," *Wall Street Journal*, May 26, 1925.

24. These estimates are based on data I compiled from the volumes prepared by the National Statistical Service (Schwarzchild, various issues).

25. Data on the date of incorporation for issuing companies were obtained from Moody's. However, precise data were available for only twenty-three of the companies that issued stock.

26. "Radio Industry in the Doldrums," *Barron's*, June 8, 1931, 13.

27. Goldsmith drew his estimates from Cox (1963).

28. This statement is based on the data provided in Gompers and Lerner (2003) on the number of initial public offerings in the United States from 1935 to 1972.

29. An issue was classified as "unseasoned" if the issuer had not registered stock previously under the Securities Act or registration A and if its stock was not listed on a national securities exchange or known to be traded over the counter. On occasion, this classification resulted in the inclusion of large, well-established companies offering their stock to the public for the first time (SEC 1963).

30. Registration A issues are those that, because of their small size, did not have to be registered with the SEC.

31. As to what constituted a "public" company, the SEC, echoing its studies in the 1940s, set the bar at 300 shareholders or more, covering 5,500 stocks then trading in the OTC market; it was reduced to companies with at least 500 shareholders and at least $1 million in assets in the Securities Acts Amendments passed in 1964. In effect, that brought about 4,000 new companies within the ambit of the Exchange Act.

References

Atje, Raymond, and Boyan Jovanovic. 1993. "Stock Markets and Development." *European Economic Review* 37, 632–640.

Baskin, Jonathan, and Paul Miranti. 1997. *A History of Corporate Finance*. Cambridge: Cambridge University Press.

Bruchey, Stuart. 1991. *Modernization of the American Stock Exchange, 1971–1989*. New York: Garland.

Bygrave, W., J. Lange, J. R. Roedel, and G. Wu. 2000. "Capital Market Excesses and Competitive Strength: The Case of the Hard Disk Drive Industry, 1984–2000." *Journal of Applied Corporate Finance* 13, 8–19.

Carosso, Vincent. 1970. *Investment Banking in America: A History*. Cambridge, Mass.: Harvard University Press.

Cox, Edwin B. 1963. *Trends in the Distribution of Stock Ownership*. Philadelphia: University of Pennsylvania Press.

Dodd, Paul A. 1933. *Financial Policies in the Aviation Industry*. Philadelphia: University of Pennsylvania Press.

Doyle, William M. 1991. "The Evolution of Financial Practices and Financial Structures among American Manufacturers, 1875–1905: Case Studies of the Sugar Refining and Meat Packing Industries." Ph.D. dissertation, University of Tennessee.

Eddy, George. 1937. "Security Issues and Real Investment in 1929." *The Review of Economics and Statistics*, 79–91.

Freudenthal, Elsbeth E. 1940. *The Aviation Business: From Kitty Hawk to Wall Street*. New York: Vanguard Press.

Garvy, G. 1944. "Rivals and Interlopers in the History of the New York Security Market." *Journal of Political Economy* 52, 128–143.

Goldsmith, Raymond. 1955. *A Study of Saving in the United States*. Princeton, New Jersey: Princeton University Press.

Goldsmith, R. 1969. *Financial Structure and Development*. New Haven, Conn.: Yale University Press.

Gompers, Paul A., and Josh Lerner. 2003. "The Really Long-Run Performance of Initial Public Offerings: The Pre-Nasdaq Evidence." *Journal of Finance* 58, 1355–1392.

Hazard, John, and Milton Christie. 1964. *The Investment Business: A Condensation of the SEC Report.* New York: Harper and Row.

Klepper, Steven. 2002. "The Capabilities of New Firms and the Evolution of the US Automobile Industry." *Industrial and Corporate Change* 11, 645–666.

Lamoreaux, N. 1985. *The Great Merger Movement in American Business, 1895–1904.* Cambridge: Cambridge University Press.

Lazonick, W., and L. Ferleger. 1994. "The Role of US Government in the Emergence of the Commercial Airline Industry." Unpublished note, University of Massachusetts, Lowell.

Leeth, John D., and J. R. Borg. 2000. "The Impact of Takeovers on Shareholder Wealth during the 1920s Merger Wave." *Journal of Financial and Quantitative Analysis* 35, 217–238.

Leffler, G. 1957. *The Stock Market.* 2nd ed. New York: Ronald Press.

Levine, Ross. 2003. "More on Finance and Growth: More Finance, More Growth?" *Federal Reserve of St. Louis Review,* July–August, 31–46.

Levine, Ross, and Sara Zervos. 1998. "Stock Markets, Banks, and Economic Growth." *American Economic Review* 88, 537–558.

Maclaurin, W. Rupert. 1971 *Invention and Innovation in the Radio Industry.* New York: Arno Press.

Maddison, Angus. 1991. *Dynamic Forces in Capitalist Development: A Long-Run Comparative View.* New York: Oxford University Press.

Morck, R., A. Shleifer, and R. Vishny. 1990. "The Stock Market and Investment: Is the Market a Sideshow?" *Brookings Papers on Economic Activity* 2, 157–202.

Navin, T., and M. Sears. 1955. "The Rise of a Market for Industrial Securities." *Business History Review* 29, 105–138.

O'Sullivan, Mary. 2004. "Finance and Innovation." In J. Fagerberg, D. Mowery, and R. Nelson, eds., *The Oxford Handbook of Innovation.* New York: Oxford University Press.

O'Sullivan, Mary. 2005. "Living with the U.S. Financial System: The Experiences of GE and Westinghouse in the Last Century." *Business History Review,* forthcoming.

O'Sullivan, Mary. 2006. "The Deficiencies, Excesses and Control of Competition: The Development of the US Stock Market from the 1930s to 2001." In Lou Galambos and Caroline Fohlin, eds., *Balancing Public and Private Control: Germany and the United States in the Postwar Era.* Cambridge: Cambridge University Press, forthcoming.

Rae, John B. 1965. "Financial Problems of the American Aircraft Industry, 1906–1940." *Business History Review* 39, 99–114.

Rajan, Raghuram, and Luigi Zingales. 1998. "Financial Dependence and Growth." *American Economic Review* 88, 559–586.

Sahlman, W., and H. Stevenson. 1985. "Capital Market Myopia." *Journal of Business Venturing* 1, 7–30.

Schwarzchild, Otto P., ed. Various issues. *American Underwriting Houses and Their Issues*. New York City: National Statistical Service.

Securities and Exchange Commission. 1963. *Special Study of the Securities Markets*. Washington, D.C.: Securities and Exchange Commission.

Seltzer, Lawrence H. 1928. *A Financial History of the American Automobile Industry*. Boston: Houghton Mifflin.

Smith, Philip H. 1968. *Wheels within Wheels: A Short History of American Motor Car Manufacturing*. New York: Funk and Wagnalls.

Sobel, Robert. 1970. *The Curbstone Brokers: The Origins of the American Stock Exchange*. New York: Macmillan.

Thomas Register. 1921. *Thomas Register of American Manufacturers*. New York: Thomas Publishing Company.

Twentieth Century Fund. 1935. *The Security Markets: Findings and Recommendations of a Special Staff of the Twentieth Century Fund*. New York: Twentieth Century Fund.

U.S. Bureau of the Census. 1976. *Historical Statistics of the United States from the Colonial Times to the Present*. Washington, D.C.: U.S. Government Printing Office.

Van der Linden, Robert F. 2002. *Airlines and Air Mail: The Post Office and the Birth of the Commercial Aviation Industry*. Lexington, Kentucky: University Press of Kentucky.

Vander Meulen, Jacob A. 1991. *The Politics of Aircraft: Building an American Military Industry*. Lawrence, Kansas: University Press of Kansas.

Zingales, Luigi. 2003. "Commentary on More Finance, More Growth." *Federal Reserve of St. Louis Review*, July–August, 47–52.

5

Stock Market Swings and the Value of Innovation, 1908–1929

Tom Nicholas

A recurrent theme in the modern literature on the economics of financial markets is the extent to which stock market swings reflect changes in the present discounted value of expected future earnings or the "animal spirits" of investors. For example, the rapid acceleration in stock prices during the 1990s can be explained by both changes in expected investor payoffs in response to the accumulation of intangible capital by firms (Hall 2001) and the behavioral phenomenon that caused a speculative bubble (Shiller 2001). Whether swings in the stock market are driven by the diffusion of new technologies or by periods of irrational exuberance is an important question in the economics of innovation and finance.

While this question is central to the debate over the causes of the recent stock market boom and bust, it is also important to a fuller understanding of another major event in the American stock market: the run-up in equity prices during the 1920s and the Great Crash of 1929. While Irving Fisher famously reported on the eve of the crash that stock prices would remain permanently higher than in past years due to the arrival of new technologies and advances in managerial organization that created positive expectations about future profits and dividend growth, retrospective analysis has indicated the presence of a bubble (DeLong and Shleifer 1991, Rappoport and White 1993, 1994). The speculative bubble hypothesis has become orthodox in the literature given that the S&P Composite Index fell by more than 80 percent from its September 1929 peak to its level in June 1932. The Great Crash is the canonical example in American financial history of market prices diverging significantly from fundamentals.

Despite the conventional wisdom that stock market prices were unrealistically high during the 1920s, we know little about the types of assets that investors are said to have been overvaluing. In particular, evidence on the relationship between innovation and the stock market is sparse. How rapid was the growth of intangible capital during this period? Did the stock market encourage investment in innovation? Did a technological revolution lead to higher stock market valuations? This chapter seeks to answer these questions using a rich data set of balance sheets, stock prices, and patent citations for 121 publicly traded corporations between 1908 and 1929. The aim is to determine whether movements in stock prices can be correlated with the intangible assets of firms and why it matters whether markets in the 1920s got valuations right.

The new data lead to at least two advances over the current literature. First, they introduce a robust measure of intangible capital based on the patenting activity of firms during the 1920s. Although patents are a noisy measure of innovation, citations to patents in the current data set in patent grants between 1976 and 2002 significantly enhance the signal-to-noise ratio. Aside from McGrattan and Prescott (2005) who present estimates of intangible capital for U.S. corporations during the 1920s, there are no systematic data on the intangible capital of firms for this period. Moreover, McGrattan and Prescott are only able to measure intangibles indirectly using equilibrium relations from a growth model. In this study, patents and their citations capture intangibles directly at the microlevel.

Second, it is well known that the predictability of U.S. stock returns is an increasing function of time (Fama and French 1992, Barsky and De Long 1990), yet most studies of the stock market in 1929 concentrate on relatively short intervals, in particular the bubble period from early 1928 to October 1929 (Rappoport and White 1993, 1994). With over twenty years of data prior to the Great Crash, this chapter is able to track firm-level innovation over major swings in financial markets and correlate these swings with changes in investor forecasts about the value of fundamentals.

The main finding of this study is that intangible capital growth was substantial in 1920s America, investors realized it, and they integrated this information into their market pricing decisions. Between 1920 and 1929 the U.S. Patent and Trademark Office (USPTO) granted the 121 firms included in this study 19,948 patents, 4,215 of which were subsequently cited in patent grants between 1976 and 2002. Insofar as these citations represent flows of knowledge from one generation of inventors to the next, this was a major epoch of technological progress. Using historical patent citations as a proxy for the intangible capital of firms, patent market value regressions reveal that a 1 percent increase in the firm's stock of cited patents is associated with a 0.26 percent increase in market value during the 1920s. The returns to intangible capital were approximately three times larger during the 1920s compared to the 1910s, reflecting large changes in the configuration of company assets between these decades. Moreover, as the ratio of the coefficient on intangibles to the coefficient on tangible capital is bigger for the 1920s, there appears to have been a major shift in investor psychology toward intangibles during the stock market boom. One implication of these findings is that investors were not only more responsive to intangible capital at this time, but through triggering large stock market payoffs for innovation, they also encouraged its growth.

Intangible Capital and the Financing of Innovation

The finding that during the 1990s stock market run-up, unmeasured intangible capital was an important element of a firm's market value (Hall, 2000, 2001) makes a historical perspective on this issue appealing. Parente and Prescott (2000) have commented that "unmeasured investment is big and could be as much as 50% of GDP." McGrattan and Prescott's (2000) calculations suggest that the growth of intangible capital may explain the postwar increase in the ratio of total market capitalization to GNP from around 0.5 in 1950 to 1.8 in the first half of 2000.

Unlike the literature on the modern period, however, intangible capital is often omitted from discussions of financial markets during the 1920s. This is a surprising omission, to the extent that intangibles are likely to be significant in stock market valuations. For example, McGrattan and Prescott (2005) estimate that the stock of intangible corporate capital was at least 60 percent of the stock of tangible corporate capital in 1929. In their model, the stock market is overvalued only if the value of intangible capital is zero. With more moderate estimates of the value of intangibles, they conclude that prices of stocks were too low in 1929, and therefore Irving Fisher was right!

Additional evidence supports the view that the 1920s was an extraordinary period of technological progress and intangible capital growth. Several firms formed during the great merger wave in American business (1897–1904) built up separate research and development laboratories, shifting innovation away from individual inventor-entrepreneurs and toward firms (Lamoreaux 1985). The centralized R&D lab became a focal point for innovation and was perhaps the most significant organizational change to influence the structure of American business in the early twentieth century. In 1921 General Electric had five labs in four states.[1] By 1927 AT&T had more than 2,000 research staff working in a 400,000-square-foot thirteen-story building on West Street in New York (National Research Council, 1927). Through vertical integration, firms avoided some of the contracting problems associated with market-based exchange, and managerial hierarchies facilitated the coordination of resources for innovation. Mowery (1983) shows that investment in R&D was positively correlated with firm survival rates between 1921 and 1946. Scientific knowledge became increasingly exploited as firms developed larger stocks of organizational capital. Within firms, star scientists played central roles in the commercialization of basic science, though unlike their counterparts in the life sciences today, few went on to start their own enterprises. Irving Langmuir spent more than four decades at General Electric, his experiments leading to the invention of the gas-filled incandescent lamp and a Nobel Prize for chemistry in 1932. At Du Pont's research center during the 1920s, Wallace Carothers's investigations into the molecular structure of polymers led to the discovery of neoprene, and nylon, which were commercialized in the early 1930s (Hounshell and Smith 1988). At Eastman Kodak, Kenneth Mees and Samuel Sheppard significantly advanced the science of photography; by reducing the width of photographic film, Kodak's research scientists permitted ever-smaller, lighter-weight cameras to be introduced. The institutionalization of innovation also extended beyond the boundaries of the firm as science in universities began to influence the direction of technological change in industry (MacGarvie and Furman 2005). Science and technology also complemented larger stocks of human capital in the economy. David (1990) reveals how falling prices for electrical capital goods after World War I encouraged electrification of the mass production economy, which in turn increased the demand for skilled, literate, and educated labor (Goldin 2001).

It is a reasonable a priori assumption that complementarities between innovation, organizational changes, and human capital had an impact on the stock market during this epoch in much the same way as researchers have discovered they do today (Bresnahan, Brynjolfsson, and Hitt 2002). The large early-twentieth-century American corporation was the principal agent of organizational and technological change according to Chandler (1990). White (1990) suggests that General Motors was attractive to investors during the 1920s because its more advanced management and organization facilitated smooth transitions from one production run to the next. Klepper and Simons (2000) show how firms with the favorable mix of innovation and complementary assets (such as marketing channels) were more likely to survive the shakeout of producers in the tire industry.

Intangible capital growth was also encouraged as markets developed to finance innovation and investors became responsive to holding equities. According to Peach (1941) the public became more willing to hold different types of securities issued by corporations following their successful experiences with Liberty Bonds during World War I. O'Sullivan (2004) documents a major financing role for the 1920s U.S. stock market as companies increasingly used external sources of finance. According to Rajan and Zingales (1998) financial development is positively correlated with the allocation of capital to areas of highest value. Nicholas (2003) shows for the 1920s that bond and stock issues by companies were positively correlated with their propensity to innovate. This finding is consistent with Schumpeter's (1942) contention that a developed and efficiently functioning capital market extends the frontier of technological progress.

Historical Balance Sheets and Patents

In order to empirically track the development of intangible capital and stock market value at the firm level during the early twentieth century, I collate data on company financials and historical patent citations. The approach is similar to studies of the modern period that, given limited disclosures by companies concerning expenditures on intangibles, used indirect measures of intangible assets. For example, Bond and Cummins (2000) use R&D and advertising outflows to proxy for investment in intangibles, while Brynjolfsson, Hitt, and Yang (2000) infer that computer-related intangibles are substantial because the coefficient on the stock of computer capital in market value regressions is much larger than other types of productive assets.[2] Although patents are an imperfect proxy for intangible assets, when combined with historical citations statistics, they provide a valuable source for tracing the dynamics of technological progress.

Company Financials

Before discussing the data on intangibles, as measured by patents and historical citations, it is helpful to describe the company financial data. The main sources on financials are *Moody's Manual of Industrials* for company balance sheets and the *Commercial and Financial Chronicle* for end-of-year share prices. The sample includes every firm with at least four years of continuous data, giving 121 firms with a time-series dimension running from 1908 to 1929. The main period of interest is the 1920s, but data going back to 1908 illustrate long-term swings in the relationship between intangible capital and the stock market. The company financial data detail major "Chandlerian" corporations of the time such as General Electric, E. I. Du Pont, Eastman Kodak, and General Motors, as well as companies that possessed a more moderate level of assets than the set of firms studied by Chandler (1990) (see figure 5.1A). While there is still a slight skew toward larger firms, by market value, my data closely approximate the population of companies collated by the Chicago Research Center in Securities Prices (see figure 5.1B).

The main financial variables used in this analysis are calculated using the methodology proposed by Lindenberg and Ross (1981). The market value of the firm is measured as the product of common equity and year-end market price, plus the book value of outstanding debt and the value of preferred stock (which is assumed to be a perpetuity discounted at the average industrial bond yield reported by *Moody's*). Capital assets (k) are estimated using the recursive formula $k_t^{rc} = k_{t-1}^{rc}[(1+i)/(1+\rho)] +$

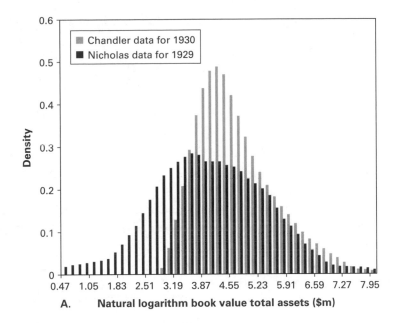

A.   **Natural logarithm book value total assets ($m)**

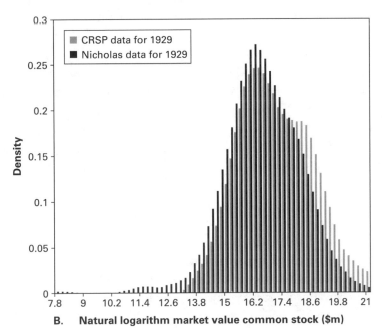

B.   **Natural logarithm market value common stock ($m)**

Figure 5.1
Benchmarking the data

$(k_t^{bv} - k_{t-1}^{bv})$ where i is the GNP implicit deflator, ρ is the depreciation rate at an assumed 5 percent, and the subscripts rc and bv denote replacement value and book value, respectively. Inventory is estimated at replacement cost by adjusting for inflation through the wholesale price index from *Historical Statistics of the United States: 1790–1950*. Tobin's q is then calculated as the ratio of the firm's market value to the replacement cost of its tangible assets. Descriptive statistics on these variables are given in table 5.1.

Historical Patent Citations

Table 5.1 also includes summary data on patenting. Patent data were assembled from the USPTO and the European Patent Office (EPO) for each firm in the sample. Over 35,000 patent grants were assigned to the 121 firms between 1908 and 1929, with 19,948 being assigned between 1920 and 1929. Figure 5.2 illustrates the level of patenting activity by firm year. High-frequency patenting firms like Westinghouse, which peaked at 564 patent grants in 1929, are included alongside lower-frequency patenting firms like Otis Elevator, which peaked at 26 patent grants in 1928. Twenty-one firms in the sample did not patent at all.

Patent counts are commonly used to proxy for innovation, but this measure is prone to error because not all inventions are patented and the quality of patents varies widely (Griliches 1990). To improve the quality of the patent measure, I use citations to 1920s patents in patent grants between 1976 and 2002. The assumption is that citations distinguish the frontier of knowledge regardless of how far back in time they go. It can be argued that inventors and patent examiners habitually cite patents from the past without regard to prior art. However, if innovation is cumulative, as suggested by Scotchmer (1991), and citations come from the frontier, these references will reflect knowledge transfers between generations of inventors.

Of the 19,948 patents granted to firms between 1920 and 1929, 21 percent are cited in patents granted between 1976 and 2002. Of the 4,215 patents cited, 2,548 receive one citation, while 1,667 receive two or more citations, with the maximum number of cites for a patent being 27. This is a notable number of citations given that citations fall off sharply a decade after the patent's grant date (Caballero and Jaffe 1993).

Table 5.1
Descriptive statistics

	Market value ($m)	k ($m)	q	Patents	Patent citations
1908	69.35	53.37	0.72	13.70	.
	(174.58)	(180.43)	(0.30)	(52.28)	.
1909	77.75	51.04	0.83	15.23	.
	(199.46)	(173.01)	(0.31)	(53.52)	.
1910	72.20	49.08	0.82	13.74	.
	(183.70)	(163.32)	(0.29)	(48.81)	.
1911	71.52	44.57	0.74	17.08	.
	(181.38)	(153.92)	(0.28)	(59.61)	.
1912	70.36	44.75	0.82	16.77	.
	(179.80)	(148.33)	(0.35)	(61.77)	.
1913	64.60	42.36	0.77	17.18	.
	(168.88)	(140.80)	(0.32)	(57.21)	.
1914	61.80	41.28	0.78	20.00	.
	(157.78)	(135.07)	(0.43)	(65.65)	.
1915	66.47	38.03	0.81	21.45	.
	(170.88)	(125.88)	(0.31)	(65.60)	.
1916	70.97	41.25	0.72	18.92	.
	(183.29)	(137.41)	(0.26)	(59.39)	.
1917	65.55	52.30	0.51	18.43	.
	(168.64)	(169.35)	(0.18)	(59.42)	.
1918	70.95	56.70	0.51	16.87	.
	(173.03)	(185.18)	(0.18)	(58.18)	.
1919	66.85	39.84	0.68	13.74	.
	(159.26)	(149.43)	(0.30)	(49.26)	.
1920	55.65	47.00	0.49	11.83	3.79
	(142.50)	(165.29)	(0.18)	(41.83)	(12.34)
1921	59.09	37.95	0.66	13.46	4.70
	(150.13)	(133.26)	(0.27)	(41.32)	(11.93)
1922	70.63	34.14	0.76	13.09	4.68
	(171.14)	(116.22)	(0.37)	(38.29)	(13.45)
1923	73.87	35.81	0.74	13.55	4.88
	(178.26)	(118.75)	(0.36)	(40.26)	(12.60)
1924	88.86	35.13	0.84	16.26	6.37
	(208.09)	(116.66)	(0.41)	(51.39)	(19.75)
1925	108.55	34.91	1.01	20.94	8.86
	(249.68)	(115.10)	(0.63)	(69.91)	(32.02)
1926	139.24	37.79	1.07	21.86	8.90
	(368.52)	(113.54)	(0.64)	(62.41)	(23.77)

Table 5.1
(continued)

	Market value ($m)	k ($m)	q	Patents	Patent citations
1927	161.94	39.43	1.25	21.46	10.54
	(401.17)	(114.28)	(0.82)	(65.02)	(32.86)
1928	250.22	44.44	1.58	28.31	10.45
	(581.75)	(116.52)	(1.13)	(82.73)	(30.69)
1929	234.27	51.16	1.25	29.61	13.30
	(569.41)	(115.19)	(0.92)	(90.50)	(36.59)

Note: Standard deviations in parentheses.

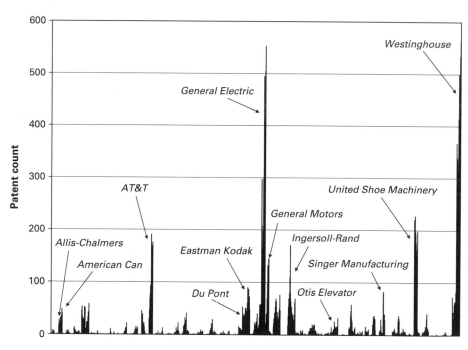

Figure 5.2
Patent counts per year for the firms in the sample, 1920–1929

To show how large this proportion is, I collected data on 132 successful grants by the USPTO between 1910 and 1930 to the great inventor-entrepreneur Thomas Edison. Great inventors were typically entrepreneurial figures who developed important inventions in response to market demand (Khan and Sokoloff 1993). Forty-two (31.8 percent) of Edison's patents are cited in patents granted between 1976 and 2002. Although this proportion may be inflated if patent examiners have a propensity to "cite the classics," I use Edison's patents as a benchmark for the upper tail of the patent quality distribution. This comparison suggests that the proportion of citations observed for the firms in the sample is both large and significant.

Patents as Intangibles

Recent research suggests that the 1930s was the most technologically progressive decade of the twentieth century. And the significance of productivity growth during this period may be attached to technological "larder-stocking during the 1920s and earlier, upon which measured advance built" (Field 2003). Insofar as the patent citation data reveal that the 1920s was a major epoch of technological innovation, how far can patent rates be used to measure the stock of intangible capital in the economy?

It is important to note at the outset that this measure is imperfect. Although twenty-one firms in the sample did not patent, their stock of intangible capital was undoubtedly greater than zero. Innovations in food processing, for instance, are much less likely to be patented compared to innovations in machinery (Moser 2005). Thus, companies like Coca Cola and Quaker Oats developed intangibles through branding, secrecy, and distribution networks even though their intangibles are effectively put at zero for this study. Nevertheless, an insight into the significance of patents as intangible assets can be gained from the financials detailed by *Moody's*. Numerous companies found intellectual property rights to be so important in their portfolio of assets that they reported patents directly in financial statements. J. I. Case Threshing Machine Company, the Wisconsin agricultural machinery manufacturer, reported $1.04 million in patents between 1922 and 1928. American Bosch Magneto Corporation, which manufactured devices for internal combustion engines, valued its patents at between $594,176 and $633,356 from 1924 to 1929, equivalent to around one-third of the total assets of the company.

More generally, patents were critical for appropriability during the 1920s. Extrapolating backward from Cohen, Nelson, and Walsh (2000), discrete industries such as chemicals in which patent protection is deemed to be important were cornerstones of industry structure in the early twentieth century. Mokyr (1990) puts innovations in chemicals at the heart of the second industrial revolution. Patents not only increased the effectiveness of research and development activities, but they also enhanced the market power of incumbents. Lerner, Strojwas, and Tirole (2003) argue that patent pools were a principal means through which firms during the early twentieth century used intellectual property rights to foster collusion. For example, the American conglomerates Du Pont, Standard Oil, and Allied Chemicals; the English firm I. C. I; and I. G. Farben of Germany captured a commanding share of the fertilizer market through 1,800 patents relating to the synthetic nitrogen process (Comer 1946).

Perhaps the most compelling evidence that patents were an important component of a firm's portfolio of intangible assets can be gained from figures 5.3, 5.4, and 5.5. Figure 5.3 plots indexes of total patents granted by the USPTO and those granted to the firms in the sample for comparable years. A striking result to emerge from this figure is the growth of patenting activity by firms over time. This growth in firm-level patenting activity is consistent with what we know about the role of the early-twentieth-century-industrial research laboratory in increasing the rate of innovation within firms (Mowery, 1995)

Using the means from table 5.1, figure 5.4 plots trends in patenting and market value over time. The logarithmic scale allows a more informative comparison of large (market value) and small (patent count) values because an equal percentage change is shown as an equal amount of space on the graph. It is notable to observe during the 1920s how close the growth rate of patents, and especially of patent citations, is aligned with the growth rate of market value.

Figure 5.5 illustrates that patents as intangible capital were important in the context of the stock market for several of the companies considered. AT&T, General Electric, and Westinghouse pushed out the frontier

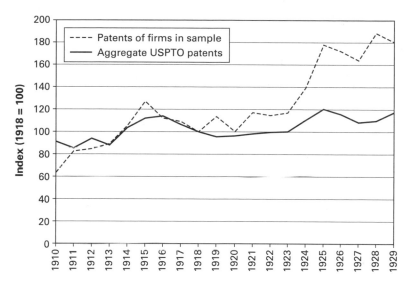

Figure 5.3
Index of patents for firms in the sample and aggregate USPTO patents

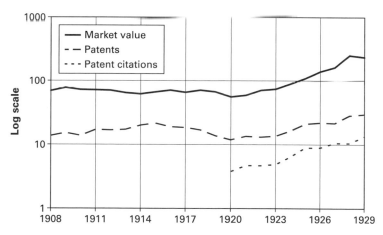

Figure 5.4
Market value and patents, 1908–1929

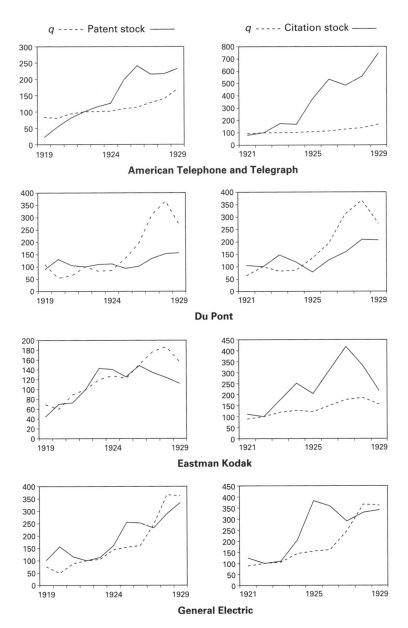

Figure 5.5
Indexes of patenting and Tobin's q, 1919–1929

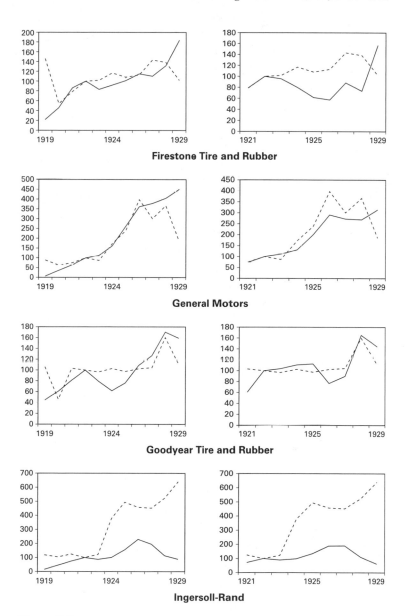

Figure 5.5
(continued)

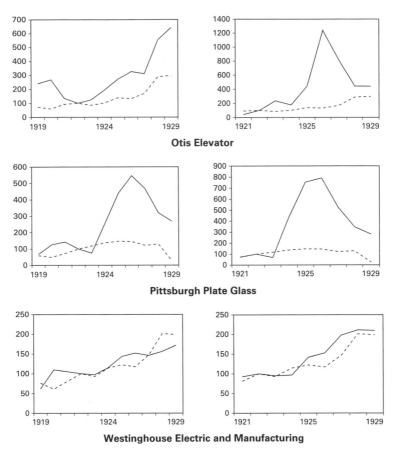

Otis Elevator

Pittsburgh Plate Glass

Westinghouse Electric and Manufacturing

Figure 5.5
(continued)

of productivity enhancing electrification technology. Eastman Kodak created a market for amateur motion pictures during the 1920s with the introduction of 16 mm reversal film on cellulose acetate. General Motors developed lighter metal casings for motor vehicles, while Firestone and Goodyear introduced advanced methods of rubber vulcanization, increasing the longevity of tires. Although the association between patenting and q is weak for some of the companies illustrated (Du Pont and Ingersoll-Rand), the fact that the correspondence is close for others warrants a more systematic investigation of the links between financial markets and the intangible assets of firms.

Estimating the Market Value of Intangibles

To analyze the data more systematically, I use an empirical approach developed from a simple model that relates patents to market value. The market value model of Griliches (1990) assumes an efficiently functioning financial market where the value of a firm (v) depends on the evolution of its cash flows, which firms attempt to maximize from their mix of tangible (k) and intangible (g) assets. This gives a value function of the form (5.1) where γ is the market premium or discount over tangible assets and π represents the relative shadow value of intangibles. Using a standard linear approximation of this value function yields equation 5.2, where q is the ratio of the firm's market value to the replacement cost of its assets. The coefficient α_1 measures the value of intangible assets relative to the tangible assets of the firm. If the value of the firm exceeds its replacement cost, the intuition behind this model is that the difference can be explained by the presence of intangibles ($\alpha_1 > 0$).

$$v_{it}(k_{it}, g_{it}) = \gamma_t(k_{it} + \pi g_{it}) \tag{5.1}$$

$$\log(q)_{it} - \log(v/k)_{it} = \alpha_0 + \alpha_1 (g/k)_{it} + X'_{it} + u_{it} \tag{5.2}$$

$$\log(v)_{it} = \theta_0 \log(k)_{it} + \theta_1 \log(g/k)_{it} + \theta_2(g = 0)_{it}$$

$$+ \theta_3 AGE_{it} + \theta_4 AGE_{it}^2 + \sum_{j}^{j-1} \theta_5 E_{tj} \cdot \log(g/k)_{it} + u_{it} \tag{5.3}$$

Although the logarithm of q is commonly used as a dependent variable, I prefer a specification (equation 5.3) with the logarithm of market value on the left-hand side because changes in stock prices, which are incorporated into the numerator of q, explain the largest component of the variation in q during the 1920s. Intangible capital is given by the firm's stock of patents, which is constructed using the declining-balance formula $g_{it} = (1 - \delta)g_{it-1} + pat_{it}$ with a depreciation rate $\delta = 0.15$. I weight each patent by the number of citations it receives, summing the total number of citations for each firm, each year. Both the stock of patents and citation-weighted patents (g_c) are normalized on the firm's capital assets. A dummy variable for when $g = 0$ is added to partial out the effect of adding one to the patent stock as a precondition of taking its logarithm. I use a logarithmic specification for g because it moderates extremes in the data and lessens the effect of outliers. Thus, θ_1 has an elasticity interpretation.

An additional variable, AGE, is calculated as time t minus the year of the firm's incorporation and included in the regressions (with a polynomial) to determine how far a firm's vintage affects its stock market capitalization in the manner of Hobijn and Jovanovic (2001). Since variation over time in the value of intangible capital is also of interest, interactions of E_{j-1} year dummies with the firm's normalized stock of patents are also included. To analyze the returns to intangible capital over major swings in the stock market and obtain more refined estimates with citation data, which I have for 1920 to 1929, equation 5.3 is estimated for two panels, 1908–1918 and 1919–1929.[3]

Results

Referring back to the descriptive statistics in table 5.1 and the plots in figures 5.3 and 5.4, it is clear that as the market value of companies grew during the 1920s, the patenting activity of these firms also increased. The aim of this section is to determine whether anything more systematic can be concluded from the relationship between these variables.

Before discussing the results themselves, it is important to consider an issue that has a wider bearing on the interpretation of the findings: the direction of causality between patenting and q. Taking a cue from the empirical literature on the q-model of investment, authors such as Barro (1990) find that the stock market predicts investment, while others such as Blanchard, Rhee, and Summers (1993) find that it does not. Theory suggests that the relationship between innovation and market value is also endogenous. Incumbents have incentives to preemptively innovate if the expected payoffs exceed the rents from maintaining the current technology (Gilbert and Newbery 1982). Equally, stock market run-ups can be driven by technology push. Equation 5.3 does not identify the direction of causality. While the standard solution to this problem is an instrumental variables estimator, there are also alternative strategies. For example, Brynjolfsson, Hitt, and Yang (2000) use a test motivated by Granger's concept of causality, finding that the lagged stock of computer capital predicts market value, while lagged market value has little predictive power for investment in computers.

For the overall interpretation of the current results, the simultaneous association between innovation and market value is not so problematic. After all, the central argument is that investors were alert to the opportunities presented by companies with stocks of intangible capital, which also encouraged further investment in innovation: both effects fed off each other concurrently, and access to external sources of corporate finance probably played a mediating role (Nicholas 2003). However, a related issue still remains: measuring and interpreting the size and significance of the relationship between patenting and market value.

To address this issue, differences in parameter estimates between periods are used. The regressions for 1908 to 1918 serve as a benchmark for the regression results of the 1920s. Changes in the stock market returns to intangible capital are captured by differences in the coefficient θ_1 between 1908–1918 and 1919–1929. Through θ_5 this effect can be measured for subperiods of the 1920s (interactions of the year dummy and $\log(g/k)$ measure changes in the relationship between stock market value and patenting relative to a baseline). Any change in investor attitudes toward intangible capital will be revealed by the ratio of the coefficient on intangibles (θ_1) to the coefficient on tangible capital (θ_0).

Table 5.2 contains the regression results. The first point to note is that the coefficient on the log of the capital stock (k) is surprisingly low with fixed effects; under the assumption of linear homogeneity in the market value model of Griliches (1990), the coefficient should be exactly one. The results from the first two columns of table 5.2 imply that for a 1 percent increase in the capital stock, market value increases by only 0.31 to 0.38 percent. In an ordinary least squares (OLS) specification without controls for unobserved heterogeneity, the estimates on (k) at the 95 percent confidence interval are larger at between 0.83 and 0.89, but they are still significantly different from unity. Possible explanations for the deviation include large, fixed capital adjustment costs or measurement error bias. A corrective approach would be to impose unity in the empirical specification, but since this has the effect of biasing downward the other coefficients in the model, the current model without parameter restrictions is preferred.

Table 5.2
Market value regressions

	Dependent variable $\log(v)$			
	I	II	III	IV
$\log(k)$	0.3817	0.3087	0.2843	0.2677
	(0.0299)	(0.0345)	(0.0354)	(0.0353)
$\log(g/k)$	0.0465	0.1245		
	(0.0286)	(0.0394)		
$g = 0$	−0.0598	0.0970		
	(0.0298)	(0.0405)		
$\log(g_c/k)$			0.2565	0.1849
			(0.0507)	(0.0553)
$g_c = 0$			0.0897	0.0777
			(0.0426)	(0.0426)
$1926 \cdot \log(g_c/k)$				0.0777
				(0.0536)
$1927 \cdot \log(g_c/k)$				0.0621
				(0.0536)
$1928 \cdot \log(g_c/k)$				0.1448
				(0.0573)
$1929 \cdot \log(g_c/k)$				0.2327
				(0.0651)
AGE	0.0307	0.1266	0.0421	0.1610
	(0.0060)	(0.0096)	(0.2405)	(0.0127)
AGE^2	−0.0006	−0.0008	−0.0009	−0.0010
	(0.0002)	(0.0002)	(0.0002)	(0.0002)
Period	1908–1918	1919–1929	1919–1929	1919–1929
Firm effects	Yes	Yes	Yes	Yes
Year effects	Yes	Yes	Yes	Yes
F	20.00	66.30	52.81	53.82
R^2 within	0.29	0.48	0.45	0.46
R^2 between	0.65	0.18	0.46	0.11
R^2 overall	0.66	0.20	0.43	0.12
Observations	781	1077	964	964

With respect to intangibles, the results reveal a strong and significant relationship between patenting and stock market value. Columns I and II report comparable estimates on the patent stock variable for the periods 1908–1918 and 1919–1929. During the 1920s, the elasticity of the firm's normalized stock of intangibles (g/k) with respect to market value is 0.12, approximately three times larger than the estimate for 1908–1918. In a comparison of columns II and III, the elasticity estimate increases from 0.12 to 0.26 when citation-weighted patents are introduced. This is important because cited patents are more likely to be commercially viable than their uncited counterparts. As the estimates show a higher value on quality patents, the results suggest that investors were integrating expectations of future growth from innovation into their assessment of market prices.

Why are the parameter estimates on the patent stock variables so different between periods? To the extent that firms were accumulating substantial stocks of intangible capital during the 1920s, this is exactly what we would expect if investors perceived that these intangibles would have a positive impact on expected future earnings. Investors would have been increasingly aware of important new inventions because widely distributed publications such as the *Scientific American* kept readers apace with technological breakthroughs. The *Official Gazette* of the USPTO published lists of patents assigned to companies, which investors could easily track.[4] If investors were uncertain about how innovation would contribute to the dividend growth rate, the effect would have been to increase the firm's fundamental value. Pástor and Veronesi (2006) incorporate uncertainty about the firm's future profitability into a stock valuation model that can reject the hypothesis of a bubble in the Nasdaq at its late 1990s peak. In their model, as uncertainty increases, the price of a stock can be justified with a significantly lower expected growth rate of earnings. The high trading volume on the New York Stock Exchange during the 1920s is consistent with high levels of uncertainty. Since no one knew what RCA's dividend trajectory looked like (because it did not pay a dividend), the firm had some probability of failing and some probability of becoming a market leader, and therefore became extremely valuable to investors in much the same way that Nasdaq stocks had some probability of failing, but also some probability of becoming the next Microsoft in the Pástor and Veronesi (2006) framework. Relative to the 1910s, the 1920s was a much more uncertain epoch concerning how technology would influence the future profitability of firms since the technological change taking place was so far reaching.

Retrospectively, we also know that technological revolutions are characterized by diffusion lags, so the large differences between the 1910s and the 1920s in the valuation of intangibles is comprehensible in this context. According to Hobijn and Jovanovic (2001), the process of innovation and delayed stock market reaction caused a lull in the stock market during the 1970s as organizations adjusted to new information and communications technology (ICT) and old capital became gradually displaced. The boom in the stock market during the 1990s was then a response to a more efficient new vintage of capital that revolved around the implementation of ICT. By this time, investors had a much clearer idea of which firms had adapted to the new technological environment and which had not. New capital became more valuable as old capital faded away; innovation did not cause a bubble in the 1990s stock market.

Inasmuch as the 1990s was a decade of ICT, the 1920s was a decade of electrification, and a similar diffusion lag was evident. David (1990) argues that the 1920s productivity acceleration occurred because manufacturing plants had developed complementary capabilities to exploit electrical power transmission inventions that dated back to the 1880s. Beyond electricity the inverted-U relationship between market value and AGE shown in table 5.2 is consistent with models of innovation and the stock market that associates new capital value creation with old capital value destruction.[5] Many dominant firms lost market share during the 1920s as rivals embraced newer technologies. For example, United States Steel overcommitted resources to producing steel for rail lines and ignored opportunities for profit created by steel skeletal construction for skyscrapers and bridges. Bethlehem Steel, in contrast, innovated in this market and by the early 1920s had gained significant market share from U.S. Steel. Old firms did not fall away altogether, but rather the threat of creative destruction encouraged preemptive innovation to prevent the dissipation of industry profits. A buoyant industrial sector (Klepper 2002 reports that more than 500 firms entered the automobile industry in its first twenty years) meant that new firms seized opportunities ignored by inefficient incumbents.

The 1920s was probably the first period in history when investors began to assess the intangible assets of companies. This would have required a major shift in investor psychology toward the bundle of assets that comprise a firm and a favorable assessment of the equity risk premium. The nuts and bolts of the railroad corporation were much easier for investors to value than the intangible assets of the 1920s technology firm. The results in table 5.2 provide an indication of the change in investor attitudes. The coefficient on intangibles (the patent stock variable) is much larger relative to the coefficient on tangible capital when comparing columns I and II. The results in the final column of table 5.2 show that the relationship between patenting and market value is stronger for years closer to the 1929 crash, contrary to what would be predicted if this was a phase of unrestrained speculation by uninformed investors. The baseline period in this regression is set at 1920–1925, where the elasticity of patenting with respect to market value is 0.18. In 1926 the elasticity rises to 0.26 ($0.1849 + 0.0777$), in 1927 to 0.25, in 1928 to 0.33, and in 1929 to 0.42. At this level of aggregation, there is not enough variation in the data to pick out the pre- and post-crash valuation of intangibles; the coefficient for 1929 is a point estimate for the entire year. Nevertheless, the coefficients for 1926–1928 provide enough evidence to support the view that stock market appreciation during the run-up to the crash was connected to expectations about the intangible capital embodied in firms.

Discussion and Conclusion

Intangible capital growth was substantial in 1920s America. Investors realized it and integrated this information into their market pricing decisions. Recall that q is computed using tangible capital, and therefore q can exceed unity as intangible assets become a larger fraction of total assets. The most important source of variation in q during the 1920s was the change in stock market prices. Table 5.1 shows that average q was very low in 1920 (0.49) but by 1929 had risen to 1.25. What explains this drastic change in the market value of firms relative to the replacement cost of their assets? While conventional wisdom suggests that unrestrained speculation created a divergence between share prices and fundamentals, this chapter has offered a new perspective: the interaction between innovation and changes in investor attitudes toward intangible capital fomented large stock market payoffs for innovating firms. Consequently, the run-up to the crash might not have been an epoch of irrational exuberance, as is often claimed; rather, investors may have been linking rational (positive) expectations about fundamentals to the market value of firms.

The basic assumption underlying this analysis is that the value of intangible assets can be inferred from the gap between q and the replacement value of capital derived from the firm's balance sheet. A common criticism of this approach is that market participants do not accurately observe intangibles and therefore the market value regression may recover an inefficient estimate of intangible value (Bond and Cummins 2000). Yet the evidence here supports the hypothesis that investors were receiving information about intangible capital during the 1920s. I measure their response to fundamentals by the value attached to citation weighted patents. The fact that cited patents have substantial explanatory power (even over and above unweighted patents) in the market value regressions shows that investors placed a premium on firms that pushed out the frontier of knowledge. Furthermore, the estimates in table 5.2 show that the parameter on the patent stock variable is much larger during the 1920s than the 1910s. Therefore, the results are consistent with a major change in the psychology of investors during the stock market boom. The 1990s run-up in the stock market has been interpreted as a response to an increase in the amount of organizational capital (Hall 2000, 2001). Similarly, the 1920s stock market boom can be seen as a response by investors to the growth of intangibles in the economy.

Technological change does trigger booms in the stock market. The 1920s was a remarkable decade of technological progress and intangible capital growth. By 1920 electricity had surpassed steam as a source of power for manufacturing and by 1929 accounted for 78 percent of total capacity (Devine 1983). A chemicals invention during this epoch—Percy Bridgman's method for growing crystals and purifying crystalline substances (patent number 1,793,672 filed February 16, 1926)—paved the way for what David (1990) describes as a breakthrough event in the computer revolution almost half a century later: Intel's silicon microprocessor. The organization of production along Taylorist lines improved workplace learning and performance (Hounshell 1984). As the threat of creative destruction in product markets increased the marginal benefits from investing in search for new technologies, firms that innovated received sizable stock market payoffs (Nicholas 2003). As capital markets became increasingly liquid due to the entry of additional investors, innovation could be financed by access to external sources of credit. Investors financed innovation directly, and indirectly they encouraged investment in technological development by inducing stock market rewards for innovation. One argument is that the change in fundamentals during this period may have caused "informational overshooting" (Zeira 1999) as market participants had different expectations about how long technological change would keep dividends growing. However, retrospectively we know that productivity growth did persist, so there was no ex ante constraint on the profits and payouts of firms. General Electric, for example, was profitable throughout the depression years (O'Sullivan 2004). Productivity growth during the 1930s was marked (Field 2003). It might be plausible therefore that investors were driving up equity values in the 1920s on expectations of productivity growth a decade later.

None of these reasons, which suggest that stock prices in 1929 were warranted ex ante, is meant to deny instability in financial markets on the eve of the crash. Neither do the findings detract from the real consequences of the precipitous decline in the stock market from 1929 to 1933. Not for the first time in the history of financial markets did investors begin to appreciate the downside of stock market risk. As Galbraith (1987) put it, "To an extraordinary degree this is a game in which there are many losers." Rather, the aim of the exercise has been to illustrate the significance of intangible capital growth during the 1920s and to highlight that the positive correlation between market value and innovation arose because of a change in investor attitudes toward the composition of assets within a firm. The evidence here suggests that soaring stock market prices during the 1920s may have made sense insofar as they were correlated with the growth of intangibles in the economy.

Determining how market participants responded to firms with large stocks of intangibles during the crash and after should further our understanding of interactions among intangible capital, investor behavior, and the stock market.

Notes

1. Schenectady, New York; Lynn, Massachusetts; Pittsfield, Massachusetts; Harrison, New Jersey; Cleveland, Ohio. See National Research Council (1921).

2. They estimate a market value specification in which the coefficient on information technology (IT) capital is around 10. According to their interpretation, the stock market does not value $1 of IT capital at $10. Rather, for every $1 of IT capital, there are $9 of related intangibles.

3. The first year of data drops off in the construction of the capital stock at replacement cost. Therefore, the panel regressions are run for 1909–1918 and 1920–1929, respectively.

4. An investor tracking Westinghouse, for example, would have been able to see that the firm was assigned more than 4,000 patents during the 1920s, approximately double the number of assignments during the 1910s.

5. However, it should also be noted that the coefficients on AGE and AGE^2 are not precisely estimated over the two time periods.

References

Barro, Robert. 1990. "The Stock Market and Investment." *Review of Financial Studies* 3, 115–131.

Barsky, Robert B., and J. Bradford DeLong. 1990. "Bull and Bear Markets in the Twentieth Century." *Journal of Economic History* 50, 1–17.

Blanchard, Olivier, C. Rhee, and Lawrence Summers. 1993. "The Stock Market, Profit and Investment." *Quarterly Journal of Economics* 108, 115–136.

Bond, Stephen R., and Jason G. Cummins. 2000. "The Stock Market and Investment in the New Economy: Some Tangible Facts and Intangible Fictions." *Brookings Papers on Economic Activity*, 61–124.

Bresnahan, Timothy F., Erik Brynjolfsson, and Lorin M. Hitt. 2002. "Information Technology, Workplace Organization, and the Demand for Skilled Labor: Firm-Level Evidence." *Quarterly Journal of Economics* 117, 339–376.

Brynjolfsson, Eric, Lorin M. Hitt, and Shinkyu Yang. 2000. "Intangible Assets: How the Interaction of Computers and Organizational Structure Affects Stock Market Valuations." Mimeo., MIT.

Caballero, Ricardo, and Adam Jaffe. 1993. "Standing on the Shoulders of Giants: An Empirical Assessment of Knowledge Spillovers and Creative Destruction in a Model of Economic Growth." In Olivier Blanchard and Stanley Fischer, eds., *NBER Macroeconomics Annual* (pp. 15–74). Cambridge, Mass.: MIT Press.

Chandler, Alfred. 1990. *Scale and Scope: The Dynamics of Industrial Capitalism.* Cambridge, Mass.: Harvard University Press.

Cohen, Wesley, Richard Nelson, and John Walsh. 2000. "Protecting Their Intellectual Assets: Appropriability Conditions and Why U.S. Manufacturing Firms Patent (or Not)." Working paper, NBER, Cambridge, Mass.

Comer, G. P. 1946. "The Outlook for Effective Competition." *American Economic Review Papers and Proceedings* 36, 154–171.

David, Paul. 1990. "The Dynamo and the Computer: An Historical Perspective on the Modern Productivity Paradox." *American Economic Review Papers and Proceedings* 80, 355–361.

DeLong, J. Bradford, and Andrei Shleifer. 1991. "The Stock Market Bubble of 1929: Evidence from Closed-End Funds." *Journal of Economic History* 52, 675–700.

Devine, Warren. 1983. "From Shafts to Wires: Historical Perspective on Electrification." *Journal of Economic History* 43, 347–372.

Fama, Eugene, and Kenneth French. 1992. "The Cross Section of Expected Stock Returns." *Journal of Finance* 47, 427–465.

Field, Alexander. 2003. "The Most Technologically Progressive Decade of the Century." *American Economic Review* 93, 1399–1413.

Galbraith, J. K. 1987. "The 1929 Parallel." *Atlantic Monthly* 259, 62–66.

Gilbert, Richard, and David Newbery. 1982. "Preemptive Patenting and the Persistence of Monopoly." *American Economic Review* 72, 514–526.

Goldin, Claudia. 2006. "The Human Capital Century and American Leadership: Virtues of the Past." *Journal of Economic History* 61, 263–291.

Griliches, Zvi. 1990. "Patent Statistics as Economic Indicators: A Survey." *Journal of Economic Literature* 28, 1661–1707.

Hall, Robert. 2000. "e-Capital: The Link between the Labor Market and the Stock Market in the 1990s." *Brookings Papers on Economic Activity*, 73–118.

Hall, Robert. 2001. "The Stock Market and Capital Accumulation." *American Economic Review* 91, 1185–1202.

Hobijn, Bart, and Boyan Jovanovic. 2001. "The Information-Technology Revolution and the Stock Market: Evidence." *American Economic Review* 91, 1203–1220.

Hounshell, David. 1984. *From the American System to Mass Production, 1800–1932.* Baltimore: Johns Hopkins University Press.

Hounshell, David A., and John Kenly Smith, Jr. 1988. *Science and Corporate Strategy: Du Pont R&D, 1902–1980.* Cambridge: Cambridge University Press.

Khan, Zorina, and Ken Sokoloff. 1993. "'Schemes of Practical Utility': Entrepreneurship and Innovation among 'Great Inventors' in the United States, 1790–1865." *Journal of Economic History* 53, 289–307.

Klepper, Steven. 2002. "The Capabilities of New Firms and the Evolution of the US Automobile Industry." *Industrial and Corporate Change* 11, 645–666.

Klepper, Steven, and Kenneth Simmons. 2000. "The Making of an Oligopoly: Firm Survival and Technological Change in the Evolution of the U.S. Tire Industry." *Journal of Political Economy* 108, 728–760.

Lamoreaux, Naomi. 1985. *The Great Merger Movement in American Business, 1895–1904.* Cambridge: Cambridge University Press.

Lerner, Josh, Marcin Strojwas, and Jean Tirole. 2003. "The Structure and Performance of Patent Pools: Empirical Evidence." Mimeo., Harvard Business School.

Lindenberg, E., and Steven Ross. 1981. "Tobin's Q Ratio and Industrial Organization." *Journal of Business* 54, 1–33.

McGrattan, Ellen R., and Edward C. Prescott. 2000. "Is the Stock Market Overvalued?" *Federal Reserve Bank of Minneapolis Quarterly Review* 24, 20–40.

McGrattan, Ellen R., and Edward C. Prescott. 2004. "The 1929 Stock Market: Irving Fisher Was Right." *International Economic Review* 45, 991–1009.

Mokyr, Joel. 1990. *The Lever of Riches.* New York: Oxford University Press.

Moser, Petra. 2005. "How Do Patent Laws Influence Innovation? Evidence from Nineteenth-Century World Fairs." *American Economic Review* 95, 1214–1236.

Mowery, David C. 1983. "Industrial Research and Firm Size, Survival and Growth in American Manufacturing, 1921–1946: An Assessment." *Journal of Economic History* 43, 953–980.

Mowery, David. 1995. "The Boundaries of the U.S. Firm in Research and Development." In Naomi Lamoreaux and Daniel Raff, eds., *Coordination and Information: Historical Perspectives on the Organization of Enterprise* (147–182). Chicago: University of Chicago Press.

National Research Council. 1921, 1927. *Industrial Research Laboratories of the United States.* Washington, D.C.: National Research Council Publication.

Nicholas, Tom. 2003. "Why Schumpeter Was Right: Innovation, Market Power and Creative Destruction in 1920s America." *Journal of Economic History* 63, 1023–1057.

O'Sullivan, Mary. 2002. "Historical Patterns of Corporate Finance at General Electric and Westinghouse Electric." Mimeo., INSEAD.

O'Sullivan, Mary. 2004. "The Financing Role of the U.S. Stock Market in the Twentieth Century." Mimeo., INSEAD.

Pástor, Lubos, and Pietro Veronesi. 2006. "Was There a NASDAQ Bubble in the Late 1990s?" *Journal of Financial Economics* 81, 61–100.

Parente, Stephen, and Edward C. Prescott. 2000. *Barriers to Riches.* Cambridge, Mass.: MIT Press.

Peach, William. 1941. *The Security Affiliates of National Banks.* Baltimore: Johns Hopkins Press.

Rajan, Raghuran, and Luigi Zingales. 1998. "Financial Dependence and Growth." *American Economic Review* 88, 559–586.

Rappoport, Peter, and Eugine White. 1993. "Was There a Bubble in the 1929 Stock Market?" *Journal of Economic History* 53, 549–574.

Rappoport, Peter, and Eugine White. 1994. "Was the Crash of 1929 Expected?" *American Economic Review* 84, 271–281.

Schumpeter, Joseph. 1942. *Capitalism, Socialism, and Democracy.* New York: Harper.

Scotchmer, Suzanne. 1991. "Standing on the Shoulders of Giants: Cumulative Research and the Patent Law." *Journal of Economic Perspectives* 5, 29–41.

Shiller, Robert. 2001. *Irrational Exuberance.* Princeton, N.J.: Princeton University Press.

White, Eugene. 1990. "The Stock Market Boom and Crash of 1929 Revisited." *Journal of Economic Perspectives* 4, 67–83.

Zeira, Joseph. 1999. "Informational Overshooting, Booms, and Crashes." *Journal of Monetary Economics* 43, 237–257.

6

Financing Fiber: Corning's Invasion of the Telecommunications Market

Margaret B. W. Graham

Optical fiber for telecommunications is a classic instance of Schumpeterian "creative destruction." A technological innovation that had been anticipated by telephone companies around the world for decades as a vitally necessary way to increase communications bandwidth paradoxically became a stumbling block for many of these same companies by the end of the twentieth century. The principal agent of this disruptive episode—what Joseph Schumpeter would have termed a "creative responder"—was the American specialty glass company, Corning Glass Works, later Corning Incorporated.[1]

As Schumpeter observed, the disruptive efforts of creative responders are destructive in part because they are economically unpredictable.[2] It follows that the financing of disruptive and radical innovations can pose major challenges for any company that has ongoing businesses to sustain and develop and that practices prudent risk management. This chapter deals with the risks and challenges that Corning faced in largely self-financing its optical fiber innovation. Corning's challenges included not only inventing and developing its critical part of the new optical fiber system (in concert and competition with other companies), but penetrating a market that was controlled by some of the most powerful telephone companies on earth. More daunting still, Corning took the substantial risk of innovating in fiber during the 1970s, a decade during which its existing businesses, like those of many other companies, were experiencing unprecedented pressures.[3]

To achieve the commanding position in fiber-optic technology that it ultimately achieved—dominant worldwide supplier of optical fiber from the 1980s on—Corning needed to solve four strategic financial problems.

Only one of these is normally recognized as a routine cost of innovation.[4] In order of their manifestation, they were: (1) funding long-term research and development for a risky project that was not expected to reach fruition for decades,[5] (2) financing the costly construction and defense of an intellectual property position for a business that did not yet exist, (3) acquiring collateral knowledge and know-how of cable and the cable business beyond Corning's own experience base, and (4) investing in costly state-of-the-art fiber-optic manufacturing capacity well in advance of predictable demand. The company addressed all four of these problems with an acceptance of risk that not only flew in the face of prevailing financial practice, but violated or avoided its own internal financial controls.

At a time when scale was believed to trump almost everything else in business, fiber optics demonstrated that special knowledge and expertise could be even more important than size and capital for an innovating company. It also demonstrated that innovating companies, and the individual entrepreneurs who ran them, should expect to assume, not avoid, personal and organizational risk if they wanted to create opportunities for superior returns.[6]

At Corning, the fiber-optics project coincided with the adoption of new, more rigorous financial controls designed to wring greater productivity from the company's operations. This regime, consisting in part of new resource allocation procedures, inevitably impinged on R&D. For many other companies in the 1960s and 1970s, the move toward greater financial rigor, measurement, and accountability often blocked, or at least weakened, long-term R&D projects while a new generation of managers adjusted to having to compete in an internationalized manufacturing arena (Liberatore and Titus 1983). A business world that regarded "management by the numbers" as the sign of a well-managed company did not consider it necessary or even desirable to seek a right balance between productivity and creativity (Lamoreaux, Raff, and Temin 2003). Yet executives in entrepreneurial companies in this financially obsessed era had to learn to make exceptions—had to decide where controls and policies had to be adhered to and where it was important to suspend the rules. Few companies' executives sought such a balance, and even fewer found it. Corning's optical fiber story deserves attention because the company's leadership took the personal risks and made the necessary exceptions. Though it would take nearly twenty years (1967–1985) for the optical fiber project to break even on an operating basis, the project managed to evade many threatening financial hurdles, receiving countless dispensations from the more adverse resource allocation policies that were being adopted at Corning. As a result, rather than putting a stop to fiber, Corning's new financial controls had positive consequences for the project. Though honored mainly in the breach in the case of fiber, the new rules helped to build and sustain the company's innovative capability,[7] improving its capacity in general to support several costly and demanding innovations at once.

The Shot Heard Round the World

Two researchers from Corning Glass Works attended a technical conference in England dealing with the future of communications technologies in the spring of 1970. Sponsored by the Institute of Electrical Engineers,[8] and titled "Trunk Telecommunications by Guided Waves," the conference represented the prevailing view of the leading communications researchers at the time that the next step in increasing the capacity of telephone and data systems would be the development of millimeter wave guides. Well along in the conference agenda, Corning's Bob Maurer shocked the assembled representatives of various national telephone companies with an announcement about the alternative to "guided waves," optical fiber. In a presentation so short and low key that it occupied only a brief technical note in the conference proceedings, Maurer revealed that a team of optical fiber researchers at Corning's Sullivan Park research center had met and surpassed the established goal of less than 20 decibels for signal attenuation that had been specified as the theoretical target for commercial success.[9] Other research teams had long since concluded that this goal was out of reach for at least the next generation of technology. Knowing this, the Corning team's revelation was not just low key; it was deliberately misleading. They wanted their presentation to accomplish two contradictory objectives: to alert technically knowledgeable parties to their breakthrough, but to attract as little business attention as possible.[10] Only one group of researchers already knew Corning's news: because Corning had a prior licensing agreement with AT&T through Western Electric, Bell Labs had been told in July.

Others who were dimly aware of Corning as a modest-sized specialty glass company were inclined to overlook the announcement. Corning was, after all, a specialty materials company, not a communications company. But to researchers from top-notch laboratories who had tried and failed to hit the goal that Corning's research team had comfortably exceeded, the announcement was a sign that a technical star to be reckoned with had appeared in the communications firmament.

Stewart Miller of Bell Laboratories was one of the few scientific experts in the audience who understood both the technical and the business significance of the Corning team's breakthrough. He recognized that the Corning researchers had accomplished something remarkable. But at first he was inclined to deny that it mattered to AT&T. Miller's laboratory was heavily committed to millimeter waveguides as the next stage of communications technology, and although the final stages of that technology were slower in coming than Miller's group had predicted, they were still confident that waveguides would achieve feasibility any day now. In any case, because AT&T had recently concluded a licensing deal with Corning that covered any developments in optics Corning might make, Bell Labs could afford to be interested but not worried. Miller's attitude would change when senior AT&T executives began asking the obvious uncomfortable question: Why had Bell Labs, with its large budget and substantial effort devoted to waveguides, allowed Corning to get ahead of it in an area that AT&T was supposed to dominate (Hecht 2004)?

Despite its relatively small size and limited resources Corning was accustomed to challenging companies that were much larger and better resourced. Since the early twentieth century, it had successfully entered or created a wide array of new businesses, each requiring special technical expertise not easily acquired elsewhere. Corning called its strategy of combining its special expertise with leading-edge technologies taking the "uncluttered path." The new technologies it developed, often in concert with others, typically allowed it to operate alone in a new market, beyond competition, for a significant period of time. Before World War II, the company had taken care to do this while limiting its own growth, by entering into "associations" (business relationships that would later be known as joint ventures or strategic alliances). Corning used these associations to leverage its specialist expertise without having to supply all the necessary resources to develop a new technology-based business on its own. Corning's associations—Pittsburgh Corning for architectural glass, Owens-Corning for fiber glass, Dow Corning for silicone, and others—shared equally both risks and rewards. Sometimes the resulting new ventures, such as Owens Corning, grew larger than Corning itself, and Corning shared in the ongoing financial returns. Because of them, Corning's owners avoided what they perceived to be the disadvantages of large size: cumbersome bureaucracy and loss of agility.

Corning's practice of forming associations ended abruptly shortly before World War II when federal government antitrust policies changed under the New Deal. President Franklin Roosevelt's antitrust enforcer, Thurman Arnold, focused on Corning and some of its technical collaborators, calling them collectively the "Glass Trust," a group of prominent glass and electrical companies (Dyer and Gross 2001, Graham and Shuldiner 2001). Under Arnold's new interpretation of monopoly law, the glass companies were not alone in being condemned as monopolists. Others were telephone, electric, and, as well, the Aluminum Company of America.[11] But Corning's penalty was especially onerous because the company was still privately owned. In addition to incurring big personal financial penalties, Corning was forced to adopt a new business model, forced to discontinue technology sharing and to adopt a more self-contained approach to innovation.

Even after Corning issued stock in 1946, in part to deal with family estate taxes, large blocks of its stock remained in the custody of family members and senior Corning employees. As a result, Corning was still shielded from the most restrictive disciplines of the financial markets and could avoid some of the practices that typically inhibited innovation in a number of companies after the war. These included selling proprietary technology for short-term returns or expecting short term payoffs on long-term projects. It was taboo at Corning to treat knowledge as a financial asset to be sold to the highest bidder. If other companies had proprietary technologies or know-how that Corning needed to succeed, then Corning might exchange proprietary knowledge of its own for other knowledge or other resources that it needed to complete an innovation. But proprietary knowledge was viewed as a resource reserved for productive use in Corning operations and ventures, not as a way of gaining additional cash.

Toward the end of the 1960s, however, it was becoming clear to Corning's leadership that more rigorous internal financial disciplines were necessary if Corning wanted to continue to fund growth in its existing businesses and still ensure that projects ready for development could get the necessary resources. In the arena of communications technology, in particular, Corning was up against powerful incumbents. Companies like AT&T enjoyed commanding technical positions, were in the main technically self-sufficient, and appeared to have enough resources to keep it that way.

Telephone's Technical Monopoly Is Challenged

For the first three-quarters of the twentieth century, AT&T, and other national telephone companies like it, had enjoyed a near-monopoly on two-way communications technologies. The size and scale of their operations, their well-resourced laboratories, their commanding intellectual property positions, their careful attention to a system that was optimized for service and for durability, if not for rapid improvements in performance, and above all their ability to negotiate public rates that covered their R&D costs—all worked together to ensure that these national companies maintained control of the rate and direction of technological change in their respective countries.[12] In the United States, this commanding position had begun to erode with a consent decree in the 1950s that required AT&T to license its patents at a reasonable rate to all companies that requested such arrangements. In view of the strength of its position in all other respects, however, AT&T still maintained control of the technologies supporting telephony in the United States (Lipartito 2002; Reich 1985; Nohria, Dyer, and Dalzell 2002). As the purchaser of four-fifths of the communications technologies in its domestic market, it had little to fear from upstarts. Many other national companies maintained equally impregnable positions in their own marketplaces (Temin and Galambos 1987, Buderi 2000).

In the 1960s corporate planning exercises revealed to many national telephone companies that the communications infrastructure, long based on cable made of copper wire or aluminum, would eventually have to be replaced by something with much greater capacity. In their usual deliberate way, organizations like AT&T and the British Post Office (otherwise known as the BPO, forerunner to British Telecom) began planning for a gradual transition to an infrastructure that would have much greater transmission capacity. The planning horizon for the technological transition they envisioned was on the order of thirty-five to forty years.

While AT&T was methodically investigating technologies that would support an orderly transition, other companies were intrigued by the prospects of a more radical shift based on the earlier exploitation of more exotic technologies. One distant candidate appeared to be optical fiber transmission—using coherent light to transmit information. The Standard Communications Laboratory (SCL) operated by ITT produced promising theoretical work that suggested an earlier and bigger transition might be economically feasible if the right materials could be found. The British Post Office, which also acted for the British military in technical matters, took note of SCL's work. It was while searching for the purest optical glass available that the British Post Office contacted Corning, which for the BPO was a little-known supplier of specialty glass located in remote upstate New York. Corning duly supplied the requested test samples of its purest conventional glass. At the same time, it turned to investigating a radical alternative, based on its own patented process for producing a pure glass alternative using vapor deposition. This chance contact triggered one of the most definitive technological incursions of modern times. In less than twenty years, this same Corning Glass Works, operating on the outer periphery of the communications industry, would establish itself as the leading supplier of optical fiber for long- and short-distance communications in the world. The BPO's initiative had catalyzed a transition that skipped most of the intermediate steps envisioned by AT&T, establishing fiber optics as the material base for a new international communications infrastructure.

Fiber Optics Technology in the 1960s

The idea of using light to transmit information was well known, but fiber optics—the combination of a laser and a waveguide (a thin glass pipe to guide and contain the light)—only became technically feasible with the discovery of the laser in 1960 (Hecht 2004). Several forms of optical transmission system then quickly came into being, including both long-distance formats and very short-distance ones. Short-distance forms of optical transmission using optical filaments included high-loss glass fiber used in devices such as endoscopes for transmitting light into hard-to-reach places. Long-distance forms were larger formats of light pipe, known as millimeter waveguides. Waveguides, while technically feasible in the broadest sense, were still a very long way from actual commercial use. As understood at the time, waveguides could retain light over distance only if laid in absolutely straight lines and buried underground. This constraint would have seemed insurmountable had fiber seemed immanent. But the tremendous light loss that occurred over even minute distances using existing fiber-optic devices made the use of it for long-distance applications such as communications seem little better than fantasy. At the same time, the need for greater bandwidth in communications, especially for military applications, was contributing to a growing impatience with the technical stranglehold that the telephone companies had on communications technology.

For their part, the telephone companies were in no hurry to move on. Although U.S. regulators had required AT&T to license its technologies at a reasonable fee to all comers, potential competitors like MCI charged AT&T with deliberately blocking speedier development of a variety of promising technologies. It had long been AT&T's standard practice to obtain access to all new technologies that pertained to communications, so that they could be controlled and shaped in ways that maintained the integrity and the returns (regulated though they were) of the existing systems. Thus, while AT&T with its incomparable Bell Laboratories facility was as capable as any other organization in the world of inventing and developing new communications-related technology, the structure of the company, regulated as it was by most states in the Union, made it unlikely that any revolutionary system would be put in place, regardless of its technical feasibility, until it was no longer revolutionary.

Corning and the Communications Business

In the public mind Corning was known best for its Pyrex brand of strong heat-resistant glasses and, starting in the 1960s, Corning Ware, ovenware made of the same material as missile nose cones that could be transferred from refrigerator to oven to table. That Corning was also a long-time supplier to the communications business was less well known, although the company had operated at the edge of the business ever since the invention of the vacuum tube.[13] Corning had long supplied the electrical companies, and the major integrated consumer electronics companies, with glass housings for their vacuum tubes, using glasses specially formulated for their electrical properties. In the early days of solid-state electronics, Corning had also supplied special glasses for electronic components, even producing a few passive components themselves. Its position as the dominant supplier of cathode ray tube envelopes to the assemblers of television picture tubes had ensured profitable growth for the company throughout the 1950s and 1960s. Other products would have contributed to Corning's fame had their very existence not been top secret in the chilliest era of the cold war: the company was the leading supplier of glasses for any number of exotic military purposes like spy satellite and telescope mirrors, missile nose cones, spaceship claddings, and radar delay lines.

Success in the television glass business after World War II changed Corning from a small specialty glass business able to take on leading problems to a force to be reckoned with in wider professional business circles. A gutsy move to concentrate most of its available resources on providing suitable glasses for television picture tubes, first black and white and then color, had led to a business that threw off such quantities of cash during the 1960s that it could well afford to self-finance its own substantial research and new-product development budgets. Its financial performance as a high-profit, high-growth company in the 1960s propelled it onto the list of "nifty fifty" high-technology companies on the stock exchange during the late 1960s and early 1970s. A forty-to-one price-earnings ratio had enabled Corning to acquire an established position in the electronics business, but it had also imposed new burdens, including even greater pressures for growth.

At the end of the 1960s, Corning's postwar formula for success in innovation based on technological self-sufficiency was losing its potency. In 1968, Corning's leading customer for CRT glass, RCA, built a plant to produce its own glass envelopes for its highest-volume television tubes.

Ironically this was the year that Corning jacked up its R&D budget to 6 percent, a number comparable to the R&D expenditures of high-tech companies rather than glass companies. Smaller than most other glass companies, Corning nevertheless by that time accounted for 10 percent of the R&D performed by the worldwide glass industry (Graham and Shuldiner 2001, Dyer and Gross 2001).

Within two years of RCA's shocking move into glass, the U.S. television business as a whole went into a nosedive, taking Corning with it: Japanese competitors had taken the television business offshore. At the same time, Corning's number one new product opportunity, safety windshields made of malleable glass, came to an abrupt and ignominious end.[14] Meanwhile Corning's other strategic initiative, its acquisition of the semiconductor company Signetics, began to require huge infusions of cash. In short, in just a few years, Corning's apparently ample innovation capacity had reversed itself. From cash-rich, R&D-rich, high-tech darling of the financial markets, capable of financing its own new businesses, Corning found itself with early-stage businesses in urgent need of continuing investment and a number of longer-term opportunities that it could not afford to support.

Disappointing Returns Lead to Financial Rigor

The performance of industrial R&D was tantamount to a patriotic duty for large technology-based U.S. companies in the aftermath of World War II. Military contractors, or any other companies that sought government favor, were expected to invest their own money in research, in support of the national goal to make U.S. industry independent of foreign sources of science, as it had not been before World War II. At first, in the aftermath of what was called the scientists' war, this obligation was not considered burdensome. Investing in R&D was naively viewed as a reasonably sure route to achieving innovation (Graham 1985, Hounshell 1996). By the mid-1960s, however, many companies had become disenchanted with throwing large sums at R&D without having found effective ways to realize its benefits. Other companies had squandered large sums on science-based products that had either failed to connect with a market or had been ordered by one military service or another, only to be cancelled before being turned into production contracts.[15] Where blockbusters had once been viewed as wise investments, the very term had become synonymous with mismanagement (Kenney and Florida 1988). Sophisticated business practice was changing toward more cautious resource allocation where R&D was concerned.

Many alluring new financial techniques sought adherents in the technocratic 1960s. Clever graduates in finance eschewed the mundane duty of finding resources for operations and looked instead to devise conglomerate strategies designed to offset risk and strip out resources by combining existing business units. Inside many large multidivisional companies, management by the numbers gave financial accountants the upper hand (Johnson and Kaplan 1987). Most well-managed companies adopted new approaches to capital budgeting and evaluating their returns, adopting hurdle rates based on internal rates of return for all parts of their businesses, and for processes within their businesses (Schall, Sundem, and Geijsbeek 1978). When held to such definitions of success at the division and department levels, managers were reluctant to base their personal performance on the outcome of uncertain ventures.

The techniques that most affected research and development related to resource allocation and project selection based on some approach to calculating future returns.[16] Using financial tools like discounted cash flow (DCF) yielding net present value (NPV), longer-term projects with uncertain future returns had to project numbers that would clear higher hurdle rates than near-term, and therefore more certain, investment opportunities. In time these calculations were applied at finer and finer levels of granularity, until by the mid-1980s, a widely admired new product innovation process, termed the stage-gate process, prescribed applying tighter financial yardsticks at each stage of a project.[17] Such practices naturally worked over time to bias managers toward the certain and the short term, where markets were well understood and uncertainties in technology were minimized. Many companies responded to the increased use of sophisticated financial techniques by focusing their efforts on investments in increased productivity rather than on investments in longer-term innovations that might eventually have major payoffs but were highly uncertain by nature.[18]

Getting Serious about Controlling R&D

In the mid-1960s Corning was also in the process of strengthening its financial disciplines and controls and making them routine. The company's ability to keep innovation alive, despite prevailing philosophies concerning financial management, was due in part to Corning management's deep familiarity with the issues inherent in managing technology. Corning's top management was not prone to adjusting its entire management philosophy to accord with the fashions of the time. Senior leadership, including several Houghton family members in top management and on the board, had inherited a strong faith in investing in R&D. The family also retained a motivating memory from an early experience of losing the company altogether, a disaster they had no desire to repeat. The company's strategy had been to follow specialty glass technology wherever it might lead. To make this work, the philosophy toward financial investment both inside the company and out was "patient money," the necessary understanding that new technology-based products and businesses could not be expected to pay off overnight.

At the same time, Corning was no ordinary family company. Professionalism in management manifested in the superior quality and preparation of its managers was a point of pride (Galambos 1970, Dyer and Gross 2001). The younger generation of managers who took over Corning's leadership in the 1960s had done their time at Harvard and Wharton, and they employed as consultants contemporaries who promoted the latest management thinking. Corning's "young Turks" were eager to apply new principles as they rotated through the various operating divisions and the newer international parts of the enterprise. They could see the need for becoming more efficient and were open to developing more stringent financial routines.

As early as the late 1950s, Corning's technical staffs had adopted the idea of the "million-dollar business" as a loose indicator of R&D success. A decade later, this effort to measure R&D performance tightened up dramatically. The target for projected annual sales revenue from any project selected to go forward increased from $1 million to $10 million. This rule of thumb was accompanied by a much more stringent requirement that any new research project receiving corporate funding had to be supported by an RFTA, a formal "request for technical assistance" from one of Corning's nine eligible business divisions (Graham and Shuldiner 2001). In 1967 as a consequence of this stricter R&D budgetary discipline, Corning pruned its corporate-funded development projects from 185 to 67. The small fiber-optics project, still in its infancy and with a highly uncertain outlook, seemed an obvious candidate for termination. Yet it survived.

It survived because Corning was led by people who understood what it took to maintain unwavering commitment to innovation. Amo Houghton had been sensitized to the critical issues by Corning's chief technology executive, Bill Armistead, in the early 1960s when Houghton was vice president of technical staffs, and he remained in close touch with Armistead by moving him to the vice chairman's office. Tom Mac-Avoy, Corning's president who had started out at Corning's laboratory, acknowledged many R&D excesses in the past and was a proponent of rigorous financial disciplines, but he remained at the same time financially supportive to those entrusted with leading Corning's innovation projects. Even Corning's chief financial officer, Van Campbell, was aware of the issues that were likely to arise if misapplied controls were allowed to undermine long-term investment. Like other Corning managers hired in the 1950s, Campbell had served in more than one function at Corning, and understood the conditions that were necessary to support innovation.

Against the Odds: Getting into Fiber Optics

Corning's entry into optical fiber was by no means a forgone conclusion. In the earliest days of optical fiber research, it had looked as though Corning would be sitting it out. As an opportunity worth exploring, optical fiber failed to meet some of the standard strategic tests. When a new scientific opportunity came along, Corning's technology strategy as a specialty glass company was to take advantage of the unique combinatorial properties of glass at several levels. It would seek commanding proprietary advantage first in the area of glass compositions, of which it had more than anyone else,[19] then at the level of unique products, and finally in the form of special process technology. But Corning's options were restricted with respect to optical fiber because the American Optical

Company had already foreclosed some of them by seizing the leading patent position for devices made of regular glass used to transmit light.[20] American Optical's patent position was broad enough to cover low-loss as well as high-loss fiber technologies as long as they were made from glass.

Corning's remaining area of opportunity, if any, was confined to unconventional glass-like compositions and new process technologies. As it happened, Corning had just such a combination of unconventional glass and new process: fused silica, an extremely pure glass-like substance, was made by vapor deposition, a process that Corning had patented as a result of work in the 1930s in organic chemistry.[21] Both the material and the process were extremely demanding to work with and to make, and at first the opportunity to use them for optical fiber was anything but obvious. Corning had acquired its experience with processes later used to produce fiber optics when it produced radar delay lines for the military. The government contract had come to Corning because of its patents covering vapor deposition, but work for the government was not high on Corning's list of priorities. Government work involved conditions and work that were not attractive to Corning's researchers at a time when attracting and retaining researchers was critically important.

Corning attracted leading researchers to its research program in part by giving them a measure of autonomy in their choice of exploratory projects. When the British Post Office showed an interest in glass-based optical communications systems in 1966, one of the researchers who had worked with delay lines at Corning, MIT-educated Robert Maurer, saw that the difficulty of producing fused silica and the challenges of vapor deposition might offer just the kind of technical advantage that Corning needed to excel against larger competitors. Maurer started a small research effort using vapor deposition to produce what he thought might be a superior form of low-loss optical fiber for long-distance communications. The project soon picked up newly minted Ph.D. physicist Donald Keck. In the early 1960s, Corning had conducted some laser research, and the laboratory retained this laser expertise in its research group. It was not a leader in lasers, however, whereas others, at AT&T and elsewhere, had far more extensive experience at working with lasers.

Unintended Consequences of Financial Controls

Maurer's fiber-optic project ran into the financially motivated drive to reduce the number of small corporate research projects almost from the start. For the strategic reasons already mentioned, Armistead, as head of Corning's laboratories, was skeptical about pursuing the BPO lead at all. Feeling the pressure to establish clearer quantifiable priorities among the many projects that were consuming Corning's corporate R&D budget (Graham and Shuldiner 2001), Armistead assessed the business case for fiber as negative. He reasoned that if American Optical held the controlling patent position on optical communications devices and AT&T accounted for most of the domestic market for telephone cable, there could be little room for Corning. Armistead warned Maurer's little group working on optical fiber in 1967 that they could continue the project only if they could secure funding from one of Corning's business units.

A long-time proponent of technologies related to optical fiber, Chuck Lucy, then head of business development for Corning's television electronics business, came to the project's rescue. Lucy had joined Corning after administering the government's contract with Corning for radar delay lines, and he retained a strong interest in the prospects for optical communications. While the new project was too long term for the television electronics business to invest in immediately, he volunteered to find funding for the project from potential international customers. Judging Lucy's willingness to trot the globe drumming up support as a valid sign of Corning business interest, if not quite an RFTA, Armistead gave the project a reprieve. In fact, it took Lucy several years to attract the right international parties, and more to get the funding in place, but meanwhile the project's technical momentum gave Armistead the technical justification to keep it going on corporate funding.[22]

This was the first instance in which the new resource allocation procedures, though honored in the breach, nevertheless had a sustaining effect on the fiber project. Lucy's proposal to seek funding outside Corning had the effect of altering the nature and the pace of the project itself. Though still within the domain of research, the work on fiber optics changed from a long-term research project—small, discovery oriented, largely theoretical, and aimed primarily at accumulating all possible knowledge about chemical formulations, their physical properties and behaviors— to an urgent results-oriented early-stage development effort aimed at producing visible, testable quantities of real salable fiber. Potential customers were unlikely to be moved by theory, but working prototypes could convince them. Until Lucy's intervention, Maurer's team had fabricated 1 kilometer of testable fiber, and they were measuring steady gradual decreases in light loss, still from a very high base. The research team's new goal was directly linked to a widely accepted measure of potential commercialization: to demonstrate in prototype the 20 decibel benchmark that ITT's Charles Kao had calculated to be economically equivalent to the full cost of installing conventional copper cable.[23] The shift in the Corning project's objectives involved abandoning the purer goals of finding just the right material in theory and of learning all there was to know about the material mechanisms. Instead, it settled on the most promising material at hand, choosing the best-known approach to producing it, then testing and improving and testing again until the desired attenuation was achieved.

AT&T's Bell Labs was pursuing a number of projects related to optical communications systems at the time, many more than Corning, but their researchers lacked a similar sense of urgency. For reasons already stated, Bell Labs researchers were working on a different timetable from the smaller Corning effort, and they were conducting their research in a more specialized and long-term way, aiming first to develop a deep theoretical understanding of all possible optical systems, including, but not limited to, fiber optics. Their main efforts to make prototypes were directed toward millimeter waveguides. From a business perspective, AT&T's technical superiority and control of the rate of adoption in the market seemed to be incontrovertible advantages. But Corning's combination of different technical objectives and a different timetable, driven by the need to seek external resources, allowed it to leap ahead. AT&T's ample internal resources were in this case an odd comparative disadvantage.

Complacency was another compelling reason for AT&T not to be unduly concerned. It had already secured licensing access to any proprietary developments Corning might achieve in the area of optical communications. Here Corning's underdog position worked to its benefit, but at the time, many Corning insiders believed that a terrible error had been committed. Corning's forfeiture of its proprietary position in fiber optics to AT&T was criticized by some Corning insiders as nothing short of a financially motivated sellout. The licensing deal in question, negotiated between 1968 and 1970, involved its senior technical leadership. Like the contemporaneous move to raise external funding for research, the AT&T licensing deal was prompted by the new emphasis on gaining early financial returns on Corning's technology. What critics did not fully appreciate was the substantial benefit Corning gained in sharing its early fiber-optics knowledge with AT&T.

Corning opened negotiations with Western Electric (AT&T's manufacturing and licensing arm) to gain an immediate reduction in royalties to AT&T for some of its semiconductor technology to benefit Corning's West Coast subsidiary Signetics. Corning had acquired this spin-off of Fairchild Semiconductors in the mid-1960s in the hopes of solidifying and developing its own small position in electronic components. Signetics urgently needed a reduction in the 4 percent royalty it was paying to AT&T for its solid-state patents. Corning anticipated obtaining a favorable cross-licensing deal for certain glass patents that AT&T had long wanted to license. But the Western Electric negotiators held out instead for Corning's future optical communications patents on a royalty-free basis to AT&T, with no cross-licensing privileges. The Corning negotiating team was concerned about giving away potential control over future fiber-optic patents. But however promising Corning's unique approach to fiber optics might be, the research team had yet to make any real breakthroughs, and the potential business opportunity, though enormous, was fuzzy at best. The immediate benefit to Signetics, by contrast, was eminently quantifiable. By early 1970, after two years of hard bargaining and much further exchange of information, Corning persuaded Western Electric that Corning's research programs were worth more than was first understood and that access to them merited a reduction of licensing fees and a less restrictive licensing arrangement. Meanwhile, AT&T came away from the negotiations confident of being able to produce its own fiber-optic products and devices without owing future royalties to Corning.[24]

During the two years of negotiation, Corning's project had made steady progress. By sustained experimentation, the group had fabricated stronger and stronger fibers of greater and greater lengths using completely proprietary techniques and formulations.[25] When in early 1970 the Corning team announced its breakthrough—an astounding 16 decibels loss of light over 1 kilometer, which handily exceeded the 20 decibel target believed to ensure economic viability for a fiber communications system—Corning was able to interact with Bell Labs's optical communications research team precisely because the licensing agreement was in place.[26]

The early interchange with the Bell Labs team headed by Stewart Miller working on optical research proved vital for Corning's progress in moving beyond its first prototype. While some Corning researchers had some earlier experience of research on lasers, its laser knowledge was far from current. Bell Labs was a key source not only for upgrading the Corning team's understanding of recent developments in laser technology but also for the systems expertise they needed to characterize the optical communications system as a whole. It was not long before Bell Labs cooled to sharing further information with the Corning team. But by then, Corning had already received the help it needed most. Bell Labs began to think of Corning's effort as serious rivalry only when it was too late. By the time they recognized the need to put fiber on a more rapid timetable, Corning could not be headed off (Hecht 2004).

Once the parameters of its basic system were specified, Corning continued its pursuit of fiber-optics technology by reverting to its prewar strategic pattern of forming active partnerships beyond the glass industry. The times were once again on its side. In the hostile environment of the New Deal, Corning's practices of association around new technology had provoked a costly and embarrassing antitrust suit against the "Glass Trust." Then an unfriendly Justice Department had portrayed Corning as a small but deadly spider controlling a web of knowledge-based agreements that had monopolized the technology of the glass industry (Graham and Shuldiner 2001). In the 1970s, however, the shoe was very much on the other foot: the Justice Department of this new day was out to cut AT&T's mighty telephone monopoly down to size and open up its technologies to more parties. In that context, Corning was no longer a malefactor but a prospective agent of change in public policy toward the communications industry.

Hurdle One: Funding Long-Term R&D

When it finally recognized that Corning could pose a threat to its control of the pace and direction of communications technology, AT&T tried to discourage Corning from further activity in optical fiber by announcing that it would produce its own optical cable and never buy it from an outside supplier.[27] According to a now seriously disgruntled Bell Labs leadership, Corning would have to explore an international market if it wanted to find a market for its product. This seemed to be a risk-free suggestion for AT&T, as there was little likelihood that such a newcomer to the communications industry would be able to produce and sell fiber-optic cable successfully. Nevertheless by 1972 Chuck Lucy had identified five companies outside North America that were willing to commit $100,000 per year for the privilege of sharing in Corning's findings, having access to test samples of the fiber, and being promised access to licensing when the technology was available. Research is cheap. At a time when Corning's entire R&D budget hovered at between $20 and $25 million per year (or about 5 percent of sales), $500,000 was a significant amount to be getting from outside sources. Half a million dollars per year would support more than ten full-time researchers, keeping a substantial group focused on fiber rather than redeployed on other more immediate priorities.

Using a form of alliance reminiscent of Corning's earlier associations, the Corning team set up what it called joint development agreements (JDAs) with companies in Britain, Japan, France, Italy, and Germany. The five-year period that these agreements covered, most beginning in 1972 or 1973, coincided with the worst downturn in sales Corning had yet experienced. Luckily for the fiber-optics project, the JDAs had to be treated as an irrevocable company commitment, and there was no attempt to end the work. The project team had five steady years of outside funding, combined with significant learning opportunities in technologies like cabling and lasers. At the end of this time, during which Corning had focused its efforts on producing and selling test samples to any potential users it could find as a way to gain knowledge and experience, much useful information had also been gathered from potential customers. Not only had it staked out a commanding patent position in its chosen version of optical fiber, but it had also identified the know-how it would need to succeed. A strategy had also been devised for the product's protection and its production, though not yet for its distribution (Graham and Shuldiner 2001, Dyer and Gross 2001, Morone 1993).

An unfolding irony of the JDA arrangement was that Corning had promised more than it needed to promise for the money it received. Working in the context of the postantitrust logic that prevailed at Corning, its patent lawyers had granted the JDA's U.S. licensing access to technologies arising out of the agreements. This might have proved inconvenient when the company wanted to settle on one partner and when it wanted to position itself against competition in the United States and abroad. Fortunately, only one company wanted to continue on another round as a full-fledged partner with Corning in commercializing optical fiber. This one company was Siemens, a German electrical giant with enormous reserves of knowledge in cabling. Siemens wanted to learn to produce fiber for the European market. Like Corning, Siemens had a long history of steady innovation and proved to be a good partner (Buderi 2000).

Recent studies of other innovating companies have suggested that Corning's experience in seeking outside funding would later be recognized as a distinguishing feature of entrepreneurial enterprises. Like Corning, they also purchased their freedom from internal constraints by seeking outside funding; they were motivated to show early practical results, and they generally worked outside but alongside the controls and rule structures of their companies (Burgelman 1993).

Hurdle Two: Funding the Intellectual Property Strategy

As project head, Robert Maurer insisted from the start that detailed records be kept for optical fiber, even for ideas that seemed of scant commercial relevance.[28] When Corning's patenting philosophy had been directed primarily at filing broad, fundamental patents, as it was after World War II, this careful attention to detail was unusual, even somewhat onerous, from the researchers' point of view. But Corning had suffered from earlier laxity in its documentation of research, and its practices had become far stricter in the postwar era. Maurer's research area was highly competitive, and even the slightest revelation of project knowledge could quickly be replicated by competing research teams. Other areas of Corning, while still mindful of security, were not generally working in such competitive fields (Graham and Shuldiner 2001). Maurer knew from the start that it was crucial that Corning take a fundamentally novel course of action. Corning could remain different only if its solution was kept secret and if it had the evidence that the novel approach was intentional from the start. The project documentation rule paid off. Corning filed for and obtained twelve very strong patents that gave it a commanding intellectual property position with respect to optical fiber technology.

Lacking the massive resources of some of its research competitors and well aware that at least some of them were determined to flout Corning's claims and enter the market anyway, Corning looked for ways to buy time. Before it had a real business in optical fiber, it set about demonstrating the strongest intent to enforce its patent position aggressively and to prosecute all infringers. While this approach was consistent with Corning's fundamental policy never to treat intellectual property as merely a financial asset, it was unusual for Corning to pursue infringers. Moreover, this policy was out of step with legal practice at the time. Patents and other forms of intellectual property were increasingly being treated as extra sources of short-term income. None of the mostly large companies that Corning sued for infringement dreamed that Corning's aggressive legal stance was motivated by anything other than the pursuit of a lucrative licensing agreement.

Instead, to the surprise of many, Corning adopted a policy of suing to win. It hired a young litigator from its patent lawyers Fish and Neave, Al Michelsen. Though almost all of its sales of optical fiber were samples, Corning set out, with the strong support of Amo Houghton and Tom MacAvoy, to pursue an aggressive campaign of litigation against infringers. The tough intellectual property strategy was the scheme of Lee Wilson, Corning's head of electronic products, who decided in 1976 to buy time for Corning and gain maximum visibility by filing suit against the two most visible targets available at the time. ITT, which had employed scientist Charles Kao to do the initial theoretical work in fiber optics, was producing optical fiber on contract to the U.S. military. Corning sued ITT first, refusing to negotiate a reasonable payoff or to settle for a license. ITT also refused to settle and countersued, charging Corning with the intent to monopolize. In defiance of conventional wisdom, Corning pressed its case, at the cost of roughly $1 million per year. The matter dragged on so long that Corning lost much of the advantage it had hoped to gain from scaring off possible infringers. Finally, ITT settled on the eve of the trial date in 1981. Even ITT's settlement, and a further settlement with the U.S. military in 1983, came nowhere near covering Corning's legal expenses for the case.

Ultimately Corning had to prosecute several large infringers around the world, its efforts culminating in a suit against Sumitomo, which was selling fiber to Canada Cable and Wire. The legal expenses, continuing at well over $1 million per year, weighed heavily on Corning at a time when there was no extra money and other Corning businesses already in their growth phase wanted legal support for their own needs.

While some of Corning's fiber settlements were more generous than others, the financial payoff from litigation never made up for the full costs expended for worldwide litigation, much less the cost of executive time in depositions and testimony or the opportunity cost for other businesses. Nevertheless, the policy of aggressive litigation served a purpose: buying Corning the time it needed to solidify a well-rounded intellectual property position for fiber beyond its patent holdings alone. By the time some of the major defendants had their day in court, Corning had surrounded its key patent holdings with a base of know-how protected mainly by trade secret that was almost impossible to imitate and that gave it a significant and sustainable lead in both performance and quality. This lead became vital when Corning's continued efforts to broaden and deepen its mastery of fiber technology allowed it to switch away from producing AT&T's preferred multimode fiber format and to supply MCI with single mode fiber, the higher-performance format that Corning preferred and had already mastered.

The strategic importance of Corning's complete intellectual property position was not at first evident to the outside world. In 1987, only three years after Corning had secured its first large orders, *Financial World* held that fiber had become a commodity and that companies like Corning would never be able to sustain the profitability levels of the early business. Fiber was without question a volume business by that time and it was crucial to maintain a large share of the total market to gain the necessary economies of scale, but Corning demonstrated that it could deliver significantly better quality and performance than its competitors. Moreover, far from allowing fiber to become a commodity, it was already working on successive generations. It could capitalize on this position only because it had become market leader as well as technology leader. Corning became a market leader by integrating downstream into making optical fiber cable, a strategic move that had been strongly advised against by consultants.

Hurdle Three: Acquiring Collateral Technology: Cable

Having decided to extend their relationship past the initial JDA, Corning and Siemens eventually formed two parallel alliances: Siecor GMBH, which produced fiber in Germany, and Siecor, Inc., which purchased fiber from Corning and manufactured cable in Hickory, North Carolina. In order to make these strategic moves, the alliance needed to acquire greater access to the North American cabling market.

It soon became clear that the alliance with Siemens alone was not sufficient to give Corning the competence in cabling it needed to integrate forward in the business. The Boston Consulting Group, when it was retained to help Corning's Lee Wilson formulate a strategy for optical fiber, advised simply selling optical glass for fiber in boule form, a commodity business if ever there was one. Key to BCG's analysis was Corning's lack of market access. Even in partnership with an international expert in producing communications cable like Siemens, how would Corning break into the tight world of cable suppliers in the United States?

Corning leaders, especially Amo Houghton and Bill Armistead, could see that the original plan, for the Siemens Corning venture, Siecor Inc., to develop a cabling business from scratch, would take far too long. There was no way around it: Corning had to acquire a real cable company with real customers. The only cable company "available," Superior Cable of Hickory, North Carolina, was on offer for $50 million. By objective standards, this was no bargain, but Corning paid. Considering the need to take the company virtually down to its foundations and rebuild, the two owners received a sizable premium. Even so, there was a going cable business with a set of customers, and it employed a workforce that knew how to make copper cable. Moreover, the substantial investment that had been made posed just the right sort of challenge for an experienced team of Corning managers headed by long-time Corning marketer Al Dawson, head of the Electronics Division after Lee Wilson.

Bolstered by the acquisition of a real cable company, the Siemens-Corning alliance, Siecor Inc., soon became a successful enterprise in its own right. Its fifty-fifty ownership arrangement, patterned on arrangements Corning had used in the 1930s and 1940s for its most successful associations, gave the new company the measure of autonomy it needed to be creative in its own right in developing the fiber-optic cable business around the world. In time, Siecor even managed the unusual feat for an American company of breaking into the Japanese market. To do this, it had to overcome challenges from companies that had the direct support of the Japanese government with its carefully constructed long-term strategy for developing its communications infrastructure (Graham and Shuldiner 2001). The joint ownership structure lasted until Corning bought out the Siemens interest altogether in 2000. The Siecor alliance proved to be one of the chief enablers of Corning's ability to innovate continuously in fiber optics (Dyer and Gross 2001).

Hurdle Four: Building Capacity in Advance of the Market

During the early building phase of its business, Corning gained experience and financed some of its development by selling specialty quantities of fiber in small lots to many different customers who were using it for research and experimentation purposes themselves. But although it could supply specialty lots on its own timetable at low cost, it was clear that Corning would not gain a viable place in the developing market for fiber and cable without investing heavily in production capacity. Companies that were already in the telecommunications industry had enough credibility to secure orders solely on the basis of previous performance. Corning had to convince potential customers that it had both operating capacity and staying power. For this reason, Dave Duke, an experienced corporate entrepreneur who had spent the time that optical fiber was in development building Corning's Celcor business (ceramic substrates for catalytic convertors) in record time, was tapped to head the fledgling fiber business. Duke brought to the fiber business not only the experience of starting a high-volume business from scratch; he also brought the necessary credibility with senior management and deep knowledge of how the Corning system worked. These endowments enabled him to take matters into his own hands. Time and again, Duke found ways to build plant capacity and have it well under way while business developers were still drumming up orders. Most of the time the necessary requisitions for Duke's projects were still winding their way through the corporate approval process.

The first major pilot plant investment came in 1975, the same year as Corning's layoff of management personnel known by the ominous label "The Guns of August." Amo Houghton's firm charge to Dave Duke had been simple: "Never run out of capacity." The ultimate sign of the CEO's unflinching commitment to the fiber project was that he signed the requisition for a pilot facility for fiber within days of giving the order for the first massive layoff of managerial-level employees Corning had ever had. Duke's pilot plant for fiber supported work on two different Corning processes for making fiber, the IV (inside vapor deposition) and OV (outside vapor deposition) processes.

The next, Wilmington, North Carolina, fiber plant started up in 1979; it was a full-scale production facility, making fiber using only the OV process. From an initial production capacity of 30,000 kilometers expandable to 100,000 kilometers per year, it ramped up to 200,000 kilometers per year. By dedicating that plant exclusively to the more challenging OV process, when AT&T and its licensees were producing fiber using the easier IV process, Corning launched itself on an aggressive cost-reduction schedule that allowed it to drop its selling price steadily from $3 to $5 per meter on the earliest production lots to ten cents twenty years later. It did this by continuously reinventing the production process and completely changing equipment generations every two years and by investing in advance of demand. Ultimately the same Wilmington plant produced millions of kilometers of fiber annually.

Corning's big break came in 1984 when MCI contracted to take a significant quantity of Corning fiber. MCI went with the untried Corning not only to avoid buying from AT&T, but because Corning offered single-mode fiber. MCI recognized single mode to be a superior format in fiber optics. AT&T, supplying its own needs, was using multimode fiber. Corning had already produced quantities of single-mode fiber in advance of getting the business. Corning secured the business because Dave Duke representing Corning and Al Dawson representing Siecor were willing go out on a limb to meet MCI's terms. They agreed to MCI's demanding schedule (one year to delivery) and the price MCI was willing to pay ($90 million for 150,000 kilometers) knowing how hard it would be to achieve. They were used to ramping up aggressively in other demanding businesses like Corning's Celcor (substrates for catalytic converters). They had had time to resolve the major technical questions, and they had worked through the necessary choices that had to be made on the business side while waiting for the business to materialize.

Outcomes for Fiber

By clearing the four strategic financial hurdles, Corning positioned itself as a leading supplier of both types of fiber (multimode and single mode). It had time to build a useful and extensive knowledge base, and it was ready to take advantage of breaks when they came. The breaks were long in coming. It was a full twelve years after Corning's first big technical breakthrough when, in 1982, it became known that AT&T would be ordered to split into different independent businesses. MCI could move quickly to challenge AT&T's technical monopoly because Corning was ready to supply the advanced form of optical fiber MCI wanted just one year from the time of the order. Based on its order and the ones that soon followed from the new Baby Bells, Corning and Siecor were launched on a high-volume fiber business at last.

In 1985 Corning's optical fiber business broke even for the first time. Other suppliers poured in to follow Corning's lead, and by 1987 the financial analysts were already issuing commodity warnings. But Corning's long run-up to a full-scale business, disciplined but not stopped by the Corning's financial controls, had prepared it well. It launched a brisk campaign to keep its version of fiber's performance ahead of rivals. The business followed a steady upward trend for the next fifteen years, with Corning continuing to hold its position as leading supplier of successive generations of fiber-optic cable. AT&T eventually reorganized, freeing the high-technology business, renamed Lucent Technologies, which included both Bell Labs and the cable business, to operate at the speed of the rest of the industry. But even with much superior resources and overall scale, AT&T and its successor companies never regained the command of the rate and direction of technology that it had enjoyed before the breakup.[29]

In the 1990s Corning was impelled by a soaring stock price reminiscent of the 1960s and 1970s, to use its own highly priced stock to acquire a place among suppliers to larger optical systems businesses. As at the time of the "nifty-fifty" stock prices of the late 1960s, when stock achieved such high price-to-earnings ratios, companies felt obliged to put the inflated stock price to use quickly by acquiring assets in the form of companies. Having purchased with stock a number of small firms that participated in fiber-optics systems and photonics, Corning attained the status of a major player in the telecommunications arena just in time to be punished by the markets when the whole telecommunications industry's house of cards collapsed. In 2001 the telecommunications equipment and systems business imploded as suddenly, and even more completely, than the domestic television business had collapsed in the early 1970s. Optical fiber was a victim of overestimated market demand all along the chain, and Corning, at the end of the supply chain as always, was one of the companies caught holding the bag.

This time, although the financial analysts had scolded Corning for not becoming simply a telecoms company, the company had other eggs in its basket. Several other promising new business opportunities—Celcor for diesel engines, substrates for computer displays and flat-screen televisions—were in their early growth phases. But survival depended on getting the money it needed to invest in these businesses from financial markets that had classified it as a telecommunications company and were intent on writing it off with the rest of its imputed industry. When the financial markets became disenchanted with technology generally, even a company as diversified and flexible as Corning was denied the capital it needed. Corning's stock plunged from $320 to less than $2 per share before starting a steady climb to recovery.

Conclusion

It has become commonplace to equate R&D with innovation, and thus to equate the financing of innovation with the allocation of resources to R&D. Corning's fiber-optics experience suggests that especially for innovations with the potential for creative destruction, the cost of innovating goes far beyond the standard R&D budget. Further, the process of innovation consumes far more than the activities of R&D, even when the R&D has to be maintained at a level and with a kind of consistency that is hard in difficult times. For Corning, staying the course with optical fiber was a matter of assuming and managing several major types of risks and uncertainties. In addition to taking on and finding ways to support long-term research that no existing Corning business unit was willing to invest in, financing the development of optical fiber involved the costly process of establishing (both patenting and defending) an intellectual property position, acquiring and reshaping a cable business outside the scope of Corning's existing competency, and building a full-scale manufacturing process in advance of demand.

At Corning, the challenges of financing fiber had to be met in the context of an ongoing business. Fiber had to compete for resources with all of the shorter-term demands for immediate resources, for which the needs were not only urgent but fully understood and comparatively easy to articulate. Moreover, the risks associated with advancing fiber's cause in the 1970s had to be assumed at a time when the company was going through an especially difficult time. Although this was a troubling period for many American companies, such periods are isolating. Companies and their leaders tend to experience them in ways that are uniquely difficult and painful to themselves. Advocating long-term investments during such a period is not for the faint of heart. Amo Houghton was often heard to declare that had he not been a member of the Houghton family, he would have been fired by the board for his dogged persistence on behalf of optical fiber. Houghton and Bill Armistead, along with a number of fiber's other executive supporters, all retired early from Corning, some before the company landed the big contract from MCI. Houghton and MacAvoy left Corning when the fog of uncertainty had only begun to dissipate, having borne the substantial personal discomfort of uncertainty for more than a decade. When they retired, the company had recovered from the lowest lows of the mid-1970s, but was still financially underperforming.[30]

Though it was a struggle for Corning to succeed with optical fiber, it is still easy to see how the company's smaller size and innovation-dominated managerial logic gave it an advantage over AT&T when pursuing disruptive innovation. It had a similar advantage over the large automotive companies with catalytic converters, even though the larger companies with their massive laboratories and enormous R&D budgets always enjoyed far more resources. Corning's clear advantage in all such cases was its ability to concentrate resources, respond quickly, and be able to share the cost and learning burdens with allies. These advantages were not merely structural; they were based on a collective competence born of past experience.

Beyond the advantages of flexibility and experience-based competence, it took risk-accepting leadership to carry the weight of Corning's innovation process, leadership that could not be removed at the first sign of setback or whiff of failure. Perhaps just as important, it took risk-accepting leadership to make the decision to set aside or circumvent newly developed financial controls, controls that were clearly necessary and justified for most of the business, when exceptions were needed. In these decisions chairman and CEO Amo Houghton had the direct benefit of the deep technical expertise of his executive colleagues with R&D backgrounds, like Armistead and MacAvoy. He also had the aid of an unusually broad-gauged financial executive, Van Campbell. But even leadership with the peculiar advantages Amory Houghton and colleagues enjoyed, such as significant family ownership with the continuity that afforded, would not have been enough to ensure Corning success in undermining of telephone company control of its own technology. In addition to a large degree of financial discretion and autonomy, Houghton relied on a combination of other factors as well, evident in the way Corning managed to clear the four financial hurdles recounted here.

Corning had both an active knowledge base of glass and related materials and an active experience base. The active knowledge base, only some of it protected by patents, was resident in processes, defined and undefined, and in the know-how surrounding them, protected by trade secret. This knowledge base could be deployed and redeployed. It was embedded not only in laboratories and experimental process facilities and carefully tended intellectual property positions. It was embedded in ways of working, and in specialized Corning equipment and capabilities like high-temperature furnaces and the ability to work with them. Corning's unique store of knowledge was seldom treated as a financial asset and rarely for sale. With rare exceptions, Corning adhered to its policy of keeping its technology for its own innovating purposes and prosecuted vigorously not only those who infringed on its patents, but also former workers who left the company and dared to use elsewhere what they had learned at Corning.

Further, Corning had an active experience base as an innovating company. This too was embedded in the company's practices and expectations. It had the experience of moving quickly into unfamiliar markets, of ramping up difficult processes in short periods of time. It had the collective memory of technical failures that had not cost employees their jobs or shut down the research enterprise, of people who had been redeployed on new projects and then allowed to return to their preferred work. Finally, it had the experience of forming alliances, ventures that endured, based on equally shared risks and shared returns. These ventures required patience and cooperation, and the ability to assess and reassess mutual goals and interests. In short, Corning had an integrated set of business practices that were not common in companies that adhered to the logic of size, scale, control, or short-term profit as the primary measure of success.

Ultimately optical fiber is a Schumpeterian story because it is shot through with the stubborn subjective rationality that Schumpeter identified as the hallmark of creative response. Like the individual creative responder as entrepreneur, Corning had deep and extensive knowledge and experience bases that amounted to institutional self-knowledge, with its attendant ability to seize sudden opportunity and exploit it. It had financial support that was not only solid but discriminating. Its resources were allocated by executives who were able and willing to make judgments, follow routines and break them, institute financial controls, and suspend or make exceptions to them. Without leaders with the ability and the autonomy to make real decisions, Corning would never have succeeded in challenging the telephone companies' technical monopoly. Fiber-optics communication systems would today be a story of an emerging twenty-first-century technology rather than a familiar maturing technology from the century just past. We can only speculate what difference that might have made, whether this particular episode of creative destruction was ultimately beneficial to consumers or society. It is possible that future historians will conclude that a more measured adoption of high-capacity communications networks with the staggering changes they brought with them would have benefited society. Either way, we can say with certainty that Corning's fiber-optic breakthrough was enabled by a set of financial practices that were adopted and carried out consciously and with intention, in a nonroutine way far different from what has generally been considered best practice for large, diversified companies in the twentieth century.

Notes

1. See chapter 10 in Swedberg (1991). For discussion of this argument, see Graham (2003). See also Rosenberg (1994).

2. Schumpeter referred to the entrepreneur who caused creative destruction as a "creative responder," one who operated under the influence of "subjective rationality." By comparison, an "adaptive responder" was one who merely reacted to foreseeable change and operated under the influence of the "objective rationality" predicted by neoclassical economics. For Schumpeter's distinction between "objective rationality," which does not have to be conscious, and "subjective rationality," which is a highly conscious working out of an individual, particular, and contextual view of economic rationality, see chapter 7 in Swedberg (1991).

3. At the time of the invention, Corning Glass Works was 192nd on the Fortune 500 list of companies. It was known in the United States as the producer of Pyrex in the form of high-tech laboratory ware, ovenware, and tableware, as well as by the Corning Ware brand. In Europe it was known as the licensor of laboratory glass, Le Pyrex.

4. For Schumpeter "invention" and "innovation" are distinct phenomena. Absent that critical distinction, there is a tendency to anticipate only the costs of invention (e.g., idea embodied in working prototype) and to expect the costs of the innovation (e.g., process technology and market development) to be borne by the new business when it is established.

5. Research alone can be relatively inexpensive, but to engage in product and market development in the iterative way that Corning adopted early in the project is a far more expensive proposition. See Morone (1993).

6. See Teece (2000), for a succinct discussion of the difference between entrepreneurial returns, surplus rents justified by the risk taken, and monopoly rents that are simply an abuse of market power.

7. For this useful phrase and the argument that accompanies it, see Carpenter, Lazonick, and O'Sullivan (2003).

8. Unless otherwise specified, the account of Corning's optical fiber innovation draws especially on Graham and Shuldiner (2001) and Hecht (2004). The British Post Office did not, as AT&T did through Western Electric, make its own equipment, but it kept in close contact with equipment suppliers like Standard Electric, an ITT subsidiary.

9. This benchmark had been set by Charles Kao of Standard Electric Laboratories, who became one of the most effective promoters of fiber-optic technology. Hecht (2004).

10. The extended note submitted by Robert Maurer and Donald Keck, both physicists, deliberately omitted the name of another significant contributor to the project, chemist Peter Schultz. Had Schultz's name appeared on the paper, knowledgeable members of the technical community would have realized that the breakthrough in attenuation involved a different form of glass, fused silica, and not just a better way of achieving purity in regular glass. See Graham and Shuldiner (2001) and Hecht (2004).

11. It was in deciding the case against the Aluminum Company of America (ALCOA) that Judge Learned Hand promulgated the new interpretation of de facto monopoly as controlling over 90 percent of the market, regardless of proof of misuse of monopoly power.

12. For AT&T's decline, see Temin and Galambos (1987). For the history of the regulated monopoly and its rationale for controlling technology in the interests of system service and for an account of the consent decree of 1956 that did not require a breakup of the company, see "Setting the Stage" (9–19). For the denouement of AT&T's integrated monopoly, see "Reflections" (336–366). Though the authors say relatively little about the emerging technologies and the forecast for steadily growing demand through the end of the century, they do suggest that one problem AT&T had with new technology stemmed from a decentralized structure that aimed to optimize local service but diminished the ability to plan centrally, which led to failure to adapt to a changing environment in 1970–1972. Owing to regulatory misunderstanding of the problems of operating under a combination of regulation and competition, AT&T found itself under increas-

ing pressure from competition at the margins from MCI and other special service providers.

13. Graham (2000). For the significance of the "active periphery," see also Graham (2003b).

14. Corning's windshield project was a ten-year project with top management support which had consumed millions of dollars of development money and had already reached the stage of investing in new production facilities. The project was a strategic initiative designed to make Corning into a major supplier of safety windshields to the auto industry. It used malleable glass manufactured using a new Corning process called fusion glass that would later be used for display glasses. The auto companies had led Corning to believe that they were interested in malleable glass because they were under pressure from the government to adopt new safety standards for windshields. They bought test quantities, and American Motors even used small amounts for the rear windows of one of its convertible models, the Javelin. But in 1969–1970 Corning discovered that Pilkington Glass had found ways to adapt its new, much cheaper float glass process for glass thin enough for safety windshields and the auto glass companies quickly embraced this less expensive, less disruptive alternative.

15. In the 1960s Corning bore the cost of massive glass structures for naval applications, only to have the navy decide not to use them. See Graham and Shuldiner (2001) and Dyer and Gross (2001).

16. Liberatore and Titus's 1983 study of research management in twenty-nine Fortune 500 Companies found that NPV/IRR was used by 74 percent of the companies; followed by cost-benefit and Gantt charts, 62 percent; payback period, 58 percent; checklists and scoring models, 47 percent; and project network diagrams, 41 percent. One study estimated that 31 percent of its subject companies required a cost and sales forecast, 30.5 percent required discounted cash flow, 28 percent required return on investment, and 18 percent used payback or breakeven analyses. Obviously a number of the companies studied employed more than one. Cooper and Kleinschmidt (1986). These suggested some diversification of technique from a study in the 1960s that found that rate of return and payback were the two most widely used techniques for R&D project selection.

17. Cooper and Kleinschmidt (1987) Corning adopted the stage-gate process for managing innovation after its internal review of its own innovation process. Many people commented, however, that fiber, which was just beginning to pay off, would never have made it to product status if it had been subjected to the new Corning innovation process. Graham and Shuldiner (2001).

18. See Hayes and Abernathy (1980). Hayes and Abernathy, professors of operations management at the Harvard Business School, challenged the prevailing financial analysis and measurement practices, arguing that overemphasis on productivity measures was making the U.S. manufacturing system less competitive with European countries like France and Germany that did not make use of these financial analytical techniques to the same extent.

19. Only a small part of Corning's vast store of glass compositions was patented. Corning relied on trade secret protection to cover much of its technology until it was ready to make use of it. This required very tight security where its glass compositions were concerned.

20. Curiously, American Optical had been drawn into research pertaining to communications using light by the film producer Mike Todd, who wanted to improve the technology for large screen projection and convinced American Optical to do the necessary research for him, Hecht (2004).

21. Graham and Shuldiner (2001). Corning's vapor deposition process was discovered by Corning researcher Franklin Hyde in the 1930s, but Corning was slow to apply for the patent, and it issued only after World War II.

22. In 1971 Armistead moved up to the position of Corning's vice chairman in charge of technology. This gave Armistead more influence over resource allocation at the corporate level than he had had as head of the technical staffs.

23. At less than 20 decibel attenuation, repeaters would have to be installed at intervals that would make the installation cost roughly the same as existing cable installations. This was one of the conclusions that Charles Kao of ITT's Standard Communications Laboratory had arrived at in his seminal theoretical work predicting the technical and financial feasibility of optical fiber. It was this study that had launched the British Post Office on the quest for pure glass that had led it to Corning. Lacking pure glass or the ability to produce it, Kao had tested his theory on the closest clear substance he could find: ice.

24. Graham and Shuldiner (2001). Signetics came out of the deal with its royalty reduced from 4 to 2 percent of sales. Unfortunately, it was too late to do much for Signetics, which had already entered a serious slump caused by the general recession in electronics.

25. Research groups at other companies were making fibers by drawing them simultaneously from a hotdog-shaped (having an inside and outside layer) configuration of two kinds of glass. Corning drew a tube and then sputtered glass on the inside using its vapor deposition method invented by its veteran researcher Frank Hyde.

26. Ken Lipartito (2003) argues that the Bell Labs organization, arranged in specialized concentrated teams focusing on specific technologies, worked against its ability to innovate. This episode supports that view. The Corning research team was integrated—smaller and more versatile—and able to use Bell Labs's deep expertise in optical components and systems much better than its own researchers were. Even in a case where the ultimate customers were its own operating units, AT&T was much less flexible.

27. Before the effort by the Justice Department to break up AT&T's monopoly this seemed like a reasonable statement to make. A decade later, it would seem far less certain.

28. When Maurer took his fiber to be tested by the BPO, despite meticulous efforts, a tiny amount of glass fiber ended up on the floor, and subsequent analysis revealed to other research teams what Corning's material composition was.

29. AT&T by this time had secured a significant share of multimode fiber, supplying its own needs and those of other companies that were performing major field tests. Because of the deal concluded in 1969–1970, it paid Corning no royalties. Hecht (2004).

30. Dyer and Gross (2001). Amo Houghton was succeeded by his brother Jamie in 1983, leaving the Corning board later to run successfully for Congress in 1988. He served eight terms in the U.S. House of Representatives representing Corning's district in New York State. Tom MacAvoy left Corning in 1988, though he remained actively involved in Corning affairs.

References

Buderi, Robert. 2000. *Engines of Tomorrow: How the World's Best Companies Are Using Their Research Labs to Win the Future*. New York: Touchstone.

Burgelman, R. A. 1993. "A Process Model of Internal Corporate Venturing in the Diversified Major Firm." *Administrative Science Quarterly* 28, 223–244.

Carpenter, Marie, William Lazonick, and Mary O'Sullivan. 2003. "The Stockmarket and Innovative Capability in the New Economy: The Optical Networking Industry." *Industrial and Corporate Change* 12, 963–1034.

Cooper, R. G., and E. J. Kleinschmidt. 1986. "An Investigation into the New Product Process: Steps, Deficiencies, and Impact." *Journal of Product Innovation Management* 3, 71–85.

Dyer, Davis, and Daniel Gross. 2001. *The Generations of Corning*. New York: Oxford University Press.

Galambos, Louis. 1970. "The Emerging Organizational Synthesis in Modern American History." *Business History Review* 44, 279–290.

Graham, Margaret B. W. 1985. "Industrial Research in the Age of Big Science." *Research in Technological Innovation*. New York: JIT Press.

Graham, Margaret. 2000. "The Threshold of the Information Age: Radio, Television and Motion Pictures Mobilize the Nation." In Alfred D. Chandler and James Cortada, eds., *A Nation Transformed by Information: How Information Reshaped the United States from Colonial Times to the Present* (137–175). New York: Oxford University Press.

Graham, Margaret. 2003a. "Corning as Creative Responder." McGill Innovation Working Paper, December.

Graham, Margaret. 2003b. "From Satellite Laboratory to Partner." Paper presented at Business History Conference, Lowell, Mass., June.

Graham, Margaret, and Alec T. Shuldiner. 2001. *Corning and the Craft of Innovation*. New York: Oxford University Press.

Hayes, Robert, and William Abernathy. 1980. "Managing Our Way to Economic Decline." *Harvard Business Review*, # 80405.

Hecht, Jeff. 2004. *City of Light, The Story of Fiber Optics.* Rev. ed. New York: Oxford University Press.

Hounshell, David. 1996. "The Evolution of Industrial Research in the United States." In Richard Rosenbloom and William J. Spencer, eds., *Engines of Innovation: U.S. Industrial Research at the End of an Era.* Boston: Harvard Business School Press.

Johnson, Thomas H., and Robert S. Kaplan. 1991. *Relevance Lost: The Rise and Fall of Management Accounting.* Boston: Harvard Business School Press.

Kenney, Martin, and Richard Florida. 1988. *The Breakthrough Illusion: Corporate America's Failure to Move from Innovation to Mass Production.* New York: Basic Books.

Lamoreaux, Naomi R., Daniel M. G. Raff, and Peter Temin. 2003. "Beyond Markets and Hierarchies: Toward a New Synthesis of American Business History." *American Historical Review* 108, 404–433.

Liberatore, M. J., and G. J. Titus. 1983. "The Practice of Management Science in R&D Project Management." *Management Science* 29, 962–974.

Lipartito, Ken. 2002. "Bell Labs in Perspective." Working paper presented to Innovation Conference, Johns Hopkins University, October.

Lipartito, Kenneth. 2003. "Picture Phone and the Information Age: The Social Meaning of Failure." *Technology and Culture* 44, 50–81.

Morone, Joseph G. 1993. *Winning in High-Tech Markets: The Role of General Management.* Boston: Harvard Business School Press.

Nohria, Nitin, Davis Dyer, and Fred Dalzell. 2002. *Changing Fortunes: Remaking the Industrial Corporation.* New York: Wiley.

Reich, Leonard. 1985. *The Making of American Industrial Research.* Cambridge: Cambridge University Press.

Rosenberg, Nathan. 1994. "Joseph Schumpeter Radical Economist." *Inside the Black Box.* New York: Oxford University Press.

Schall, L. D., G. L. Sundem, and W. R. Geijsbeek. 1978. "Survey and Analysis of Capital Budgeting Methods." *Journal of Finance* 33, 281–287.

Swedberg, Richard, ed. 1991. *Joseph A. Schumpeter: The Economics and Sociology of Capitalism.* Princeton, NJ: Princeton University Press.

Teece, David J. 2000. *Managing Intellectual Capital: Organizational, Strategic and Policy Dimensions.* New York: Oxford University Press.

Temin, Peter, and Louis Galambos. 1987. *The Fall of the Bell System: A Study in Prices and Politics.* Cambridge: Cambridge University Press.

7

The Federal Role in Financing Major Innovations: Information Technology during the Postwar Period

Kira R. Fabrizio and David C. Mowery

No summary of major postwar innovations in U.S. economic history can ignore the information technology (IT) sector. Advances in electronics and related technologies created three new industries— electronic computers, computer software, and semiconductor components—and these three industries combined to give birth to the Internet, a general-purpose technology spanning many industrial sectors. Technological change supported the growth of new firms in these industries and revolutionized the operations and technologies of more mature industries, such as telecom munications, banking, and airline and railway transportation.

In considering the sources of finance for these innovations in IT, it is impossible to overlook the role of the federal government. Indeed, the federal government accounted for a substantial share of overall U.S. national R&D spending for much of the postwar period. Beginning in 1953, the first year for which reliable historical data are available, federal sources accounted for more than 50 percent of total national R&D investment, a share that increased to nearly 67 percent by 1964. Federal sources accounted for more than 50 percent of total national R&D spending through 1978 and dropped below 40 percent only in 1991 (figure 7.1). Through most of the 1953–2005 period, more than 50 percent of this federal R&D budget was devoted to defense purposes.

The IT sector, which scarcely existed in 1945, was a key focus of federal R&D and defense-related procurement spending for much of the postwar period. Moreover, the structure of these federal R&D and procurement programs exerted a powerful influence on the pace of development of the underlying technologies and the structure of the industries that developed these technologies for defense and civilian applications.

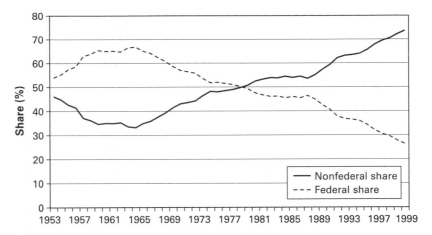

Figure 7.1
Federal and Nonfederal Shares of national R&D Spending, 1953–2000
Source: National Science Foundation, Division of Science Resources Studies.
2001. *National Patterns of R&D Resources: 2000 Data Update*. Available at
⟨http://www.nsf.gov/sbe/srs/nsf01309/start.htm⟩.

Indeed, the structures of the U.S. semiconductor, computer hardware,
and computer software industries all differ in varying degrees from those
of their counterparts in other industrial economies. Along with other
scholars, we believe that the scale and structure of federal R&D and procurement programs in the United States are responsible in part for these
structural contrasts.

This chapter reviews the history of federal R&D and procurement
support for the development of the key industries in the IT sector. We include procurement with R&D programs for two reasons: (1) in many
cases, the most significant effects of federal spending on industry structure and technology diffusion resulted from procurement, rather than
R&D programs, and (2) a discussion of the sources and financing of innovation cannot divorce technology development from technology diffusion through the progressive expansion of markets for innovations. New
technologies undergo a prolonged period of debugging, performance and
reliability improvement, cost reduction, and learning on the part of users
and producers about applications and maintenance (Mowery and
Rosenberg 1998). The pace and pattern of such progressive improvement affect the rate of adoption, and the rate of adoption in turn affects

the refinement and improvement of these innovations. Government procurement allowed companies in these industries to benefit from learning by increasing the scale of production for early stages of the technology. As these industries matured, however, and as their civilian markets expanded to outstrip defense and other government markets, the influence of federal R&D and procurement declined. Nonetheless, these federal policies were critical in establishing the initial conditions for industry evolution that influenced the subsequent development of firm strategies and industry structure.

Our discussion of the federal role in financing innovations in the IT industry is essentially descriptive. Credible tests of the obvious counterfactual, What would have happened without federal funding? are virtually impossible. But this descriptive analysis highlights the limitations of the conventional market failure justification for public funding of R&D. Although the market failure framework is applicable to these industries, it omits many of the important channels through which public funding of R&D in IT affected industry growth, dynamism, and market structure.

The Economics of Public R&D Investment

The classic analyses of market failure in innovation are Arrow (1971) and Nelson (1959), both of whom argued that government funding of research and development was especially important in technologies where private companies would not allocate sufficient resources to innovation. This is particularly true when the results of R&D investments are characterized by indivisibility, limited appropriability, and uncertainty (Arrow 1971), all of which discourage private firms from investing in R&D. These conditions often characterize industries in early stages of development, when new technologies differ radically from existing technologies or when the technology in an industry requires considerable investment in basic research, results from which are uncertain and often produce economic returns that cannot be captured by the company performing the research (Nelson 1959). Government funding thus is necessary for the fundamental research that often is essential to the development of innovations, and the absence of public funding may slow or prevent the development of such innovations. Arrow's original analysis pointed out that intellectual property rights protection may improve the appropriability of the returns from innovation, but the associated temporary monopolies created by patents introduce significant costs. Indeed, Arrow argued that direct funding of such projects was preferable to private financing because of the welfare costs of the temporary monopolies created by patent protection.

Federal financing of IT-related R&D in the postwar United States supported longer-term research on a broader range of technological alternatives and approaches in this nascent field than private funding alone could have done. This body of basic science became an important input to many IT-sector innovations. But the effects of federal financing of innovation in IT went well beyond amelioration of the market failures highlighted by Arrow and Nelson. The IT sector was also a major beneficiary of large flows of private investment during the 1960s, 1970s, 1980s, and 1990s, and it is arguable that in the absence of federal support, private sources would have supported some level of R&D and innovation in this sector. Moreover, in some important instances (e.g., the invention of the transistor and integrated circuit), private rather than public sources supported the R&D that led to the key technical breakthroughs. But the timing of commercialization, the speed of adoption of the technologies resulting from these processes, and the structure of the industries that emerged to commercialize these innovations all would have been very different. Federal funding supported the exploration of a wider range of technological alternatives than exclusive reliance on private funding would have, largely as a result of the large scale of these public investments.

One of the most important long-term consequences of federal financing of innovation in IT was the creation of a relatively weak intellectual property rights environment and, in some cases, the direct encouragement of high levels of interfirm technology diffusion by federal agencies funding R&D or procurement. Federal funding for procurement of the products of these new industries also encouraged the entry of new firms and interfirm technology diffusion. In addition, federal procurement supported the rapid attainment by supplier firms of relatively large production runs, enabling faster rates of improvement in product quality and cost than otherwise would have been realized. Finally, federal support for innovation in IT contributed to the creation of a large-scale R&D infrastructure in federal laboratories and, especially, in U.S. universities, which became important and highly productive sources of innovations. Indeed, federal support for this R&D infrastructure led to the early deployment of the large-scale computer networks in the United States, which led to a wave of innovation, much of it driven by users of these networks, that produced the Internet.

These effects of federal funding for innovation in IT are at best only hinted at in the classic market failure justification for federal support. And we do not claim that the positive results of federal investments in IT can or will be reproduced in other fields. At least some of the catalytic effects of federal support for innovation in IT were enhanced by the general-purpose characteristics of information technology, the enormous capacity of this technology for rapid improvement in price-performance ratios, and the tendency for these reductions in the price-performance ratio to accelerate adoption in a widening array of applications. In other respects, the history of federal support for innovation in IT is a history of a specific chapter in U.S. political and economic development dominated by the cold war.[1] Moreover, as we noted earlier, these federal policies exercised considerable influence over industry development precisely because they were present at the creation of the industry; indeed, they accelerated the creation of these industries. It is unlikely that comparable intervention in other industries at later stages of development of IT would have such a powerful long-term influence.

Federal Funding and IT Innovation

Federal policy played a central role in the development of all four of the technologies that we are including in our definition of the IT sector. The military applications of semiconductors and computers meant that defense-related R&D funding and procurement were important to the early development of these industries. The R&D infrastructure created in U.S. universities by defense-related and other federal R&D expenditures made significant contributions to technical developments in semiconductors, computer hardware, and computer software. The Internet itself emerged from federal programs, largely motivated by national defense, that developed a national network linking the far-flung components of the academic and industrial R&D infrastructure created with federal funds.

In all of these technologies, the direct influence of federal R&D and procurement policies was strongest in the early years of their development, when federal expenditures on R&D or procurement accounted for the majority of such funding. These federal programs in turn focused primarily on the development of defense-related applications. The semiconductor, computer hardware and software, and Internet industries now encompass many markets and applications beyond national defense, which now accounts for a much smaller share of demand or applications in all of these industries. Indeed, the technological spillovers that once flowed from defense-related technologies to civil applications now frequently move in the opposite direction, and the ability of Defense Department policymakers to influence the direction of technological change has diminished considerably. Nonetheless, the substantial role of federal support programs in the earliest stages of development of many of these industries means that the influence of these programs on intellectual property policies, interfirm technology flows, entry, and overall industrial structure remains significant today.

The electronics revolution that spawned the semiconductor and computer industries, as well as the Internet, can be traced to two key innovations: the transistor and the computer. Both appeared in the late 1940s, and the exploitation of both was spurred by cold war concerns over national security. The creation of these innovations also relied on domestic U.S. science and invention to a greater extent than many important U.S. innovations of the pre-1940 era. The following sections briefly survey the development of each of these four technologies, describing key aspects of their industrial and technological evolution and highlighting the role of the federal government in financing major innovations.

Semiconductors

The transistor, invented at Bell Telephone Laboratories in late 1947, marked one of the first tangible payoffs to an ambitious program of basic research in solid-state physics that Mervin Kelly, Bell Labs's director, had launched in the 1930s. Facing increasing demands for long-distance telephone service, AT&T sought a substitute for the repeaters and relays that would otherwise have to be employed in huge numbers, greatly increasing the complexity of network maintenance and reducing reliability. Kelly felt that basic research in the emergent field of solid-state physics might yield technologies for this purpose.[2]

The transistor had important potential military applications in military electronics and computer systems, but the inventing firm, AT&T, was not producing it in commercial quantities. Considerable process R&D and trial-and-error experimentation were needed, and by 1953, the U.S. Defense Department was funding pilot transistor production lines operated by AT&T, General Electric, Raytheon, Sylvania, and RCA (Tilton 1971). As figure 7.2 shows, federal development contracts with these and other industrial firms were initially dominated by production engineering, but by 1959, R&D spending accounted for more than 80 percent of federal support for semiconductor-related technology development within these firms. According to Tilton (1971), federally supported R&D accounted for nearly 25 percent of total industry R&D spending in the late 1950s.

Interestingly, the bulk of this federal R&D spending during the 1950s was allocated to established producers of electronic components, includ-

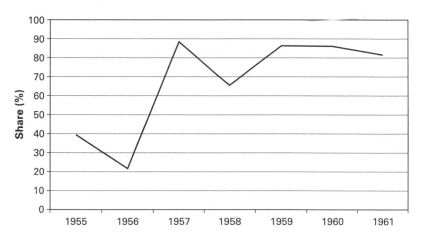

Figure 7.2
R&D Share of Federal Semiconductor Development Contracts to Firms, 1955–1961
Source: Tilton (1971).

290 Kira R. Fabrizio and David C. Mowery

ing those listed in the preceding paragraph, during the 1950s. Indeed, Tilton (1971) shows that new firms, including Texas Instruments, Shockley Laboratories, Transitron, and Fairchild, received only 22 percent of federal R&D contracts in 1959, although these firms accounted for 63 percent of semiconductor sales in that year. The major corporate recipients of military R&D contracts were not among the pioneers in the introduction of innovations in semiconductor technology, while the pioneering firms did so without military R&D contracts (Kleiman 1966). Defense procurement contracts proved to be at least as important as public funding of industry R&D in shaping this nascent industry.

The first commercially successful transistor was produced by Texas Instruments, rather than AT&T, in 1954.[3] Texas Instruments' silicon junction transistor was quickly adopted by the U.S. military for use in radar and missile systems. The next major advance in semiconductor electronics, in 1958, was the integrated circuit, which combined a number of transistors on a single silicon chip. The integrated circuit (IC), invented by Jack Kilby of Texas Instruments, drew on TI's innovations in diffusion and oxide masking technologies that had initially been developed for the manufacture of silicon junction transistors.

Kilby's search for the IC was motivated by the desirability of a device that could expand the military (and, eventually, the commercial) market for semiconductor devices. Little of Kilby's pathbreaking R&D was supported by the U.S. military, but defense-related procurement dominated TI's early shipments of integrated circuits. Figure 7.3 demonstrates the significant share of IC shipments accounted for by government purchases, as well as the decline in this share as commercial markets for the IC grew. A longer time series for the government share of semiconductor shipments (figure 7.4) shows the importance of government procurement in the early years of the broader semiconductor industry, as well as the decline in the share of demand represented by federal procurement after the 1960s. By the 1990s, military demand accounted for less than 10 percent of integrated circuit sales (figure 7.5).

One result of the substantial presence of the federal government in the early postwar semiconductor industry as a funder of both R&D and procurement was the emergence of an industry structure that contrasted with those of pre-1940 technology-intensive U.S. industries, such as

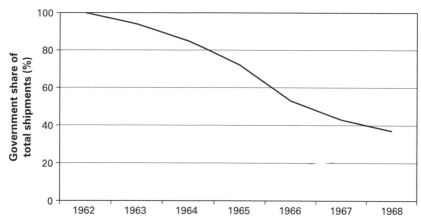

Figure 7.3
Government Purchases of Integrated Circuits as a Percent of Total Shipments
Note: Includes circuits produced for Department of Defense, Atomic Energy
Commission, Central Intelligence Agency, Federal Aviation Agency, and Na-
tional Aeronautics and Space Administration.
Source: Levin (1982).

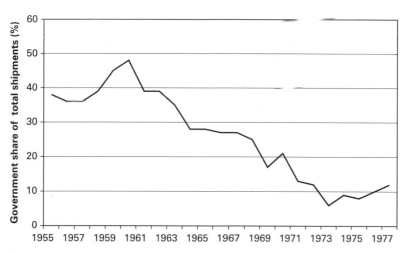

Figure 7.4
Government Purchases of Semiconductor Devices as Share of Total Shipments
Note: Includes devices produced for Department of Defense, Atomic Energy
Commission, Central Intelligence Agency, Federal Aviation Agency, and Na-
tional Aeronautics and Space Administration equipment.
Source: Levin (1982).

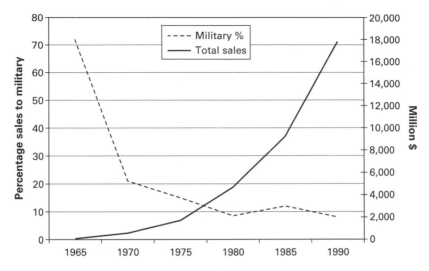

Figure 7.5
Total and Military Share of U.S. Integrated Circuit Sales
Source: Alic, Branscomb, Brooks, Cater, and Epstein (1992).

chemicals or electrical machinery, as well as the postwar semiconductor industries of such nations as Germany and Japan. In a virtual reversal of the prewar situation, the R&D facilities of large U.S. firms provided many of the basic technological advances that new firms commercialized. Entrants' role in the introduction of new products, reflected in their often dominant share of markets in new semiconductor devices, significantly outstripped that of established firms. Moreover, the role of new firms grew in importance with the development of the integrated circuit.

Although the military market for ICs was rapidly overtaken by commercial demand, military demand spurred output growth and price reductions that expanded commercial demand for ICs. The large volume of ICs produced for the military market allowed firms to move down firm- and product-specific learning curves, reducing component costs and expanding commercial applications.[4] Table 7.1 traces the growth between 1962 and 1978 in total shipments of ICs and the concurrent growth in shares of shipments to industrial and commercial uses.

Military procurement policies also influenced industry structure by promoting competition and the intraindustry diffusion of technological

Table 7.1
End use shares of total U.S. sales of integrated circuits and total market value
1962–1978

Markets	1962	1965	1969	1974	1978
Government	100%	55%	36%	20%	10%
Computer	0	35	44	36	38
Industrial	0	9	16	30	38
Consumer	0	1	4	15	15
Total U.S. domestic shipments (millions)	$4	$79	$413	$1,204	$2,080

Note: Total percentages may not equal 100 because of rounding.
Source: Langlois and Steinmueller (1999).

knowledge. In contrast to Western European defense ministries, the U.S. military awarded substantial procurement contracts to new entrants such as Texas Instruments with little or no history of supplying the military.[5] The U.S. military's willingness to purchase from untried suppliers was accompanied by requirements that mandated substantial technology transfer and exchange among U.S. semiconductor firms. To reduce the risk that a system designed around a particular IC would be delayed by production problems or the exit of a supplier, the military required its suppliers to develop a second source for the product, that is, a domestic producer that could manufacture an electronically and functionally identical product. Compliance with second-source requirements meant that firms had to exchange design and sufficient process knowledge to ensure that the component produced by a second source was identical to the original product.

By facilitating entry and supporting high levels of technology spillovers among firms, defense-related procurement policy (along with other federal policies, such as the 1956 AT&T consent decree) increased the diversity of technological alternatives explored by individuals and firms within the U.S. semiconductor industry during a period of significant uncertainty about the direction of future development of this technology. Extensive entry and interfirm technology diffusion also intensified competition among U.S. firms. This highly competitive industry structure enforced a rigorous selection environment, weeding out less effective firms and technical solutions. For a nation that was pioneering in the semiconductor industry, this combination of technological diversity and strong selection pressures proved to be highly effective.

As nondefense demand for semiconductor components grew and came to dominate industry demand, defense-civilian technology spillovers declined in significance and actually reversed in direction. By the 1970s, military specification semiconductor components often lagged behind their commercial counterparts in technical performance, although these "milspec" components could operate in much more hostile environments of high temperatures or vibration. Nonetheless, concern among U.S. defense policymakers over this technology gap grew and resulted in the creation of the Department of Defense (DOD) Very High Speed Integrated Circuit program (VHSIC) in 1980. Federally funded VHSIC projects linked merchant semiconductor firms largely devoted to commercial production, semiconductor equipment manufacturers, and defense systems houses in development projects intended to produce advanced, high-speed milspec components.

Originally planned for a six-year period and budgeted at slightly more than $200 million, the VHSIC program lasted for ten years and spent nearly $900 million. Nonetheless, the program failed to meet its objectives, demonstrating the limited influence of the federal government within a U.S. semiconductor market that by the 1980s was dominated by commercial applications and products. Rather than seeking to redirect the course of innovation within the U.S. semiconductor industry, defense policymakers now seek to change procurement policies to enable more rapid incorporation of commercial innovations (Alic et al. 1992).[6]

Computers

The U.S. computer industry also benefited from cold war military spending, but in other respects, the origins and early development of this industry differed from semiconductors. Although they were at best peripheral actors in the early development of semiconductor technology, U.S. universities were important sites for the government-funded development and research activities that produced the earliest U.S. computers. In addition, federal spending during the late 1950s and 1960s, which was funded largely from military sources, provided an important basic research and educational infrastructure for the development of this new industry.

During the war years, the American military sponsored a number of projects to develop high-speed calculators. Many of the first-generation computer projects were funded by the U.S. military (See table 7.2). The ENIAC, generally considered the first fully electronic digital computer, was funded by Army Ordnance, which was concerned with the computation of firing tables for artillery. Developed by J. Presper Eckert and John W. Mauchly at the Moore School of the University of Pennsylvania, the ENIAC did not rely on software but was hard-wired to perform each set of calculations. In 1944, John von Neumann began advising the Eckert-Mauchly team, which was working on the development of a new machine, the EDVAC. This collaboration developed the concept of the stored-program computer: instead of being hard-wired, the EDVAC's instructions were stored in memory, facilitating their modification.

From the earliest days of their support for the development of computer technology, the U.S. armed forces supported wide diffusion of technical information on this innovation. This attitude, which contrasted with that of the military in Great Britain or the Soviet Union, appears to have stemmed from the U.S. military's concern that a substantial industry and research infrastructure would be required for the development and exploitation of computer technology.[7] The technical plans for the military-sponsored IAS computer developed by von Neumann at Princeton's Institute for Advanced Study were widely circulated among U.S. government and academic research institutes and spawned a number of clones (e.g., the ILLIAC, the MANIAC, AVIDAC, ORACLE, and JOHNIAC; see Flamm 1988).[8] Public funding supported research on many problems that might not have been supported from private sources, consistent with the market-failure analysis discussed above, but equally important was the relatively liberal dissemination of the results of this publicly supported research.

By 1954, the ranks of the largest U.S. computer manufacturers were dominated by established firms in the office equipment and consumer electronics industries. The group included RCA, Sperry Rand (originally the typewriter producer Remington Rand, which acquired Eckert and Mauchly's embryonic computer firm), and International Business

Table 7.2
Early U.S. support for computers

First generation of U.S. computer projects	Estimated cost of each machine (thousands)	Source of funding	Initial operation
ENIAC	$750	Army	1945
Harvard Mark II	840	Navy	1947
Eckert-Mauchly BINAC	178	Air force (Northrop)	1949
Harvard Mark III	1,160	Navy	1949
NBS Interim computer (SEAC)	188[a]	Air force	1950
ERA 1101 (Atlas I)	500	Navy/NSA[b]	1950
Eckert-Mauchly UNIVAC	400–500	Army via census; air force	1951
MIT Whirlwind	4,000–5,000	Navy; air force	1951
Princeton IAS computer	650[a]	Army; navy; RCA; AEC	1951
University of California CALDIC	95[a]	Navy	1951
Harvard Mark IV	NA	Air force	1951
EDVAC	467	Army	1952
Raytheon Hurricane (RAYDAC)	460[a]	Navy	1952
ORDVAC	600	Army	1952
NBS/UCLA Zephyr computer (SWAC)	400	Navy; air force	1952
ERA Logistics computer	350–650	Navy	1953
ERA 1102 (3 built)	1,400[c]	Air force	1953
ERA 1103 (Atlas II, 20 built)	895	Navy/NSA	1953
IBM Naval Ordnance Research Computer (NORC)	2,500	Navy	1955

Source: Flamm (1988).
[a] Estimated cost in 1950 in "Report on Electronic Digital Computers by the Consultants to the Chairmand of the Research and Development Board," June 15, 1950, app. 4, cited by Redmond and Smith (1980, 166).
[b] The National Security Agency (NSA) includes army and navy predecessor agencies.
[c] Cost for three machines.

Machines, as well as Bendix Aviation, which had acquired the computer operations of Northrop Aircraft. These firms' early computers were sold primarily to federal government agencies, particularly the defense and intelligence agencies. The National Security Agency, the Atomic Energy Commission, and the Defense Department all supported the development of advanced computer systems for specialized applications in air defense, cryptography, and nuclear weapons design.

IBM's technology development efforts benefited from the firm's experience as supplier of more than fifty large computers for the SAGE air defense network that was developed under the supervision of MIT's Lincoln Laboratories in the 1950s, and the firm also was awarded a contract by the Atomic Energy Commission for an advanced computer (referred to as the Stretch project) for use by Los Alamos National Laboratories. Other U.S. computer firms, including Sperry Rand and ERA, produced advanced computers in small quantities for federal intelligence and defense agencies during the 1950s. According to Flamm (1987), federal funds accounted for 59 percent of the combined computer-related R&D spending of General Electric, IBM, Sperry Rand, AT&T, Raytheon, RCA, and Computer Control Corporation between 1949 and 1959.

Business demand for computers gradually expanded during the early 1950s, and the most commercially successful machine of the decade, with sales of 1800 units, was the low-priced IBM 650 (Fisher, McKie, and Mancke 1983). Even in the case of the 650, however, government procurement was crucial: the projected sale of 50 machines to the federal government (a substantial portion of total projected sales of 250 machines) influenced IBM's decision to initiate the project (Flamm 1988). Sales to the government made up a substantial portion of IBM sales during the 1950s, but declined through the following two decades as private sector sales grew (see figure 7.6). The ability of the major producers to use extensive federal R&D funding to penetrate commercial markets varied. IBM benefited from these programs, but also successfully exploited its lengthy experience in business equipment in designing and manufacturing both the computers and the peripherals that were so important to business computers.[9]

Even after the emergence of a substantial private industry dedicated to the development and manufacture of computer hardware, federal R&D

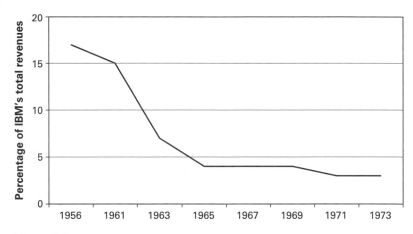

Figure 7.6
IBM Sales of Special Products and Services to U.S. Government Agencies
Source: Flamm (1978, 108).

spending aided the creation of the new academic discipline of computer science. Universities were important sites for applied as well as basic research in hardware and software, and contributed to the development of new hardware and networking technologies. The training by universities of engineers and scientists active in the computer industry also was extremely important. U.S. universities provided important channels for cross-fertilization and information exchange between industry and academia, but also between defense and civilian research efforts in software and in computer science generally.

In 1963, about half of the $97 million spent by universities on computer equipment came from the federal government, while the universities themselves paid for 34 percent and computer makers picked up the remaining 16 percent (Fisher, McKie, and Mancke 1983). Federal monies for computer-related research accounted for a significant portion of total research expenditures in industry and academia through the 1980s (see figure 7.7). During the 1970s and 1980s, roughly 75 percent of the mathematics and computer science research performed at universities was funded by the federal government (Flamm 1987). According to a recent report from the National Research Council's Computer Science and Telecommunications Board, federal investments in computer science

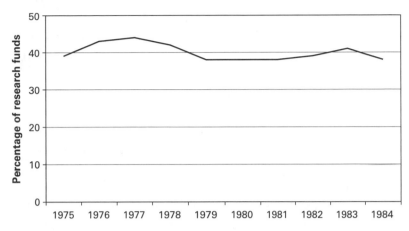

Figure 7.7
Federal Math and Computer Science Funds as a Percent of All Computer-Related
Research Funds
Source: Flamm (1987).

research increased fivefold during the 1976–1995 period, from $180 million in 1976 to $960 million in 1995 in constant (1995) dollars. Federally funded basic research in computer science, roughly 70 percent of which was performed in U.S. universities, grew from $65 million in 1976 to $265 million in 1995 (National Research Council 1999a). Defense-related R&D spending in software appears to have declined somewhat in the 1980s, even as civilian agencies such as the National Science Foundation increased their computer science research budgets. The defense share of federal computer science research funding declined from almost 60 percent in fiscal 1986 to less than 30 percent in fiscal 1990 (Clement 1987, 1989; Clement and Edgar 1988), and defense funding of computer science research in universities appears to have been supplanted somewhat by the growth in funding for quasi-academic research and training organizations.

As was true of semiconductors, the relationship between federally supported innovation and civilian innovation in computers changed as the U.S. computer industry matured. In both the hardware and software areas, the government's needs differed from those of the commercial sector, and the magnitude of technological spillovers from military R&D and procurement to civilian applications appears to have declined as the computer industry moved into the 1960s. Just as had been the case in semiconductors, however, military procurement demand attracted new firms to enter the industry, and many such enterprises entered the fledgling U.S. computer industry in the late 1950s and 1960s.

The Computer Software Industry

By the 1980s, the development of the semiconductor and computer industries had laid the groundwork for the expansion of another new postwar industry, the production of standardized computer software for commercial markets (as opposed to the commercial production of custom software or user-developed custom software). The growth of the U.S. computer software industry has been marked by at least four distinct eras. During the early years of the first era (1945–1965), covering the development and early commercialization of the computer, software as it is currently known did not exist. The concept of computer software as a distinguishable component of a computer system was effectively born with the advent of the von Neumann architecture for stored-program computers. But the development of a U.S. software industry began only when computers appeared in significant numbers. The large commercial market for computers that was created by the IBM 650 provided strong incentives for industry to develop standard software for this architecture.

Along with the development by IBM and other major hardware producers of standard languages such as FORTRAN, widespread adoption of a single platform contributed to substantial growth of internal software production by large users. But the primary suppliers of the software and services for mainframe computers well into the 1960s were the manufacturers of these machines. In the case of IBM, which leased many of its machines, the costs of software and services were bundled with the lease payments. By the late 1950s, however, a number of independent firms had entered the custom software industry. These firms included the Computer Usage Company and Computer Sciences Corporation, both founded by former IBM employees. In the late 1950s, the Computer Usage Company secured contracts with NASA, and the following year, it went public. Other start-ups followed during the early 1960s (Campbell-Kelly 1995, 2003).

Procurement of products and services by the federal government was an important factor in the early development of the software industry. As we previously noted, IBM was the primary supplier of computers for the SAGE air defense project, but the RAND Corporation was the contractor responsible for the bulk of the huge amount of software required for SAGE. RAND in turn created the Software Development Division to produce the software. This division separated from RAND, forming the Systems Development Corporation (SDC), in 1956. Since large-scale software development projects of this sort were well beyond the technological or scientific frontier of academic computer science (a discipline that itself scarcely existed in the early 1950s), the SAGE software development project acted as a "university" of sorts for hundreds of software programmers, laying the foundations for the software industry's future development within the United States (Campbell-Kelly 1995). In part to facilitate this role as a training ground and in part because SDC was restricted by air force pay scales, the company encouraged turnover of employees, and these "SAGE alumni" in turn contributed to the development of the broader software industry (Langlois and Mowery 1996).[10]

In the late 1950s and early 1960s, defense contractors, including TRW, MITRE Corporation, and Hughes, began to produce large-scale systems software for military applications under federal contracts. IBM and other mainframe computer manufacturers also produced large one-of-a-kind software applications for customers and became important suppliers in the software-contracting industry. Much of the software-related know-how developed from defense contracts, and the *Apollo* manned space flight program spilled over to commercial applications. For example, IBM's collaboration with American Airlines to develop the SABRE reservation system drew on IBM's background from the SAGE development (Campbell-Kelly 1995).

Federal procurement programs influenced the evolution of specific programming languages as well. A Department of Defense effort to establish a standard programming language resulted in the widely used common business-oriented language, COBOL. The DOD required that general-purpose computers purchased by the military support COBOL and that any business-related applications for defense programs be written in the language. Since the DOD accounted for such a large share of the market for custom software, its procurement requirements facilitated the development and diffusion of COBOL (Flamm 1987).

The second era of the software industry's development (1965–1978) witnessed the first entry of independent producers of standard software. Although independent suppliers of software began to enter in significant numbers in the early 1970s in the United States, computer manufacturers and users remained important sources of both custom and standard software during this period. Some consulting firms and other suppliers that had provided users with operating services and programming solutions began to unbundle their services from their software, providing yet another cohort of entrants into the independent development and sale of traded software. Sophisticated users of computer systems, especially users of mainframe computers, also developed expertise in the creation of solutions to their applications and operating system needs. A number of leading U.S. suppliers of traded software were founded by computer specialists formerly employed by major mainframe users and by the Systems Development Corporation.

During the third era of development of the software industry (1978–1993), the appearance of the desktop computer produced explosive growth in packaged software markets. Once again, the United States was the first mover in this transformation, and the U.S. market quickly emerged as the largest single one for such packaged software. Rapid adoption of the desktop computer in the United States supported the early emergence of a few dominant designs in desktop computer architecture, creating the first mass market for packaged software. The independent software vendors (ISVs) that entered during this period were largely new to the industry. Few of the major suppliers of desktop software came from the ranks of the leading independent producers of mainframe and minicomputer software, and mainframe and minicomputer ISVs are still minor factors in desktop software.

The large size of the U.S. packaged software market, as well as the fact that the United States was the first large market to experience rapid growth (reflecting the earlier appearance and rapid diffusion of mainframe and minicomputers, followed by the explosive growth of desktop computer use during the 1980s), gave the U.S. firms that pioneered in the domestic packaged software market a formidable first-mover advantage that was exploited internationally. U.S. firms' market shares in their home market exceed 80 percent in most classes of packaged software and exceed 65 percent in non-U.S. markets for all but applications software.[11]

Much of the rapid growth in custom software firms during the 1969–1980 period reflected expansion in federal demand, which was dominated by DOD demand. But like the semiconductor industry, defense markets gradually were outstripped by commercial markets, although this trend occurred more gradually in software than in hardware or semiconductors. There exists no reliable time series of DOD expenditures on software procurement that employs a consistent definition of software (e.g., separating embedded software from custom applications or operating systems and packaged software). Nevertheless, the available, imperfect data suggest that in constant dollar terms, DOD expenditures on software increased more than thirtyfold during the 1964–1990 period (see Langlois and Mowery 1996, and figure 7.8). Throughout this period, DOD software demand was dominated by custom software, and DOD and federal government markets for custom software accounted for a substantial share of the total revenues in this segment of the U.S. software industry. By the early 1990s, however, defense demand accounted for a declining share of the U.S. software industry's revenues.

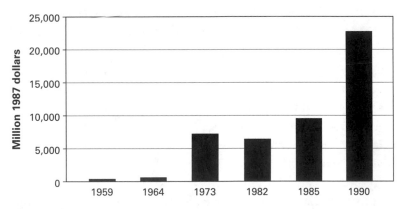

Figure 7.8
Department of Defense Software Procurement, 1959–1990
Source: Langlois and Mowery (1996, 69).

Its declining share of total demand meant that the defense market no longer exerted sufficient influence on the path of R&D and product development to benefit from generic academic research and product development; defense and commercial needs had diverged. The tangled history of the DOD's generic software language, Ada, unveiled in 1984, illustrates the declining influence of federal procurement on the rapidly growing software industry. Billed as a solution to the severe problems of system maintenance and software development resulting from the bewildering variety of software languages in use within defense systems, Ada was designed to be employed in all defense applications. By standardizing all DOD programs around a single language, Ada proponents argued, the commercial developers that no longer were interested in serving the military market would be motivated to produce software that could be used in both civilian and military applications. But these aspirations were largely unrealized. Partly because of the huge difficulties associated with inserting Ada into the enormous installed base of defense-related software, the language failed to attract the attention of commercial developers. The contrast between the failure of this DOD-supported language to take hold and COBOL's rapid diffusion into military and commercial applications underscores the points made earlier concerning the tendency for the influence of defense-related R&D and procurement demand to decline as commercial markets expand.

The fourth era in the development of the software industry (1994–present) has been dominated by the growth of networking among desktop computers, both within enterprises through local area networks linked to a server and among millions of users through the Internet. Networking has opened opportunities for the emergence of new software market segments (e.g., the operating system software that is currently installed in desktop computers may reside on the network or the server), the emergence of new dominant designs, and, potentially, the erosion of currently dominant software firms' positions. Although the Internet was in most respects created by public R&D funding, private spending on R&D and commercial applications has dominated the path of development of the U.S. software industry during this fourth Internet-driven era. Like the previous eras of this industry's development, the growth of network users and applications was more rapid in the United States than in other industrial economies during the 1990s, and U.S. firms have maintained dominant positions in these markets.

The Internet

The Internet was developed and commercialized first in the United States, although scientists and engineering in other industrial economies, especially France and the United Kingdom, made important contributions to computer networking technologies during the 1970s, and the key advances behind the creation of the World Wide Web were invented at CERN, the European nuclear physics research facility. Nonetheless, U.S. entrepreneurs and firms led the transformation of these inventions into components of a national and global network of networks and were early adopters of new applications.

Federal agencies such as the DOD and National Science Foundation played a critical role in funding the development and diffusion of early versions of the technology. Federal spending on R&D and procurement was complemented by the R&D investments of large corporations and the many start-ups that populated Internet-related industries. These small firms often drew on expertise developed in U.S. research universities or in large corporations and benefited from the regulatory and antitrust policies of federal agencies such as the Federal Communications Commission and the Justice Department. But the explosive adoption and commercial exploitation of the Internet during the 1990s built on a foundation of computer-networking R&D and investment, much of which was from federal sources, and experience in the use of this networking infrastructure, that had developed during the previous thirty years.

During the early 1960s, several researchers, including Leonard Kleinrock at MIT and Paul Baran of RAND, developed various aspects of the theory of packet switching.[12] The theoretical work and early experiments of Baran, Kleinrock, and others led the DOD's Advanced Research Projects Agency (DARPA) to fund the construction of a prototype network. In December 1968, DARPA granted a contract to the Cambridge, Massachusetts–based engineering firm of Bolt, Beranek and Newman (BBN) to build the first packet switch.[13] The switch, called an interface message processor (IMP), linked computers at several major computing facilities over what is now called a wide-area network. A computer with a dedicated connection to this network was referred to as a host. The resulting ARPANET is widely recognized as the earliest forerunner of the Internet (National Research Council 1999a). By 1975, as universities and other major defense research sites were linked to the network, ARPANET had grown to more than 100 nodes.

In 1974, two DARPA-funded engineers, Robert Kahn and Vinton Cerf, published the first version of the TCP/IP protocol suite.[14] The new data networking protocol allowed physically distinct networks to interconnect with one another as "peers" and exchange packets through special hardware called a gateway. By publishing TCP/IP and placing the standard within the public domain, Kahn, Cerf, and their collaborators strengthened its position in a two-decade-long competition among a variety of networking protocols, including proprietary standards such as IBM's SNA and Digital Equipment's DECNET, open alternatives such as Datagram (UDP) networking and the Unix to Unix Copy protocol (UUCP), and standards supported by established telecommunications firms, such as X.25 and Open Standard Interconnect (OSI). These other networking protocols were largely developed with private funds, although they benefited from the R&D infrastructure in universities and laboratories that was supported in large part by federal funds.

TCP/IP's origins in a federally funded research project (and its development at a time during which the results of federally funded research rarely were patented) were crucial to the eventual victory of this open, nonproprietary standard as the foundation for the architecture of the Internet. Had a proprietary protocol established dominance, the evolution of computer networking and eventually the Internet might have been very different (see Mowery and Simcoe 2002 for further discussion). The weak intellectual property protection for TCP/IP and related networking innovations reflected the network's academic origins, the DOD's support for placing research in the public domain, and the inability of proprietary standards to compete with the open TCP/IP standard. The resulting widespread diffusion of the Internet's core technological innovations lowered barriers to the entry by networking firms in hardware, software, and services.

Although U.S. scientists and engineers made important contributions to packet-switching and computer-networking technologies and protocols, they were by no means alone. French and British computer scientists also contributed important technical advances during this period, and publicly supported prototype computer networks were established in both France and the United Kingdom by the early 1970s. U.S. dominance in computer networking did not result from a first-mover advantage in the invention or even the early development of a packet-switched network. The factor that seems to distinguish ARPANET from these simultaneous projects was its funding for large-scale deployment of the network. Its size and inclusion of a diverse array of institutions as members appear to distinguish the ARPANET from its British and French counterparts and accelerated the development of supporting technologies and applications.

We lack the necessary data to estimate the total federal investment in Internet-related R&D. Even were such data available, the complex origins of the Internet's various components would make construction of such an estimate difficult. Nevertheless, federal investments in the academic computer science research and training infrastructure that contributed to the Internet's development were substantial. The large scale of the U.S. defense-related programs in computer science research and networking distinguished them from those in the United Kingdom and France. In addition to their size, the structure of these substantial federal R&D investments enhanced their effectiveness. In its efforts to encourage exploration of a variety of technical approaches to research priorities, DARPA frequently funded similar projects in several different universities and private R&D laboratories. Moreover, the DOD's procurement policy complemented DARPA's broad-based approach to R&D funding. As had been true of semiconductors, major development and procurement contracts were awarded to small firms such as BBN, which received the contract to build the first IMP. This policy helped foster entry by new firms into the emerging Internet industry, supporting intense competition and rapid innovation.

In 1985, the National Science Foundation, by then one of several federal government agencies managing the backbone of the U.S. national network, made the first in a series of policy decisions that encouraged the standardization of Internet infrastructure and promoted expansion and utilization of the network. Beginning in 1985, any university receiving NSF funding for an Internet connection was required to provide access to all qualified users and to use TCP/IP on its network. Standardization around TCP/IP encouraged interoperability and supported the creation of a large pool of university-trained computer scientists and engineers skilled in use of the protocol.

The process of infrastructure rationalization concluded with the decommissioning of the original ARPANET in 1990 and the transfer of its users and hosts to the new NSFNET. In spite of growing private sector participation in the management of the Internet, the NSF maintained an acceptable use policy (AUP) throughout this period that prohibited use of NSFNET for commercial purposes. As more commercial users attached to the network, on their own or in partnership with academic institutions, they lobbied the NSF to abandon the AUP. By 1991, the NSF had abandoned the AUP, and Internet backbone management was privatized in 1995. Once again, rapid growth in nongovernmental applications led to a reduction in direct governmental influence over the evolution and adoption of this technology, although in this case, governmental "withdrawal" might be more accurately depicted as a purposive rather than an inadvertent outcome.

In May 1991, Tim Berners-Lee and Robert Cailliau, two physicists working at the CERN laboratory in Switzerland, released a new document format, hyper-text markup language (HTML), and an accompanying document retrieval protocol, hyper-text transfer protocol (HTTP).[15] Together, HTML and HTTP turned the Internet into a vast cross-referenced collection of multimedia documents, dubbed by these collaborators the World Wide Web (WWW). In order to use the WWW, a computer needed a connection to the Internet and the application software that could retrieve and display HTML documents. Although it was not the first functional Internet browser, Mosaic, a free program written by a group of graduate students at the University of Illinois National Center for Supercomputing Applications that included Marc Andreesen, was widely adopted and accelerated the growth of the Web.[16] During 1993, the first year that Mosaic was available, HTTP traffic on the Internet grew by a factor of 3,416. By 1996, HTTP traffic was generating more packets than any other Internet application. The "gold rush" of Internet commercialization and hype had begun, and in this field as in other parts of the IT sector, private funding of R&D and technology deployment vastly outstripped public R&D investment.

HTML and HTTP were not invented in the United States, but decades of previous federal and private sector investments in R&D and infrastructure supported their rapid domestic adoption and development. By the early 1990s, the basic protocols governing the operation of the Internet had been in use for nearly twenty years, and their stability and robustness had improved considerably. As Greenstein (2000a) has pointed out, the explosive growth of the Web during the 1990s benefited from the lengthy period of gestation and refinement of the network infrastructure. This prolonged adolescence was supported in large part by public funds.

Conclusion

The postwar U.S. economy has benefited from the creation and innovative dynamism of a series of new industries, including those discussed in this chapter. Although the development and commercialization of innovations by firms in all of these industries have relied heavily on private sources of finance, it is difficult to overstate the importance of the federal support for R&D, and especially in industries such as computer hardware or semiconductors, procurement. We have argued that these public R&D and procurement programs accelerated the growth of U.S. firms in these industries. The structure of these public programs also enhanced the innovative dynamism and competitive strength of the IT sector. But the influence of public funding for R&D and procurement was enhanced by the fact that such public intervention occurred at the inception of these industries. As the IT sector developed, federal R&D and procurement contracts exercised less influence over the semiconductor, computer hardware, and computer software industries in the United States.

Our discussion of the importance and effects of federal financing of innovation in IT is broadly consistent with the market failure analyses of Arrow and Nelson, but it highlights some of the limitations of these analyses. In particular, the historical overview of these industries' development emphasizes the dynamic effects of public financing of innovation, as well as the path-dependent nature of its influence on industry evolution. The structure of federal financing and procurement programs was no less important than their magnitude in affecting industry evolution. Moreover, because federal financing of innovation was most significant in the earliest years of the IT sector's development, the effects of these policies were unusually substantial and long lived. The initial conditions surrounding the evolution of these industries were significant, and they were affected by the structure of federal policies for financing innovation.

Although the influence of federal R&D funding on industry innovativeness and development is readily apparent in the IT sector, federal R&D programs have played an important role in the development of other postwar high-technology industries in the United States. The most obvious examples are the pharmaceuticals and biotechnology industries, which have benefited from the large-scale basic research programs sponsored by the National Institutes of Health. In the biomedical sciences no less than in information technology, the presence of these large-scale federal R&D programs is almost entirely a post-1945 phenomenon. Postwar public sector R&D programs in the United States have few pre-1945 counterparts, which may be one reason that the pattern of growth and structural evolution of the industries benefiting from these public R&D programs differs in some respects from the prewar industries of chemicals and electrical equipment, among others.

The influence of federal R&D and procurement on the technological and structural development of the IT sector was amplified by other policies in such areas as antitrust policy and intellectual property rights. Remarkably, in the light of the consistently procompetitive posture of all of these disparate policies during the 1945–1970 period, there is little, if any, evidence of systematic coordination among them. Nonetheless, their powerful and largely beneficial effects on industry dynamism and growth are apparent in retrospect. It is also apparent in retrospect that this influence was greatest by far in the early development (roughly the first twenty years) of each of the IT-related industries discussed in this chapter. Moreover, federal programs rarely attempted to directly affect the development of commercial applications or technologies, an area in which the U.S. government, no less than other industrial economy governments, has compiled a very undistinguished record (see Eads and Nelson 1971 for one of the earliest and still one of the best summaries).

At least some of the beneficial effects of these federal R&D and procurement programs appear to have flowed from their encouragement for relatively liberal disclosure and interfirm diffusion of technology. This posture stands in sharp contrast from today's conventional wisdom about the desirability of patenting the intellectual property resulting from federal R&D, conventional wisdom now embodied in such policies as the Bayh-Dole Act of 1980 (see Mowery et al. 2002). Although the evidence provided by these histories of selected high-technology industries must be treated with great caution, the discussion in this chapter suggests that relatively weak intellectual property rights regimes may prove beneficial, or at least not harmful, to the development and innovative performance of industries in their early stages of development.

Notes

An earlier version of this chapter was presented at the SSRC conference on "Financing Major Innovations," Irvine, California, March 22, 2003, and benefited from comments from the participants at that meeting, especially Linda Cohen, Steve Usselman, and Zorina Khan. Portions of this chapter draw on Mowery and Rosenberg (1999) and Mowery and Simcoe (2002).

1. Scientific research in nanotechnology, much of which has important defense applications, may be an example of another field in which federal R&D funding could play an important role in the development of a general-purpose technology. A prominent San Francisco venture capitalist, Steve Jurvetson, recently characterized the role of private investment in the development of nanotechnology applications as follows: "We [venture capitalists] have to come in much later, after [something that] was an incredibly risky proposition bears some fruit in terms of a prototype of some products. That's when we first invest.... every one of our nanotech and related companies was first federally funded in a university setting or in a government lab" (Tansey, 2004).

2. "As early as 1936, Kelly felt that one day the mechanical relays in telephone exchanges would have to be replaced by electronic connections because of the growing complexity of the telephone system and because much greater demands would be made on it. As this is hardly technically feasible using valves, it seems that Kelly was thinking not simply of a radically new valve technology, but perhaps of radically new electronics.... It seems most likely that Kelly saw the logical progression from a semiconductor rectifier in copper oxide to be a semiconductor switch" (Braun and MacDonald 1982, 36).

3. Commercial exploitation of Bell Labs's discovery was influenced by the antitrust suit against AT&T filed in 1949 by the U.S. Department of Justice. In 1956, the antitrust suit was settled through a consent decree, and AT&T restricted its commercial activities to telecommunications service and equipment. The 1956 consent decree also led AT&T, holder of a dominant patent position in semiconductor technology, to license its semiconductor patents at nominal rates to all comers, seeking cross-licenses in exchange for access to its patents. As a result, virtually every important technological development in the industry was accessible to AT&T, and all of the patents in the industry were linked through cross-licenses with AT&T.

4. From the beginning, TI and other firms were aware of the commercial potential of the IC. As one of its first demonstration projects, TI constructed a computer to demonstrate the reductions in component count and size that were possible with ICs.

5. "European governments provided only limited funds to support the development of both electronic component and computer technology in the 1950s and were reluctant to purchase new and untried technology for use in their military and other systems. European governments also concentrated their limited support on defense-oriented engineering and electronics firms. The American practice was to support military technology projects undertaken by industrial and business equipment firms that were mainly interested in commercial markets. These firms viewed their military business as a development vehicle for technology that eventually would be adapted and sold in the open marketplace" (Flamm 1988, 134).

6. The history of federal involvement in the SEMATECH manufacturing R&D consortium further illustrates the shifting roles of defense-related and civilian R&D and innovation in semiconductors. Founded in 1987, SEMATECH was devoted to collaborative R&D on semiconductor manufacturing processes, in an effort to improve U.S. civilian semiconductor firms' manufacturing performance in the face of intense competition from Japanese producers. It initially received half of its $200 million operating budget from the federal government, based on a series of influential reports from the Defense Science Board (U.S. Department of Defense, 1987; see also Alic et al. 1992) that argued that the civilian semiconductor "industrial base" in the United States was essential to the nation's defense establishment (see Grindley, Mowery, and Silverman 1994 for a more extensive discussion of the changing structure of SEMATECH).

7. Goldstine, one of the leaders of the wartime project sponsored by the army's Ballistics Research Laboratory at the University of Pennsylvania that resulted in the Eckert-Mauchly computer, notes that "a meeting was held in the fall of 1945 at the Ballistic Research Laboratory to consider the computing needs of that laboratory 'in the light of its post-war research program.' The minutes indicate a very great desire at this time on the part of the leaders there to make their work widely available. 'It was accordingly proposed that as soon as the ENIAC was successfully working, its logical and operational characteristics be completely declassified and sufficient be given to the machine...that those who are inter-

ested...will be allowed to know all details'" (1993, 217). Goldstine is quoting the "Minutes, Meeting on Computing Methods and Devices at Ballistic Research Laboratory," October 15, 1945 (note 14).

8. Other computers developed for or initially sold to federal agencies during this period include the Whirlwind, developed at MIT with funding from the navy and the source of important technical advances that were incorporated into the SAGE strategy air defense system of the 1950; Remington Rand's UNIVAC, based on the Eckert-Mauchly technology and sold to the Census Bureau and other government agencies, as well as private firms; and the IBM 701, developed as a scientific computer in 1953 for the Defense Department.

9. "Like most 'new' technology, the computer had important antecedents, and existing firms had developed capabilities related to those antecedents. None had done so more thoroughly than IBM. Despite some superficial differences between computers and the earlier tabulating equipment that had formed the core of IBM's business, computers involved a mix of knowledge and capabilities that matched those existing at IBM extraordinarily well" (Usselman 1993, 5). Usselman also emphasized the skills at IBM's Endicott, New York, factory that had long been the source of many of the firm's tabulating machines: "The Endicott facility also produced a series of input-output devices that helped develop the market for both large and small computers. Though these products made use of electronics, they also drew on the mechanical skills available at Endicott. Printers and disk storage devices in particular were as distinguished [sic] as much for their rapid, precise mechanical motions as for their logical design" (12). Chandler (1997) makes a similar argument.

10. One such programmer noted in the early 1980s, "The chances are reasonably high that on a large data processing job in the 1970s you would find at least one person who had worked with the SAGE system" (Benington 1983, 351).

11. Most analyses of packaged software markets distinguish among operating systems (referred to by the U.S. Commerce Department in the most recent *US Industry and Trade Outlook* as "systems infrastructure software"), the software used to control the operations of a given desktop, mainframe, or minicomputer; applications, software designed to support specific, generic functions such as word processing or spreadsheets; and development tools, which include programming languages and application development programs (see U.S. Department of Commerce 2000 for further detail).

12. Packet switching is fundamentally different from circuit switching, the technology that connects ordinary telephone calls. On a packet-switched network, information is broken up into a series of discrete "packets" that are sent individually and reassembled into a complete message on the receiving end. A single circuit may carry packets from multiple connections, and the packets for a single communication may take different routes from source to destination.

13. Bolt, Beranek and Newman, an MIT spin-off founded in 1948, was an early example of the new firms that played an important role in the Internet's development. The firm was started by MIT professors Bruce Bolt and Leo Beranek in

partnership with a graduate student, Robert Newman. Populated as it was in its early years by a mixture of recent graduates, professorial consultants, and other technical employees with close links to MIT research, BBN is a good example of the quasi-academic environment within which many Internet-related innovations were developed (Wildes and Lindgren 1985).

14. "Transmission Control Protocol (TCP)," 1974.

15. The development of these important technical advances was motivated by Berners-Lee and Caillau's interest in facilitating the ability of physicists to archive and search the large volumes of technical papers being transmitted over the Internet as it then existed.

16. NCSA was an NSF-funded facility devoted to research on supercomputing architecture and applications. By the early 1990s, networking technologies and powerful desktop computers had reduced the need of academic researchers for access to supercomputers. As a result, Andreesen and colleagues at the NCSA focused on developing new technologies to support expanded use of computer networking (Abbate 1999). Federally funded "excess capacity" in the research computing infrastructure thus contributed to an important innovation in networking.

References

Abbate, J. 1999. *Inventing the Internet*. Cambridge, Mass.: MIT Press.

Alic, John A., Lewis M. Branscomb, Harvey Brooks, Ashton B. Cater, and Gerald L. Epstein. 1992. *Beyond Spinoff: Military and Commercial Technologies in a Changing World*. Boston: Harvard Business School Press.

Arrow, K. 1971. "Economic Welfare and the Allocation of Resources for R&D." In K. Arrow, ed., *Essays in the Theory of Risk Bearing*. New York: American Elsevier.

Benington, H. D. 1983. "Production of Large Computer Programs." *Annals of the History of Computing 5*, 350–361.

Braun, E., and S. MacDonald. 1982. *Revolution in Miniature*, 2nd ed. New York: Cambridge University Press.

Campbell-Kelly, M. 1995. "Development and Structure of the International Software Industry, 1950–1990." *Business and Economic History 24*, 73–110.

Campbell-Kelly, M. 2003. *From Airline Reservations to Sonic the Hedgehog: A History of the Software Industry*. Cambridge, Mass.: MIT Press.

Chandler, A. D., Jr. 1977. *The Visible Hand*. Cambridge, MA: Harvard University Press.

Clement, J. R. B. 1987. "Computer Science and Engineering Support in the FY 1988 Budget." In Intersociety Working Group, ed., *AAAS Report XII: Research and Development, FY 1988*. Washington, D.C.: American Association for the Advancement of Science.

Clement, J. R. B. 1989. "Computer Science and Engineering Support in the FY 1990 Budget." In Intersociety Working Group, ed., *AAAS Report XIV: Research and Development, FY 1990*. Washington, D.C.: American Association for the Advancement of Science.

Clement, J. R. B., and D. Edgar. 1988. "Computer Science and Engineering Support in the FY 1989 Budget." In Intersociety Working Group, ed., *AAAS Report XIII: Research and Development, FY 1989*. Washington, D.C.: American Association for the Advancement of Science.

Eads, G. C., and R. R. Nelson. 1971. "Government Support of Advanced Civilian Technology: Power Reactors and the Supersonic Transport." *Public Policy*, 405–428.

Fisher, Franklin M., James W. McKie, and Richard B. Mancke. 1983. *IBM and the U.S. Data Processing Industry*. New York: Praeger.

Flamm, Kenneth. 1987. *Targeting the Computer*. Washington, D.C.: Brookings Institution.

Flamm, Kenneth. 1988. *Creating the Computer*. Washington, D.C.: Brookings Institution.

Goldstine, H. H. 1993. *The Computer from Pascal to Von Neumann*, 2nd ed. Princeton, N.J.: Princeton University Press.

Greenstein, S. 2000a. "Framing Empirical Work on the Evolving Structure of Commercial Internet Markets." In E. Brynjolffson and B. Kahin, eds., *Understanding the Digital Economy*. Cambridge, Mass.: MIT Press.

Grindley, P., D. C. Mowery, and B. Silverman. 1994. "Sematech and Collaborative Research: Lessons in the Design of High-Technology Consortia." *Journal of Policy Analysis and Management* 13, 723–758.

Kleiman, H. 1966. "The Integrated Circuit Industry: A Case Study of Product Innovation in the Electronics Industry." DBA dissertation, George Washington University.

Langlois, R. N., and D. C. Mowery. 1996. "The Federal Government Role in the Development of the U.S. Software Industry." In D. C. Mowery ed., *The International Computer Software Industry: A Comparative Study of Industry Evolution and Structure*. New York: Oxford University Press.

Levin, Richard C. 1982. "The Semiconductor Industry." In Richard R. Nelson ed., *Government and Technical Progress: A Cross-Industry Analysis*. New York: Pergamon Press.

Mowery, D. C., R. R. Nelson, B. N. Sampat, and A. A. Ziedonis. 2002. "The Growth of Patenting and Licensing by U.S. Universities: An Assessment of the Effects of the Bayh-Dole Act of 1980." *Research Policy* 30, 99–119.

Mowery, D. C., and N. Rosenberg. 1998. *Paths of Innovation: Technological Change in Twentieth Century America*. Cambridge: Cambridge University Press.

Mowery, D. C., and T. Simcoe. 2002. "The History and Evolution of the Internet." In B. Steil, R. Nelson, and D. Victor, eds., *Technological Innovation and Economic Performance*. Princeton, N.J.: Princeton University Press.

National Research Council. 1999a. *Funding a Revolution: Government Support for Computing Research*. Washington, D.C.: National Academy Press.

Nelson, R. R. 1959. "The Simple Economics of Basic Scientific Research." *Journal of Political Economy* 67, 297–306.

Redmond, Kent C., and Thomas M. Smith. 1980. *Project Whirlwind: History of a Pioneer Computer*. Bedford, Mass.: Digital Press.

Tansey, B. 2004. "An Insider's View of Nanotechnology." *San Francisco Chronicle*, February 1.

Tilton, J. 1971. *International Diffusion of Technology: The Case of Semiconductors*. Washington, D.C.: Brookings Institution.

"Transmission Control Protocol (TCP): Request for Comments 0675." Available at: www.faqs.org/rfcs (accessed July 15, 2006).

U.S. Department of Commerce. 2000. *U.S. Industry and Trade Outlook 2000*. Washington, D.C.: U.S. Government Printing Office.

U.S. Department of Defense. Office of Undersecretary for Acquisition. 1987. *Defense Semiconductor Dependency: A Report of the Defense Science Board*. Washington, D.C.: U.S. Department of Defense.

Usselman, S. 1993. "IBM and Its Imitators: Organizational Capabilities and the Emergence of the International Computer Industry." *Business and Economic History* 22, 1–35.

Wildes, K. L., and N. A. Lindgren. 1985. *A Century of Electrical Engineering and Computer Science at MIT, 1882–1982*. Cambridge, Mass.: MIT Press.

8

Learning the Hard Way: IBM and the Sources of Innovation in Early Computing

Steven W. Usselman

Accepted wisdom in current scholarship holds that modern computing is a product of massive public investment. Virtually all histories of computing technology place significant emphasis on the role of government funding in general and on that of military expenditures in particular. Pioneering analyses by Kenneth Flamm written some two decades ago documented the hefty investments made by government and the military in early mainframe electronic computing through direct subsidies of research and procurement of advanced data processing equipment.[1] A more recent study sponsored by the National Academy of Science traced similar influences over the essential technologies that have given us the current world of networked personal computers operating with graphic user interfaces and linked by servers, routers, protocols, and other components that constitute the World Wide Web.[2] Important work by David Mowery and his colleagues (including chapter 7 in this volume) summarizes these perspectives while adding further examples of government influence in promoting software development and the flow of talent into computing.[3]

This chapter offers a moderate adjustment to this important body of work. My goal is not to dispute fundamentally the proposition that government has exerted a profound influence over computing. Rather, I wish to suggest some ways in which private enterprise and private capital have also figured in the story. In particular, I seek to comprehend how one firm, IBM, contributed to the emergence and refinement of the stored program electronic computer during the two decades from the close of World War II until its production of System/360, a comprehensive line of such machines whose basic architecture would dominate American and world markets for mainframe computing for another two decades or more.

We can perhaps best understand the place of IBM in early computing by considering the firm as a key element in a broad-based effort at learning that occurred across many fronts and included many participants. This exercise in learning focused on two overarching objectives: (1) use of electronic computing for advanced processing and display of data and for control and (2) mastery of solid-state electronic components of decreasing size and cost and increasing capability. Each line of development mobilized an array of experts located in diverse institutions: in university laboratories and computing facilities distributed across multiple academic disciplines; in government agencies ranging from civilian bureaucracies such as the Social Security Administration to defense installations and laboratories; and in numerous corporate settings, including data processing departments of most Fortune 500 companies as well as research laboratories located at both large, established electronics firms such as AT&T and RCA and small start-ups such as Fairchild and Texas Instruments.

IBM played an essential role as a critical intermediary in these multinodal and multifaceted efforts at learning. Reaching across many domains, IBM linked diverse researchers with one another and connected them to a broad range of consumers. Perhaps most important, the firm offered opportunities for participants in the emergent fields of computing and solid-state electronics to consolidate lessons learned and to meld them in a series of stable designs that would be produced in significant volume. The interludes of stability made possible by these dominant designs in turn facilitated further learning not only within IBM, but also among suppliers of components and software.

In performing these functions, which it pursued, of course, for reasons of self-interest rather than altruism, IBM effectively mobilized a great deal of capital toward early computing and solid-state electronics. Some of this capital, to be sure, came in the form of government defense contracts granted to IBM and its customers. But much of it also came through earnings derived from commercial exchanges in the private civilian economy, which had a vast appetite for data processing and electronic communications, and from private investors seeking to capitalize on opportunities presented by that commercial demand. In reinvesting its earnings and attracting such investment, IBM effectively leveraged its established technical and business capabilities—and the profits and goodwill they generated—to funnel massive investment toward the emergent field of electronic digital computing.

In bridging the military and commercial realms, IBM managers naturally hoped that developments would flow from the former to the latter. Like many analysts to follow them, they looked for military contracts to generate research and learning that would diffuse into products aimed at the commercial market. Much to their frustration, however, the flows were actually far more turbulent. Military projects frequently drew IBM down highly specialized cul-de-sacs whose arcane lessons did not transfer readily to the commercial line. As often as not, the real learning flowed in the opposite direction, as IBM contributed assets and capabilities it had developed in the commercial realm to defense projects. The grand development effort that culminated in System/360, which melded significant new capabilities in solid-state electronics with a comprehensive approach to the market for computing, emerged from a complex brew of experience gained through both military and commercial endeavors. In the end, it involved a massive mobilization of private capital in support of a project aimed overwhelmingly at commercial users.

Overview

Accounts stressing the importance of government and the military to computer development at IBM typically emphasize two large projects launched at the time of the Korean War. One was the Semi-Automatic Ground Environment (SAGE) system, which deployed twenty-three pairs of massive electronic computers operating in real time to monitor American airspace and deploy antiaircraft fire in response to potential Soviet invasion. Growing out of Project Whirlwind, a computer development project sponsored by the Air Force and coordinated by MIT scientist Jay Forrester, SAGE offered IBM an opportunity to manufacture electronic circuits in high volume and to deploy and maintain machines built from them and operating in highly demanding circumstances.[4]

The second project was the IBM 701, an advanced programmable computer of standard design intended primarily for use in engineering calculations and other tasks performed by large defense laboratories and military contractors. Its nickname during development—the Defense Calculator—made clear its ties to the military market. During the early 1950s, IBM placed approximately twenty of these machines in the field. No previous computer of such power had been produced in such numbers.[5]

There is no disputing that IBM pursued these military projects in hopes of gaining critical advantages over its competitors in the market for commercial computing. Thomas Watson, Jr., who as executive vice president had taken over responsibility for product development and strategy from his CEO father, made this perfectly clear at the time and in later recollections. "I knew if you got the SAGE contract," he once told an interviewer, "you got the computer business."[6]

In diverting his most talented people from ongoing projects aimed at commercial markets, Watson looked to gain experience in electronic component design and manufacture that would soon be transferred into commercial products of more advanced capabilities. Watson closely monitored efforts to build circuits for SAGE using automated equipment for inserting electronic components into printed circuit boards and for wiring and soldering the boards themselves. He toured IBM facilities showing films of the equipment and urging his engineers to make use of them in commercial projects.[7] Watson expressed severe disappointment that Forrester and his MIT colleagues decided against using solid-state transistors, a technology he and others at IBM regularly referred to as an "automation technique" because in their minds it promised above all to drive down the cost of building electronic circuits.[8] SAGE did, however, take advantage of another mass-produced solid-state component when it used arrays of small magnetic doughnuts for its core memories. The doughnuts were formed using pill-pressing equipment modified by IBM mechanics and were wired with inserters also designed by those skilled holdovers from the firm's heyday as a manufacturer of electro-mechanical machines. Such arrays remained the basic memory technology of IBM computers through System/360, until the advent of silicon chip memory in the early 1970s.[9]

Watson also intended for the 701 to serve as a learning exercise that would yield benefits for the commercial sector. IBM conducted a parallel development project, the 702 Business Calculator, that would borrow techniques developed for the Defense Calculator and apply them to a machine targeted at large commercial customers such as insurance companies and banks. When development costs for the 701 soared, effectively doubling the promised monthly rental charge, Watson decided against declaring the project a loss leader and absorbing the costs as the price of learning. He wanted the development engineers and the sales and maintenance force to operate within the constraints imposed by a balance sheet showing true expenses and profits, as they would have to do in the commercial sector. When all eighteen customers agreed to the hefty price increase, Watson later recalled, "I knew we had the bull by the horns."[10]

The record, then, seems quite clear. IBM not only benefited from public investment in the form of military contracts; it consciously viewed these defense projects as springboards to commercial dominance. The fact that System/360 incorporated magnetic core memories and electronic circuits built using descendants of the assembly equipment developed for SAGE appears only to confirm the overriding influence of government on early computing. Add in that System/360 also used silicon transistors, a technology whose development was supported in large measure by military projects, and that influence looms still larger.[11]

In reality, however, the story is not quite so straightforward. Probing more deeply into events at IBM, several complicating features come to light. First, efforts to integrate electronics into advanced calculating machines were underway at IBM even before the advent of stored program computing during World War II. Second, in the years immediately following the War, as its fortunes soared from returns on investment in its traditional product line, IBM had embarked on several development projects aimed at bringing electronic computing to the commercial sector; Watson's play for military contracts at the start of the Korean conflict diverted vital resources and temporarily derailed those commercial ventures. Third, IBM won the military contracts because it had established capabilities in marketing, manufacturing, and maintenance that government and the military recognized as lacking in other potential contractors. The defense programs thus served as a vehicle through which essential knowledge and practices accumulated in the commercial sector were transferred to the emergent field of electronic computing. Fourth, techniques mastered for SAGE and the 701 did not move easily into commercial operations, and efforts to force the transfer often proved highly disruptive; much of the critical learning at IBM during the 1950s actually occurred in the course of producing more modest machines for the commercial market, which generated earnings that IBM invested in subsequent developments in electronics. Fifth, when circuitry built from solid-state components assembled by highly automated processes did make their way into IBM commercial computers, demands of subsequent military contracts threatened to draw circuit designers away from the objectives of simplicity and low cost that were critical for success in the commercial sector. In tying silicon technology to these more modest objectives and mobilizing enormous amounts of financial resources toward the effort, IBM helped push the emergent field in new directions that would yield enormous benefits.

Laying a Foundation

Before examining these propositions in more detail, it will be useful to become familiar with some critically important features of IBM's business practices during this period. Most of these were inherited from methods instilled by the elder Thomas Watson, who took over the firm in the early teens after a long apprenticeship at John Patterson's National Cash Register Company.[12] IBM thrived under Watson by cultivating an integrated organization aimed at providing specialized data processing services to businesses and other large organizations. The strategy hinged on IBM's dominance in accounting machinery built from complex assemblies of mechanical gears, chains, springs, and ratchets and a smattering of relays and other electromechanical devices. The machines read data stored on punched cards, performed any of a variety of calculations on it, printed the results in various formats, and stored the results again on cards. Customers leased the equipment from IBM on a monthly basis and purchased cards from IBM, which until a consent decree in 1936 had exclusive rights to the card market. Field engineers and service representatives from IBM spent considerable time with customers, maintaining the machines, devising new ways to manipulate and present data, and installing new components and other refinements that made these novelties possible.

These arrangements lent an enviable financial stability to IBM. As observers such as historian Robert Sobel have noted, IBM had many attributes of a bank or insurance company.[13] While those institutions converted the funds they raised into loans or term-life policies, IBM turned its capital into leasing agreements. Equipment placed with customers under these agreements constituted an installed base of capital that generated revenue at fairly predictable levels. In 1946, for instance, IBM valued its cumulative investment in land and buildings at $21 million. Reserves for depreciation lowered this value by $10 million, leaving a net cumulative investment in land and buildings of just $11 million. IBM's cumulative stock of rental machines and equipment, meanwhile, was valued at nearly $136 million. Even with $71 million reserved for depreciation, net cumulative investment in rental machines and equipment stood at $65 million. Of IBM's roughly $76 million in cumulative net investment, then, more than 85 percent was in the form of rental machines and equipment.[14] Payments for goods and services associated with those machines generated some $120 million in revenue for IBM in 1946. Net income after expenses stood at $31.3 million; after deducting for taxes, IBM claimed profits of $18.8 million. At the time, IBM had only $30 million in outstanding debt, even after taking out an additional $13 million in loans from Prudential Insurance in 1946 alone.[15]

Of course, rental agreements were not quite identical to loans or insurance policies. For one thing, their terms were generally much shorter. IBM customers could cancel in a matter of months. More important, the lease agreements were mediated by physical assets—the array of data processors and input-output devices and storage media and printers and displays that constituted any IBM installation. Like all other physical assets, the installed machines deteriorated over time, and they were subject to competition from newer products offering enhanced features and performance.

Several factors helped mitigate concerns about cancellations and obsolescence. The nature of the physical assets themselves went a long way to dampen such fears. Customers who stored much of their critical data on punched cards and generated their routine billing statements in standard form from IBM printers and associated equipment had strong incentives to stick with IBM systems. Those tempted to change course met substantial resistance from IBM, which did everything in its power to keep its installations exclusively proprietary and prevent customers from incorporating competitive alternatives incrementally on a piecemeal basis. Meanwhile, Watson took care to ensure that IBM engineers and designers devised new models offering increased speeds and novel features. Large customers such as Prudential provided test-beds in which IBM technicians advanced the state of the art, while new users such as those in the overseas markets Watson cultivated often made do with older refurbished machines. Watson and his vaunted sales force built the business by finding additional customers and convincing established ones to manipulate and present data in new ways.[16]

This mode of doing business enabled IBM to pursue innovation in a highly deliberate manner. The last thing IBM wanted was to introduce change so rapidly as to render its installed base obsolete. The elder Watson steered clear of the speculative investment in research that came to characterize firms such as RCA, Du Pont, and Kodak. At IBM, the vast majority of research and development expenditures occurred in the course of particular product development efforts. Many of those efforts, moreover, took the form of ongoing refinement and modification rather than breakthrough new designs. IBM honed sophisticated routines for linking field engineers at customer sites with designers and manufacturers working at its facilities in isolated Endicott, New York.

Responsibility for introducing entirely new machines or thoroughly updated models of old ones fell to a group known variously as Future Sales or Product Planning. Blending elements of sales, engineering, finance, and market forecasting, this group monitored ongoing innovation occurring at customer sites and decided when to incorporate new functions and capabilities into regular commercial products to be built in volume and leased at a standard price. Product planners set specifications for price and performance and then turned to a crack team of mechanical inventors in the design laboratory at Endicott. Often Watson set teams in this design shop against one another in a race to come up with a suitable machine. The victorious team then turned the product over to a group known as manufacturing engineering, which looked for ways to build the designs economically in volume. A manager from Product Planning, generally reporting directly to Watson, supervised the entire process. It was his job to get the product out the door and generating rental income. The task was demanding, for IBM expected each major new product development effort of this sort to pay for itself in five years. Delays ate into the projected revenue stream and pushed back the day when all associated costs would be written off and the installed machines would pump pure profit into IBM.[17]

This style of innovation, which tied research and development to particular product programs that were expected to generate revenues sufficient to cover all costs, persisted long into the 1950s as IBM attempted first to introduce electronics into the world of data processing and then to master the emerging world of solid-state components and circuitry. Both Watsons were reluctant to spend money on centralized research that was divorced from the sort of accountability provided by product development programs aimed at commercial markets.[18] Learning at IBM throughout the early computer era was thus tightly joined with established functions and routines operating across the firm and with the expanding commercial market for data processing that IBM already dominated. This strategy, largely the residue of inherited habit and financial conservatism, proved fortuitous. The sort of integrated learning and technical compromise it fostered turned out to be highly suitable for the emergent new field of electronic computing.

Venturing into Electronics

IBM's migration from electromechanical calculating equipment to electronic data processing began during the late 1930s. For some time, inventors in the design laboratory at Endicott had been incorporating increasing numbers of relays and other electrical components into their machines. Such components operated with greater speed and reliability and lower power requirements than the strictly mechanical components they replaced. Sensing that tube-based electronic components might offer benefits along similar lines, designers tentatively explored the possibility of incorporating some of them as well. In the years prior to World War II, IBM hired several college graduates with degrees in electronics to help assist the mechanical designers in these efforts. Several of these newly hired employees, including future director of engineering Ralph Palmer and his assistant Byron Phelps, would go on to play prominent roles in IBM's computer design and development efforts throughout the 1950s.[19]

Palmer and Phelps had trouble persuading designers at Endicott of the virtues of electronics. Accomplished in the mechanical realm of gears and ratchets and accustomed to exercising great authority over their projects, established designers generally stuck with what they knew. Such resistance may ultimately have worked to IBM's advantage, however, for it left Palmer and Phelps free to experiment with calculators built almost entirely from electronic components. By the eve of the war, they had nearly completed a version of IBM's standard multiplier that used electronic circuits rather than mechanical counters to perform the calculations.

This project still languished in the design laboratory when Palmer and Phelps were drafted into the military and sent to work on wartime electronics projects. Meanwhile, demand for traditional IBM equipment soared, as the Selective Service Administration and many government procurement programs used punched cards for record keeping and reporting.[20] Revenue and net income for 1946 stood at two-and-a-half times those of 1940. Net income after taxes had doubled over the course of hostilities, and IBM had added a second design and manufacturing facility located at Poughkeepsie, New York.[21] Palmer and Phelps took up residence there after obtaining their releases from the service and quickly resurrected their project. In 1946, IBM introduced it as the 603 Electronic Multiplier.[22]

This machine and a more general-purpose electronic calculator, the 604, made their way into the market in significant numbers during the late 1940s and early 1950s. By September 1952, IBM had leased over 1,500 of the 604s and had placed another 250 in the field as part of an arrangement known as the card-programmed calculator (CPC).[23] With this arrangement, users could instruct, or "program," the calculator to perform various sequences of calculations. A marketing team headed by IBM engineer Cuthbert Hurd helped to build enthusiasm for this machine among aerospace firms and other customers with extensive engineering and scientific applications. Hurd's team fostered active exchanges among such users and IBM personnel, who effectively formed a cooperative learning community dedicated to advanced electronic calculation.[24]

These early experiences with electronics point to several complexities that make it difficult to generalize about the learning process and the relative contributions of government and business. Government certainly influenced the course of events. Many early users of the CPC were engineers and scientists who worked for companies such as Northrup Aviation and other primary defense contractors. Hurd and his team cultivated this niche; they stood apart from the established sales force that marketed IBM equipment to the corporate world. Palmer and Phelps, who returned from their wartime assignments flush with new ideas about how to use electronic circuits to store and manipulate information, had never truly melded their work into the established punched card operations, which continued to thrive without relying on electronics. (Buoyed by the enormous demand for this equipment spurred by the war, the elder Watson in 1947 authorized designers at Endicott to launch a new round of product developments based on the old technology, including a mammoth chain printer and new calculators based on a larger punched card.[25]) One might even argue that the work on electronic calculators was a luxury made possible by wartime profits and other legacies of the military effort, since the facility at Poughkeepsie had been constructed under military certificate of necessity in order to produce Norden bomb sights.

Yet for all of this, it seems a stretch to conclude that the push toward electronic calculation was merely the product of military endeavors. Palmer and Phelps were hired to make contributions to the commercial line. They intended their electronic multiplier for the established business market, and they had nearly completed it when the war interrupted their efforts. Most of the 604s produced at Poughkeepsie made their way into precisely that market, where institutions used them to perform standard accounting functions. Palmer developed the electronic machine, moreover, in a setting infused with the established culture and routines of IBM. Though frustrated by the resistance of older designers steeped in the mechanical technology, he thoroughly inculcated their concept of pursuing innovation through specific machine development programs aimed at a large commercial market. With the 603, as with nearly all of his subsequent efforts in electronic calculation, Palmer did not look to learn through laboratory experiment or by building machines that pressed against the frontiers of calculation. Rather, he sought merely to introduce electronics into established domains and trusted that the resulting products would yield enhanced performance that would open new possibilities and new markets.[26]

In going after certain of those markets, as Hurd did with the CPC, IBM did find it necessary to cultivate some distinctive practices suitable for the emergent field of scientific calculation. But Hurd and his team also drew heavily on established procedures in areas such as field engineering, manufacturing, publicity, and customer service. The team modified established functions; it did not devise new ones out of whole cloth. Product engineering teams at rival firms such as Engineering Research Associates and Eckert-Mauchly foundered in trying to reach these same markets in large part because they lacked comparable institutional support.[27]

All the while, moreover, key figures in top management embraced Palmer's original vision. They pushed for the new electronic machines to permeate the established commercial market. The most important figure championing this cause was the younger Thomas Watson. Returning to IBM in 1946 after a long stint as an army pilot, Tom Watson quickly established himself as the heir apparent to his father. Soon he began exerting a strong influence over areas such as product strategy, especially that involving electronics. From the moment he saw Palmer demonstrate the 603, the younger Watson later recalled, he was convinced that electronics held the future for IBM.[28] His actions over the next decade and a half testified to the depths of this conviction. Watson looked whenever possible to get electronics into IBM products, and when the world of electronics was transformed by the rise of new techniques of solid-state components and printed circuitry, he pressed hard to get those new techniques incorporated into IBM products as well.

The younger Watson's enthusiasm for electronics was motivated to considerable degree by conspicuous developments occurring outside

IBM. World War II had fostered some dramatic developments in highly sophisticated equipment capable of cranking through the sorts of routine calculations desired by scientists and engineers. IBM itself had contributed significantly to such a project located at Harvard. The most dramatic accomplishments, however, were those of the University of Pennsylvania professors J. Prespert Eckert and William Mauchly. Their Electronic Numerical Integrator and Calculator (ENIAC), developed in an effort to generate firing tables for new guns being tested at the Aberdeen Proving Ground, used arrays of vacuum tube circuits to perform routine calculations at unprecedented speeds.[29]

Although the public attention ladled on Eckert and Mauchly irritated the Watsons, neither father nor son paid the inventors much heed until they announced plans in 1947 to launch a new company and market a modified version of ENIAC for use by insurance companies and other commercial enterprises that processed large amounts of financial data. Their list of prospective customers included some of IBM's most lucrative accounts. Even then, IBM showed little concern until two years later, when the Watsons got word that at least one major client, Prudential Insurance, had entered an agreement with Eckert and Mauchly to use an electronic calculator that substituted tape or other magnetic media for punched cards. This move by a leading customer who also served as IBM's sole lender signaled the willingness among vital commercial customers to abandon established storage media and liberate themselves from IBM.[30]

The news regarding Prudential galvanized the younger Watson. Quick reviews of all IBM product development programs convinced him the trouble lay with the old mechanical engineering culture of Endicott. Inventors there had concentrated on building a new generation of punched card equipment while allowing IBM's own work on magnetic tape and drums to languish. Watson transferred the work on a tape-based calculator to Poughkeepsie and would have moved the drum-based project there as well if not for the intervention of his new director of engineering, Wallace McDowell, who had spent his entire career at Endicott. Meanwhile, Watson told IBM's sales force, assembled in Endicott for its annual meeting, that within a decade, all IBM products would be based entirely on electronic components.[31] He also quietly explored the possibility of acquiring Eckert and Mauchly, though he quickly dropped the idea because of antitrust considerations.[32] Watson could only look on in frustration as the computer pioneers instead cut a deal with Remington-Rand, itself a recent merger that had combined one of IBM's major competitors in the commercial data processing industry with a firm that was highly regarded in scientific and engineering circles.[33]

Detouring toward Defense

These events in the half-decade following World War II left a curious legacy at IBM. In retrospect, we can see that the firm had in fact laid much of the groundwork for its later success in electronic computing. With the 603, the 604, and the CPC, it had transformed the prewar experiments with electronics into viable commercial products. These programs melded established business routines with the new technology of electronics. IBM also had launched a series of product development efforts tied to the new magnetic storage technologies, which would be crucial for input-output devices and for use as short-term memory. During the 1950s, the tape and drum machines that came under scrutiny in 1949 would emerge as the cornerstones of IBM's commercial computing products. Another venture underway in a small facility in San Jose, California, would yield machines based on random-access disc storage devices that could substitute for the large tubs of punched cards used to store data.

Yet at the time, Tom Watson was convinced IBM had fallen behind. Accustomed to seeing innovation occur at the top of the product line, with new machines designed for large business customers such as Prudential, he failed to appreciate that electronics might spread instead from more modest technical successes aimed at a broader range of customers. The high glamour associated with work on one-of-a-kind large-scale computers, which the press celebrated as "electronic brains," reinforced his thinking. Both Watsons were embarrassed that IBM had lost the initiative in this high-profile realm to small groups affiliated with universities and research laboratories. Meanwhile, the continuing success of IBM's traditional accounting equipment saddled the firm with antitrust concerns that limited its ability to counter the upstarts through merger and acquisition. To make matters worse, new placements of punched card equipment had slowed, prompting the elder Watson to fret about IBM's lingering debt and mounting costs. In meetings with plant managers, Watson even raised the possibility of layoffs.[34]

It is in this context that we can comprehend Tom Watson's decision in 1950 to turn work on electronics away from ongoing commercial development projects and toward new possibilities opened by the Korean War and by growing concerns about the Soviet nuclear threat. In short order, Watson secured the SAGE contract and a series of commitments for the Defense Calculator, as well as other large contracts to build a sophisticated computer for work on naval ordnance and to pursue work on navigational computers for bombers.[35] These defense contracts would, Watson believed, fund investment in new facilities and force the transition to electronics at the high end. From there, they would trickle down into all IBM products. All of this would occur in the context of specific product development programs, so that IBM would not need to depart significantly from its established traditions and build an extensive independent research effort. This was especially true in the case of the Defense Calculator, where the military effectively created a market for a computer of advanced design.[36]

On the surface, the strategy appears to have worked. The 701 gave IBM the largest installed base of sophisticated computers of standard design. Updated versions introduced later in the decade, known as the 704 and 705, reached still more customers in the scientific and commercial realms, respectively. The new machines incorporated a critical new electronic component, ferrite core memories, that IBM had developed in collaboration with engineers at MIT working on SAGE.[37] The SAGE project would also bequeath to IBM a highly automated manufacturing facility for assembling electronic circuits. Machines inserted components into laminated printed circuit boards containing sophisticated wiring patterns that had been laid down using new techniques of photolithography. Built between 1954 and 1956 at Kingston, several miles up the Hudson River Valley from Poughkeepsie, this facility also contained equipment that automatically wrapped the wires of the components and soldered them together by dipping the exposed ends in a bath of solder and chemically removing the excess. Virtually all of the techniques brought together at Kingston had been developed by small firms funded by the air force and other branches of the military.[38] In the closing years of the decade, circuits in similar form would make their way into most IBM commercial products. Most would be assembled using modified versions of the equipment at Kingston.

Looking more closely at the actual course of events at IBM during this period, however, we see a picture far more complex, and a legacy of Watson's military-led strategy far more ambiguous. The balance sheets for the early 1950s hint at some of the difficulties. Indebtedness, for instance, spiked sharply upward with the turn toward military projects. IBM had ratcheted up its debt with Prudential from $30 million to $85 million during the years immediately following World War II, as the company rapidly expanded its installed base of punched card machines. Debt then held steady while IBM built its base further through retained earnings, which mounted from $42 million to $102 million during the four years from 1947 through 1950. With the onset of the military projects, debt leaped upward during the early 1950s, increasing by $50 million in 1951 alone and by another $115 million over the course of the next three years. Retained earnings, meanwhile, remained flat (though IBM did build its net current assets by an impressive $63 million during this period).[39] The situation prompted Watson to write a somber note to executive vice president Albert Williams in September 1954 telling him to hold debt steady and pace expansion accordingly.[40]

Along with the increased indebtedness came a significant shift in investment. During the immediate postwar years, IBM had funneled most of its capital into rental machines and equipment. Net investment in these areas, after deducting for reserves, jumped from $65 million at the beginning of 1947 to $164 million at the close of 1950.[41] Investments in land and buildings, meanwhile, grew by just $7.5 million. At the close of 1950, net investment in such resources stood at just $18.5 million. Four years later, after doubling the size of its facilities at Poughkeepsie to accommodate work on the navigational computer, breaking ground on the SAGE manufacturing facility at Kingston, and constructing a new lab and plant near Endicott for work on bomb sights, net investment in land and buildings had reached $52.1 million. Facilities now accounted for 14 percent of total net investments, compared to just 10 percent before the defense buildup.[42]

The lion's share of investment during the early 1950s still went toward rental machines and equipment, whose net value again doubled over the course of four years, reaching $325 million by the end of 1954. But this investment did not yet flow toward a new generation of digital electronic computers. The only primarily electronic product to reach significant production volumes during these years was the 604 calculator, which remained the bread-and-butter product of Poughkeepsie. Together with the traditional punched card equipment, it accounted for in excess of 95 percent of the revenue generated from accounting equipment and data processing machines.[43] The 701 defense calculator and its companion 702 were not announced until the spring of 1953 and did not reach customers until late 1954 and early 1955. With total volumes in the teens, at the time of announcement they accounted for just 100 of the 6,000 employees at Poughkeepsie and occupied just 20,000 square feet of the factory floor.[44] Work on electronics did occupy a new 179,000 square foot laboratory completed at Poughkeepsie in 1954, and with monthly rental prices ranging from $12,000 to $18,000, the 700 series machines that did reach the field represented a substantial investment.[45] But most of that investment occurred after 1954. The same was true of the 650, the new drum-based computer built at Endicott. Finally announced in July 1953 with an average monthly lease price of $3,500, it did not reach the field until December 1954.[46]

In the short run, at least, the play for defense contracts thus left IBM in a significantly weakened position. The promised move into electronic computing remained unfulfilled. Gathering to assess the situation in the spring of 1955, management drew into question the entire effort to respond to Univac. Even with a rental price of $30,000 per month, the 705 could never generate substantial revenue, because few customers were ready to make such a revolutionary break in their data processing routines. IBM "could easily be caught in a trap trying to win the 'horsepower' race." Its "major emphasis should be placed on lower priced equipment... designed to allow thousands of companies to get into electronic data processing on a bread-and-butter basis." Yet IBM appeared vulnerable in that niche, with competitors such as NCR poised to move "very rapidly."[47] Despite IBM's hefty investments and healthy increases in gross sales, after-tax income actually slipped below 1950 levels for two years running and barely surpassed them in 1953. In its annual report for that year, IBM management felt compelled to defend its dividend practices, under which it paid out just a third of its profits in cash while offering the rest as shares. It explained that share dividends helped preserve funds for investment. IBM's annual statements during these years also featured an entry for development and engineering expenses, which the firm listed as totaling $53.4 million for the years 1951 through 1954, more than three times the level of the previous four years.

Achieving Commercial Success

Such assurances soon appeared justified, as IBM's financial picture brightened considerably during the mid-1950s. By 1958, annual gross sales had climbed to nearly $1.2 billion, more than two-and-a-half times the level of 1954.[48] Net income after taxes grew even more impressively, by a factor of 2.7, to more than $126 million annually. Such growth enabled IBM to satisfy stockholders with substantially larger dividend payments (and increased share values) while diverting a larger percentage of its income into retained earnings and investment. Of the $340 million in net income after taxes earned from 1955 through 1958, IBM paid out less than $93 million in cash dividends, or just 27 percent. During the previous four years, cash dividends had absorbed nearly 38 percent of posttax net income. Share dividends, which had covered the remaining 62 percent during the 1951–1954 period, claimed only 35 percent during the 1955–1958 period. IBM thus built its retained earnings by $129 million to more than $221 million. Its stock of net current assets soared as well, to nearly $400 million, quadruple the level of four years earlier.[49]

While this performance certainly provides impressive testimony to IBM's success in bringing electronic computing to the commercial market, it does not necessarily confirm Watson's adage about the importance of securing defense contracts. A great deal of the earnings retained during the mid-1950s came from products not associated with the defense projects, and much of the investment they made possible went toward building the installed base of those products. Prominent among these were the 650 drum computer and the 305 RAMAC (Random Access Memory Accounting Calculator). By November 1955, when not a single 704 or 705 had yet left the plant at Poughkeepsie, IBM had installed 123 of the 650s.[50] Within another year, production had surpassed 500 units, and the 650 was well on its way to becoming the workhorse computer of many universities and businesses. Nearly 2,000 would eventually enter the market.[51] The 305 RAMAC, announced in 1955, got off to a slower start and achieved only 20 percent of its anticipated production run of 5,000 units.[52] This was still an impressive number in its day, however, and meanwhile the disk storage systems associated with RAMAC became a valuable peripheral device when used in conjunction with a 650 or another electronic computer.

The success of such products set IBM scrambling to expand capacity. By the end of 1958, net investment in land and buildings exceeded $146 million, an increase of more than $94 million in just four years. Some 22 percent of IBM's net investment was then in the form of land and buildings. Some of this investment went toward the plant at Kingston and another at Owego, New York, intended to build navigational computers for bombers, but IBM also added some 2 million square feet of laboratory and plant space to support initiatives such as RAMAC, the 650, and other segments of its product line aimed at users in the middle and lower range of performance.[53] A report issued in late 1955, when IBM authorized the Owego facility and created the separate Military Products Division, indicated that of the 33,000 IBM employees projected to be located at plants and laboratories in upstate New York when the facilities at Owego and Kingston came on line, just 25 percent would be assigned to military contracts.[54] Recently constructed plants in California, Kentucky, and Minnesota focused entirely on commercial products.

As in previous years, net investment in rental machines and equipment grew less rapidly on a percentage basis but absorbed greater absolute sums of capital. At the close of 1958, this figure stood at $528 million, an increase of $213 million over four years. The amount might have grown more dramatically if not for two important changes in business practices. Under the terms of a 1956 consent decree, IBM began selling machines as well as leasing them.[55] At about this same time, it switched from straight-line depreciation of installed equipment to a years-digits method, effectively accelerating depreciation rates.[56] In conjunction with these changes, IBM significantly increased the funds it reserved for depreciation. In 1954, the company had maintained a reserve of $233 million against its installed stock of $549 million in rental machines and equipment, a ratio of 42.6 percent. At the end of 1958, the reserve ratio stood at 52.7 percent ($589 million against $1,117 million). These figures reflected a shift in IBM's business model toward more rapid turnover of more expensive machines.

The combined burdens of expanding its facilities, installing costly new machines in the field, and bolstering its reserves drastically exceeded what even IBM's impressive retained earnings could cover. Despite his earlier vow to hold the line on debt, Watson turned once more to Prudential, increasing the credit line another $175 million over four years. At the close of 1958, IBM's long-term debt stood at $425 million. Even this failed to cover the massive expenditures of 1957, when net investment in rental machines jumped by $150 million, or 38 percent, and that in facilities grew by $50 million, or a staggering 63 percent in a single year. For the first time since 1925, IBM issued a new public offering. The slightly more than 1 million new shares tendered to existing stockholders, a 10 percent increase in total shares outstanding, pumped $226 million in new capital into IBM.[57]

The burst of 1957 proved something of a watershed for IBM. For several subsequent years, large write-offs of obsolete rental machines significantly offset new expenditures, leaving the pace of additional net investment flat or in retreat. As the dust settled, Watson and other top managers struggled to assess where the whirlwind had left them. Reluctantly, Watson came to accept a cold reality: Endicott had carried the day. As he explained to a large gathering of management personnel in 1959, products designed and built at Endicott and its satellite facilities accounted for some 70 or 80 percent of IBM revenues. The more glamorous activities at Poughkeepsie and vicinity contributed less than 30 percent, while absorbing over 70 percent of funds invested in product development efforts.[58]

Although an outside observer might have taken this as a sign of success in transferring knowledge from the military side to the commercial, Watson and his management team interpreted it far differently. In fact, as Watson confessed to the assembled crowd, the situation left them feeling "scared." Watson and his colleagues understood that considerable knowledge had in fact flowed in precisely the opposite direction. Workhorses such as the 650 drum computer and the disk files, with their mobile electronic "heads" floating just above the surfaces of rapidly rotating magnetic surfaces, were essentially mechanical marvels. Skilled manufacturing personnel steeped in the mechanical culture of Endicott, working in collaboration with programmers and materials scientists, had played a large role in making them a reality. All the while, these technicians continued producing updated versions of the associated peripheral equipment, such as card punches and readers and printers, which remained essential parts of any data processing installation even after the rise of electronic central processors and memory units.

Nothing demonstrated the persistent importance of Endicott more vividly to Watson than his ongoing struggles to transfer circuit manufacturing techniques from the SAGE project to the commercial line. As is often the case with military projects, SAGE funneled resources toward novel techniques of arcane design. The project called for huge volumes of circuits built in identical format. Computers aimed at the various segments of the commercial market required a much more diverse array of circuits. This was especially true in the early decades of computing, when the high cost of memory severely limited the ability of programmers to tailor computers to perform different functions. In order to make use of the automated production techniques developed at Kingston, IBM engineers had to modify the equipment so that it could be programmed to build circuits of many different variations. Even then, the techniques met with strong resistance from computer designers, especially those working on high-end machines that emphasized high performance along narrow lines rather than low cost and enhanced flexibility. Watson had to issue a blanket order in late 1957 compelling all designers to use circuits built in the standard format.[59] Within months, designers engaged in a military-funded project known as STRETCH had abandoned the basic form and produced a circuit design that defied automated production.[60]

The critical modifications in circuit assembly technology emerged not from Kingston or Poughkeepsie, the home of military-funded projects, but from a small group at Endicott known as manufacturing engineering research. Its roots went back to the old electromechanical days when products moved from laboratory prototype to factory floor through a team of experts who looked for ways to manufacture the new model as economically as possible. Under the influence of Palmer, the manufacturing engineering force had squirreled away some funds from its work on specific product development programs in order to support research into techniques that might prove beneficial across the product line. Ed Garvey, a mechanical engineer who had helped design equipment to build circuits for the 604 and the 701, headed the group. Garvey added a cohort of more highly educated employees familiar with chemical process manufacture and with programming techniques that might be used to build greater flexibility into automated assembly equipment.[61] Palmer convinced Watson to embrace Garvey's approach to circuit manufacture, rather than some alternatives being explored at Poughkeepsie, because he believed people steeped in the manufacturing culture of Endicott were far more likely to bear down and make the critical compromises necessary to transform prototypes into reality.[62] Several years later, Garvey would ride to prominence once more when IBM summoned him from Endicott to help instill order and discipline over the new silicon chip manufacturing facility it had opened near Poughkeepsie in conjunction with System/360.[63]

Incorporating Solid State

The critical decision to integrate backward and invest in solid-state design and production capabilities flowed directly from the management meeting of 1959. On that occasion, Watson announced a thorough reorganization designed to right the balance and funnel development funds more directly toward the modest commercial area. Plans for System/360, which had coalesced by late 1961, essentially manifested that aim.[64]

In a broader sense, however, the move into solid state capped a decade of frustration for Watson and other managers, as their visions of defense projects providing a conduit for electronics innovations repeatedly ran afoul. The 702 computer, a commercial machine developed in companion with the 701, languished at Poughkeepsie as engineers focused on meeting the pressing deadlines of the defense machine. The successor

705, at last pushed out the door in January 1956, eventually gained a significant foothold in the commercial market, but its designers balked at incorporating the techniques Watson had identified as critical to the future success of IBM. They shunned printed circuits and agreed to use ferrite core memories developed in conjunction with SAGE only after Watson installed a product "czar" who dictated the change from above.[65] Most frustrating of all, engineers at Poughkeepsie and Kingston failed to use the latest sensation in solid-state electronics: the transistor. On one memorably humbling occasion, an exasperated Watson assembled his principal design engineers around a conference table and bombarded them with circuit boards removed from transistorized radios. Why could IBM engineers not make use of such techniques, Watson demanded to know, in machines that rented for thousands of dollars per month?[66] In a calmer moment Watson secured the services of Mervin Kelly, the research director at Bell Labs who had supervised work on transistors, and asked him to conduct a thorough examination of IBM's research and development efforts.[67]

The person who bore the brunt of such outbursts and scrutiny was Ralph Palmer, who had taken responsibility for all research and development efforts connected with electronics. Palmer had not in fact neglected printed circuitry or the transistor and other solid-state components. Within the constraints imposed by IBM's established procedures, which funded little research outside specific product development programs, he had managed to pursue both printed circuitry and the transistor. A small group at Poughkeepsie had obtained point-contact transistors from Bell Labs in 1948, within months of their invention. In characteristic fashion, Palmer urged the group in 1951 and 1952 to build an experimental version of the 604 calculator using the new devices. When Bell Labs announced the alloy-junction transistor in 1953, Palmer quickly shifted these efforts toward this device, which promised greater stability. In 1954 he coupled the work on transistors with that taking place in a small facility known as the printed circuits laboratory. Created by Palmer in 1950, it had for several years explored the possibility of building a small accounting machine from printed circuits.[68]

Palmer's hope of mastering transistorized circuitry by modifying an existing product foundered during the early and mid-1950s under conflicting forces within IBM. His initial experiments suffered during the early 1950s as military contracts drained away talent and energy. Designers responsible for high-end machines such as SAGE rejected transistors as too temperamental and risky. Product planners responsible for new commercial machines at IBM were similarly reluctant to hinge the success of their development programs on unproven technology. Why saddle your project with such risks, they asked, and then have others capitalize on the learning you had supported? Product planners also strongly resisted any efforts to direct money away from their self-contained development projects.[69]

Meanwhile, Kelly and others told Watson that he needed to foster a centralized research function capable of generating components that designers could pull down off-the-shelf and incorporate into new designs. Trying to develop both the hardware and the logical design of new computers simultaneously was hopelessly difficult, Kelly insisted, and would eventually lead to disaster. Firms such as RCA and GE, which built their own advanced components, would take over the computer business. Influenced by this advice and by current fashions within corporate America, Watson in late 1955 launched a central research venture, headed temporarily by Palmer until a suitable administrator from outside could be found. A year later, the Ph.D. physicist Emmanuel Piore arrived from the Naval Research Laboratories to relieve Palmer.[70]

In Palmer's view, neither product planning nor Mervin Kelly had the right idea. Product planners remained overly concerned with the needs of the most sophisticated customers and with making sure that machines placed with those customers would cover the cost of development. Their approach did not give designers significant latitude to embrace new departures in basic components. Centralized research of the sort advocated by Kelly and Piore, on the other hand, was too remote from commercial endeavors. Contrary to the advice of Kelly, Palmer believed in coupling component work closely to machine design. When Piore arrived, he branded the work Palmer had supported at the laboratory as development rather than research. Piore took the new laboratory at Yorktown Heights off in dramatic new directions tied much more closely to the physical sciences.[71]

Palmer's preferred solution to this impasse was to consolidate product development efforts in a comprehensive, modular line that would be built from circuits of standard design.[72] This was the vision that would eventually take hold at IBM between 1959 and 1961 under the leadership of product planning czar T. Vincent Learson and would culminate in early 1964 with the announcement of System/360. Failing in that objective in the mid-1950s, Palmer instead embraced the standard circuit format developed by Garvey, which he hoped possessed sufficient flexibility to satisfy designers in the separate development programs. Nor would designers flinch at incurring the cost of acquiring the new circuits. With purchased components projected to account for 80 percent of total costs, IBM could launch the standard circuit program with little capital investment. Garvey received $1.4 million at the time of announcement in November 1957, and budgets for subsequent years called for expenditures of approximately $1 million annually.[73]

For the transistors that would be included in those circuits, IBM entered into a pioneering joint venture with Texas Instruments, which along with Fairchild Semiconductor had recently made some pioneering breakthroughs with transistors made from silicon.[74] Most groups working on transistors, including those within IBM, had focused on germanium devices. Germanium was cheaper to obtain in pure form than silicon. Recent developments, however, suggested that the oxidizing qualities of silicon opened the possibility of creating a glass seal on its surface that protected the temperamental material and made it easier to manipulate during manufacture. For engineers and scientists at TI and Fairchild, such "glass passivation" opened opportunities to produce transistors (and, ultimately, entire circuits) by treating the surfaces of silicon. For engineers and scientists at IBM who had been struggling to assemble transistors using automated equipment, the surface phenomenon presented an immediate opportunity to do away with the cumbersome cans and hermetic seals used to enclose existing transistors.[75]

The contract IBM and TI entered into in December 1957 reflected these two orientations toward solid-state components. TI had assembled significant knowledge of the physics and chemistry of semiconductors and had developed sophisticated methods for manipulating those materials.

IBM had little to offer on that front. The agreement called for complete sharing of patents and knowledge of germanium devices, but guarded exchanges regarding silicon technology more closely, in part by expressly banning IBM from selling silicon devices on the open market. IBM could, however, provide two things. On the technical side, the computer maker gave TI use of the automated production technology it had developed in an effort to drive down the cost of assembling transistors and transistorized circuits. TI, which up to this time had aimed its products at small niches, including many provided by the military, had not yet cultivated production expertise of this sort. IBM thus once again found the legacy of the mechanical roots it had cultivated in pursuit of its business customers paying dividends in the world of electronics.[76]

A still larger benefit from that inheritance involved the business end of the agreement. Above all else, IBM enticed TI by offering the upstart access to its large, established market in commercial data processing equipment. IBM agreed to purchase 600,000 transistors from TI in 1959 and to buy all of its requirements in excess of 2 million from the Texas supplier during the subsequent two years.[77] IBM agreed not to acquire silicon devices from any alternative suppliers and to limit use of its own in-house silicon products to its own data processing equipment. The arrangement was projected to build TI's existing annual production of 100 million transistors by approximately 20 percent. By guaranteeing TI this highly profitable revenue stream, IBM effectively leveraged its investment in the commercial sector to gain critical skills in silicon technology.

The partnership formally lasted just over two years, until IBM decided to develop its own capabilities in silicon device manufacture.[78] Watson made this choice reluctantly. He much preferred to leave the risks and expense of mastering the complex domain of solid-state technology to others. In creating the new components division, Watson authorized expenditures of $7.2 million in 1961 and another $30 million in January 1962, and costs would continue to soar after that.[79] He was also reluctant to make individual product development managers dependent on a single internal source of supply, especially a captive one that could not sell its products on the open market and be held accountable to market-based measures of performance.[80] But Mervin Kelly and influential advisers within IBM, including long-time corporate treasurer and confidant Albert Williams, convinced Watson that with continuing technical development in solid-state, component manufacture and circuit packaging would merge into a single field. Without integrating backward, Williams told Watson, IBM might "lose its game to TI."[81] Watson then agreed to couple the new consolidated product line, which had already committed to using a new generation of circuit packaging developed by Garvey and his group, with a parallel effort to manufacture its own silicon transistors in a new facility near Poughkeepsie.[82]

Massing the Resources

This series of decisions in 1961 effectively committed IBM to an enormous, consolidated exercise in learning that would encompass virtually all of its worldwide operations and resources. IBM proposed to displace its entire existing product line, including the vast installed base, with new machines built from new technology.[83] By the time the initiative had run its course, not only had IBM regained its 70 percent share of the rapidly expanding market for computing, it had also emerged as by far the largest manufacturer of semiconductors in the world.[84] Looking back on the crucial decisions a few years later, when the success of the venture remained somewhat in doubt, a reporter dubbed it the $5 Billion Gamble. The figure did not reflect a careful assessment of dollars expended on research and product development. Rather, the reporter simply reasoned that in replacing its entire product line with System/360, IBM had gambled its entire net worth.[85]

The notion is exaggerated. IBM possessed assets that would have survived even if System/360 had failed, and the firm had laid plans for other products that did not draw on the same single source even before some of the new machines had reached the market. Still, IBM did with System/360 trade on much of its assembled assets—most important, its installed commercial base and its esteemed reputation in business and financial circles—and staked them on a single innovative venture of unprecedented scale and scope. By exploiting its assets in this comprehensive fashion and concentrating them on a single exercise in learning aimed at developing one common approach to circuit technology, the firm pushed solid-state electronic computing to new levels of performance while also substantially driving down its cost.[86]

The job of orchestrating the financing for this transition fell largely to Williams, the corporate finance officer who had insisted IBM could not avoid the strategic move into component manufacture. In a gesture intended to give Williams greater weight on Wall Street, Watson promoted him to the IBM presidency in early 1961.[87] By then, Williams and IBM had begun hoarding cash. While revenue and earnings grew at about 17 percent annually for several years after 1959, from that year through 1963, new investment lagged write-offs and depreciation by some $348 million. Nearly half the deficit came during 1963 alone, even as the firm began to acquire expensive new production equipment for silicon transistors and circuitry. To help support those activities, IBM trimmed its expenditures on product development and engineering in traditional areas. Within data systems, outlays in these categories fell from 24.4 percent of division revenue in 1959 to 13.6 percent in 1963.[88]

One cause of this austerity was the rapidly diminishing activity under military contracts. With the SAGE contract cancelled and other defense programs running out their course, IBM revenue from government contracts fell to $144 million in 1963. This constituted just 7 percent of IBM's total revenues, compared to the 77 percent generated by data processing. Four years earlier, federal contracts had accounted for more than 17 percent of total revenue, while data processing claimed a 65 percent share.[89] The plant at Kingston, once the apple of Watson's eye, now hung like an albatross around his neck. Executive conferences during the early 1960s dwelled at length on the difficulties of absorbing the plant and its workforce of several thousand into IBM's commercial operations, which "don't need it or want it."[90] When Learson inquired in mid-1961 whether remaining work under government contracts might help underwrite development of the new circuit technology, the accounting staff rebuffed him unequivocally.[91]

During this period of retrenchment, IBM funded its very modest additions to its net stock of capital entirely through retained earnings. It took on no additional debt and after 1959 curtailed its practice of issuing share dividends, claiming that it could not afford to absorb the costs while the stock value remained so high.[92] Though IBM did boost its cash dividends back up toward the 33 percent range, these still fell well below customary levels for a Fortune 25 firm. The new dividend policy left substantial amounts of cash in IBM's accounts. Of the $852 million in net income earned after taxes during 1959–1962, IBM paid out just $238 million in cash and $121 million in shares. Not surprisingly, its net current assets grew dramatically. IBM began to acquire large sums of U.S. treasury bonds and notes. By the end of 1963, its balance sheet listed nearly $1 billion in such assets.

While IBM hunkered down within the United States during the late 1950s and early 1960s, it did pump nearly $200 million of investment funds into its international subsidiary, IBM World Trade (WT), whose business was growing at a pace resembling that achieved by IBM's domestic operations during the mid-1950s. By 1963, WT gross sales totaled slightly more than 38 percent of those generated by domestic operations. Five years earlier, foreign sales had reached just 21 percent of domestic levels.[93] By the close of 1963, foreign net investment had reached $612 million, compared to $973 million for domestic operations. With more than 38 percent of IBM's worldwide net capital investment located overseas, up from less than one-quarter five years before, foreign operations stood prepared to participate as full partners in the implementation of System/360.

As that initiative went into full swing in 1964, IBM's financial pendulum swung dramatically with it. Net investment within the United States during the four years from 1964 through 1967 grew more than 140 percent, reaching over $2.3 billion. Foreign net investment nearly doubled during the same period, to just under $1.2 billion. All told, IBM increased its global net investment by more than $1.9 billion during these four years. Much of this came during 1966 and 1967, when IBM placed large numbers of System/360 installations with customers. During those two years, the net stock of capital held in the form of rental machines jumped by $1 billion worldwide. The net stock of installed rental machines in 1967, valued at $2.6 billion, was roughly two and a half times as large as in 1963. The value of plant, buildings, and equipment increased on the order of 50 percent during the same period.[94]

Funds for this massive investment came primarily from retained earnings and cash reserves, not borrowing. Though IBM dramatically restructured its debt during this period and shifted much of it overseas, in the end these maneuvers left the firm with worldwide debt in 1967 of $521 million, a slight reduction from the 1963 level of $550 million. IBM avoided taking on more debt in part because increasing numbers of consumers now purchased their computers or leased them from a third party that had bought from IBM. Buoyed by such sales, worldwide posttax income surged impressively even before the new System/360 machines reached the field. Over the course of the four years from 1964 to 1967, IBM cleared more than $2 billion in posttax profit. The firm paid out $850 million, or about 40 percent, in cash dividends. A share dividend in 1967, the first since 1959, eventually claimed another $525 million. Even setting aside sufficient funds to cover the share dividend, IBM retained $710 million in earnings over the four years.

While rapidly investing this new income, IBM for a time also spent down its cash reserves. Net current assets fell steeply during the first two years of the period and did not climb back to their 1963 levels until the close of 1967. Earlier that year, cash reserves went so low that IBM found it necessary to take out a short-term bridge loan for $200 million.[95] IBM further stretched its resources during the early years of the buildup by taking small write-offs for obsolete equipment and rental machines and declaring relatively modest depreciation expenses. In 1965 and 1966, its ratio of depreciation reserves to net stock of rental equipment dropped to 48 percent, down considerably from the 55 percent ratio that typically prevailed before System/360.[96] For a brief period, IBM even tried selling equipment and licenses connected with its automated circuit manufacturing technology.[97]

For all its cleverness in manipulating such cash flows, IBM in the end could not finance such a monumental expansion without turning once more to the capital markets. In 1966, it made its first public stock offering since early 1957. The sale of more than a million and a half shares generated some $371 million in new capital. In addition to this public offering, IBM brought increasing numbers of employees under its stock option plan. In 1967, a year prior to a two-for-one split in the stock, in excess of 300,000 shares were sold under the plan. In the throes of its transition to solid state, IBM thus began to tap a financial instrument that would later come to characterize a new generation of computing firms centered in Silicon Valley.[98]

Conclusion

The story of System/360 belongs most assuredly to an earlier era, at the height of what some have called the American Century. In that context, it has assumed a deservedly prominent spot in the annals of modern corporate enterprise.[99] By any account, the venture entailed enormous financial risks and generated impressive returns. It sparked feverish imitation at the time, among both private domestic competitors and nations that worried about American dominance of a vital industry.[100]

For all that, it must be acknowledged that IBM and System/360 never lacked for critics. Even as its new machines took hold, many viewed the company as a technical dinosaur that impeded more visionary approaches to computing.[101] The criticisms permeated IBM's own ranks. A new generation of managers, left to dig out from the messes caused by numerous delays and mistakes that accompanied the project, vowed "never again" to the idea of committing the firm to such a single comprehensive endeavor.[102] Meanwhile, Watson and some influential figures among the engineering corps ridiculed System/360 for tying the company to an archaic approach that did not take full advantage of silicon technology.[103] As it plotted to catch up, IBM found itself having to pay license fees to start-up firms such as TI and Fairchild.[104] Another small upstart, Control Data Corporation (CDC), quickly captured the high-end market for computing by using new integrated circuits obtained from the silicon pioneers. IBM's clumsy attempts to counter CDC soon landed it on the short end of a private antitrust settlement, and by 1967, its conduct had sparked what would become a monumental confrontation with the U.S. Department of Justice.[105] In tying its efforts in solid-state circuitry so closely to a single product, IBM may also have imparted an enduring conservatism to its own work on solid state. The firm soon found it difficult to seize new opportunities such as the MOS transistor, a technology exploited much more successfully by Fairchild and its successor, Intel.[106]

Yet it is also true that System/360 introduced a useful measure of stability into computing at a time when the industry might have suffocated under a staggering array of independent approaches.[107] The resulting uniformity set clear targets for competitors such as Digital Equipment Corporation, which soon made significant inroads in major segments of the market.[108] System/360 also established a vast market for computer products developed by more specialized firms, such as software and service providers and optional equipment manufacturers. Bolstered by strong antitrust interventions aimed at curbing IBM, such firms had emerged as a powerful force on the computing scene by the late 1960s.[109] Even the chip maker Intel, the famed entrepreneurial start-up, achieved its first major commercial success by building solid-state memories for the successors to System/360. The learning associated with System/360 thus extended far beyond the confines of IBM itself.

Whatever the ultimate impact of System/360, the convoluted path that brought it into existence clearly suggests that far more was involved in the establishment of solid-state electronic digital computing than the linear transfer of military-sponsored research to commercial products. The extended exercise in learning that culminated in a new prevailing platform for computing occurred across many fronts, and the lessons did not flow in a single direction. Certainly IBM benefited from the activities of government, especially its procurement of advanced calculating equipment and its more broad-based efforts in training and research among a large community of suppliers in electronics. Yet government could also exert a distorting influence, one that drew IBM away from opportunities in the commercial market that might well have blossomed without significant contribution from the public sector. Seldom did lessons learned in a military context transfer readily to the commercial sector. Often the most important learning took place when skills and routines developed for the commercial sector were brought to bear on the undertaking. The critical transition to solid state occurred only when IBM marshaled both its market power and the acquired expertise of its entire organization toward a single objective. Getting the computer business, as it turned out, was not quite as simple as getting the SAGE contract.

Notes

1. Kenneth Flamm, *Targeting the Computer: Government Support and International Cooperation* (Washington, D.C.: Brookings Institution, 1987), and *Creating the Computer: Government, Industry, and High Technology* (Washington, D.C.: Brookings Institute Press, 1988).

2. Thomas P. Hughes et al., *Funding a Revolution: Government Support of Computing Research* (Washington, D.C.: National Research Council, 1999). See also Arthur Norberg and Judy O'Neill, *Transforming Computer Technology: Information Processing for the Pentagon, 1962–1986* (Baltimore, Md.: Johns Hopkins University Press, 1996).

3. David C. Mowery, "Innovation, Market Structure, and Government Policy in the American Semiconductor Electronics Industry: A Survey," *Research Policy* 12 (1983): 183–197; David C. Mowery, ed., *The International Computer Software Industry* (New York: Oxford University Press, 1996), especially the chapters by W. Edward Steinmuller and by Richard N. Langlois and David C. Mowery; and David C. Mowery and Richard R. Nelson, eds., *Sources of Industrial Leadership: Studies of Seven Industries* (Cambridge: Cambridge University Press, 1999), Chaps. 1–4, 9. For another perspective on the software industry emphasizing the role of private enterprises, see Martin Campbell-Kelly, *A History of the Software Industry: From Airline Reservations to Sonic the Hedgehog* (Cambridge, Mass.: MIT Press, 2003).

4. In addition to Flamm, see Paul N. Edwards, *The Closed World: Computers and the Politics of Discourse in Cold War America* (Cambridge, Mass.: MIT Press, 1996).

5. Charles J. Bashe et al., *IBM's Early Computers* (Cambridge, Mass.: MIT Press, 1986), 135–164; and Emerson W. Pugh, *Building IBM: Shaping and Industry and Its Technology* (Cambridge, Mass.: MIT Press, 1995), 167–174, 185–186.

6. Thomas J. Watson, Jr., interview with Steven W. Usselman and Richard Wight, March 27, 1986, Armonk, N.Y. See also Thomas J. Watson, Jr., and Peter Petre, *Father, Son & Co.: My Life at IBM and Beyond* (New York: Bantam Books, 1990), 230–233.

7. Watson, Jr., interview with Usselman and Wight, March 27, 1986, and Richard Whalen interview with Richard Wight, Boulder, Colo., November 20, 1985.

8. On references to the transistor as an automation technique, see especially "Report on the Skytop Conference, December 1955," IBM Corporate Management Papers, IBM Corporate Headquarters, Armonk, N.Y., reporting on an executive conference held at Skytop, Pa., November 8–11, 1955, and related letters in the Thomas J. Watson, Jr., Papers, IBM Corporate Archives, Somers, N.Y. See especially letters from L. H. LaMotte to T. Vincent Learson, December 13, 1955, W. McDowell to G. E. Jones, January 3, 1955, McDowell to LaMotte, January

12, 1956; E. E. Witter to LaMotte, January 13, 1956; and McDowell to T. J. Watson, Jr., January 19, 1956.

9. Emerson W. Pugh, *Memories That Shaped an Industry: Decisions Leading to IBM System/360* (Cambridge, Mass.: MIT Press, 1984).

10. Watson, Jr., interview with Usselman and Wight, March 27, 1986. Discussion about pricing the 701 can be traced in the Watson, Jr., Papers. See esp. C. Benton, Jr., to Watson, Jr., April 4, May 2, June 13, October 9, 1951, February 18, 1952; W. W. McDowell to Watson, Jr., November 19, 1951; McDowell to J. G. Phillips, February 6, November 28, 1951; and McDowell to A. L. Williams, February 18, 1952.

11. Thomas J. Misa, "Military Needs, Commercial Realities, and the Development of the Transistor, 1948–1958," in Merritt Roe Smith, ed., *Military Enterprise and Technological Change: Perspectives on the American Experience* (Cambridge, Mass.: MIT Press, 1985), 253–287, and Stuart W. Leslie, "The Biggest 'Angel' of Them All: The Military and the Making of Silicon Valley," in Martin Kenney, ed., *Understanding Silicon Valley: The Anatomy of an Entrepreneurial Region* (Stanford, Calif.: Stanford University Press, 2000), 48–67. On government and the emergence of solid-state electronics more generally, see Richard C. Lewin, "The Semiconductor Industry," in Richard R. Nelson, ed., *Government and Technical Progress: A Cross-Industry Analysis* (New York: Pergamon Press, 1982), and Mowery, "Innovation, Market Structure, and Government."

12. Two useful recent works on the elder Watson are Kevin Maney, *The Maverick and His Machine: Thomas Watson, Sr., and the Making of IBM* (New York: Wiley, 2003), and Richard S. Tedlow, *The Watson Dynasty: The Fiery Reign and Troubled Legacy of IBM's Founding Father and Son* (New York: Harper Business, 2003). See also Pugh, *Building IBM*, and Robert Sobel, *IBM: Colossus in Transition* (New York: Times Books, 1981).

13. Sobel, *IBM*, 211–212.

14. Because this figure included the value of equipment used in IBM's own facilities, not all of this 85 percent constituted an installed base of machines generating a stream of leasing revenue for IBM. But the vast majority was in that form. When the company rearranged its accounts in 1965, creating a distinct category for rental machines, it shifted some $761 million worth of equipment from the rental machine and equipment account to the land and building account. While this may seem like a massive sum when compared to the figures from two decades before, the $761 million actually constituted less than 17 percent of the $4,552 million net stock of capital IBM reported for 1965 (not adjusted for reserves for depreciation, which lowered total net investment by roughly half). The year before this reallocation, rental machines and equipment together accounted for more than 88 percent of the total capital stock. With equipment shifted to land and buildings and new additions made across all categories, the share attributed to rental machines in 1965 still stood at more than 73 percent.

In the decade or so following World War II, rental machines in all likelihood accounted for an even greater share of accumulated capital. In 1965, the stock of equipment included expensive new plants for fabricating silicon devices and packaging them automatically into printed circuits. Such massive investments were a significant departure for IBM, which historically had assembled its products from purchased parts. Since signing a consent decree with the Department of Justice in 1956, moreover, IBM had sold as well as leased its products. Though rentals continued to generate the vast majority of IBM's revenue for many years following the decree, by the early 1960s, sales had reached levels sufficient to prompt IBM management to warn stockholders about subsequent declines in the rental stream. With the swing toward sales, capital shifted inexorably from rental machines to land, building, and equipment. Indeed, as sales mounted over the course of the 1960s, the share of net investment in rental machines drifted steadily downward. By 1971, rental machines accounted for less than 69 percent of the net stock of capital. With deductions for reserves, which were proportionally larger for rental machines than for other forms of capital, the share of net investment held in the form of rental machines had fallen to less than 64 percent. All figures come from IBM *Annual Reports*.

15. IBM *Annual Report* (1946).

16. My discussion of IBM practices regarding innovation during the 1940s and 1950s is based on my reading of Bashe, *IBM's Early Computers;* Pugh, *Building IBM;* Maney, *The Maverick and His Machine;* and sources in the IBM Corporate Archives and Management Papers, cited in more detail below. For a look at these practices from the perspective of Prudential and other consumers in the insurance industry, see JoAnne Yates, *Structuring the Information Age: Life Insurance and Technology in the Twentieth Century* (Baltimore, Md.: Johns Hopkins University Press, 2005).

17. In addition to sources cited below, perspective on these processes comes from various interviews in the IBM Archives, including Richard Wight interview with Clarence Frizzel, February 20–22, 1985, Saratoga, Calif., and IBM archival sources on product planning, product managers, and other topics.

18. Planning documents for 1954 indicate that IBM proposed to spend just $1.4 million and employ seventy-seven professional personnel in its three research laboratories. An additional $2 million and eighty-eight professional personnel would go toward work on electronic components independent of specific product development efforts. "IBM Development Engineering Manpower and Estimated Expense for Year 1954," accompanying J. W. Birkenstock to A. L. Williams et al., February 12, 1954, Williams Papers, IBM Archives. The IBM *Annual Report* for that year listed total engineering and development expenses in excess of $14 million. The number of employees involved in research and development, including product engineering, totaled approximately 3,000 in 1954. Bashe, *IBM's Early Computers*, 542.

19. Bashe, *IBM's Early Computers*, 34–46,

20. On the wartime explosion of demand for data processing, see James W. Cortada, *Before the Computer: IBM, NCR, Burroughs, and Remington Rand and the Industry They Created, 1865–1956* (Princeton, N.J.: Princeton University Press, 1993), and Thomas K. McCraw, *American Business: How It Worked, 1920–2000* (Wheeling, Ill.: Harlan Davidson, 2000), 73–94.

21. IBM *Annual Report* (1946) contains revenue data and details on the Poughkeepsie plant, which was purchased for $3 million under military certificate of necessity.

22. Bashe, *IBM's Early Computers*, p. 46. IBM produced approximately 100 of these machines.

23. T. V. Learson to L. H. LaMotte, September 3, 1952, LaMotte Papers, IBM Corporate Archives. Watson authorized production of the first twenty-three IBM 604s on December 19, 1947. "Summary of Products Committee Meeting Held on October 30th and 31st, 1947," T. J. Watson, Sr., Papers, IBM Corporate Archives. The first machine reached the field in fall 1948. All told, some 5,600 of the 604s went into the field in various arrangements. Bashe, *IBM's Early Computers*, 59–68.

24. Bashe, *IBM's Early Computers*, 68–72, 84–86, and Atsushi Akera, "Voluntarism and the Fruits of Collaboration: The IBM User Group, Share," *Technology and Culture* 42 (2001): 710–746.

25. "Summary of Products Committee Meeting Held on October 30th and 31st, 1947," T. J. Watson, Sr., Papers, IBM Corporate Archives.

26. On Palmer's philosophy, see Bashe, *IBM's Early Computers*, 541–542, 545, and Palmer's July 1967 interview with L. M. Saphire, IBM Corporate Archives.

27. Arthur L. Norberg, *Computers and Commerce: A Study of Technology and Management at Eckert-Mauchly Computer Company, Engineering Research Associates, and Remington Rand, 1946–1957* (Cambridge, Mass.: MIT Press, 2005).

28. Watson, Jr., interview with Usselman and Wight, March 27, 1986, and Watson, Jr., and Petre, *Father, Son & Co.*, 136–137.

29. Norberg, *Computers and Commerce*.

30. Yates, *Structuring*, 113–150, and Norberg, *Computers and Commerce*, 73–116, 167–208.

31. Bashe, *IBM's Early Computers*, 102–103, 114–116.

32. Watson, Jr., interview with Usselman and Wight, March 27, 1986.

33. Norberg, *Computers and Commerce*, 205–208.

34. "Highlights of Meeting of Manufacturing held Saturday, April 14, IBM Homestead, Endicott," April 30, 1951, Watson, Sr., Papers, IBM Archives.

35. On the naval ordnance machine, approved in October 1950, see Bashe, *IBM's Early Computers*, 132–133, 181–183. The navigational computer contracts were approved in early 1951. IBM Board Minutes, January 10, 1951, Armonk, N.Y.

36. On decisions to go forward with the 701 and charge for it as a commercial product, see Bashe, *IBM's Early Computers*, 133–136, and n. 42 below. On the notion of the military creating a market for computing, see Steven W. Usselman, "IBM and Its Imitators: Organizational Capabilities and the Emergence of the International Computer Industry," *Business and Economic History* 22, 2 (Winter 1993): 1–35, reprinted in David E. H. Edgerton, *Industrial Research and Innovation in Business* (London: Edgar Elgar, 1996), and Steven W. Usselman, "Fostering a Capacity for Compromise: Business, Government, and the Stages of Innovation in American Computing," *Annals of the History of Computing* 18, 2 (Summer 1996): 30–39.

37. Pugh, *Memories That Shaped an Industry*.

38. Richard Whalen interview with Richard Wight, November 20, 1985. See also E. J. Lorenz, "A History of Printed Wiring in IBM" and "History of Automatic Component Assembly," fall 1955, IBM Legal Papers, Printed Circuit Capabilities, PT 07838/0001.

39. This and other financial data come from IBM *Annual Reports*.

40. T. J. Watson, Jr., to A. L. Williams, September 15, 1954, Williams Papers, IBM Archives. At that point, IBM's debt stood at $250 million, but Watson had already secured an additional $100 million from Prudential. He sought to hold the line at $350 million.

41. Reserves for depreciation grew more modestly, from $70 million to $110 million.

42. Summary data on plant expansion can be followed in IBM *Annual Reports*. For more specific details regarding Kingston, see IBM Board Minutes, September 29, 1953, March 9, December 7, 1954, June 5, September 11, 1956, and IBM *Poughkeepsie News*, November 20, 1953, January 28, 1962. On the expansion at Poughkeepsie, see IBM Board Minutes, May 8, June 6, 1951, and IBM *Poughkeepsie News*, May 22, 1954.

43. "IBM—Electric Accounting Machine Division—Estimated Income Statement Segregating Regular EAM and EDPM Business—Ten Months ended October 31, 1955," L. H. LaMotte Correspondence, IBM Corporate Archives.

44. When their successors the 704 and 705 moved into production in late 1955, with volumes that together totaled about seventy installed machines by the end of 1956, they still engaged just 650 workers and took up only 80,000 square feet of factory space. Information on personnel, space, and production volumes comes from T. E. Clemmons to T. J. Watson, Jr., November 7, 1955; A. L. Williams to K. L. Snover, January 18, 1956; W. J. Mair to K. L. Snover, January 20, 1956; and Snover to Williams, January 25, 1956; all in Watson, Jr., Papers; and IBM *Poughkeepsie News*, March 16, 1956.

45. IBM *Record*, June 8, 1953. Consternation over appropriate charges for the 701 had occupied Tom Watson and his associates for two years, as engineering design costs soared. See W. W. McDowell to J. G. Phillips, February 6, 1951; T. V. Learson to T. J. Watson, Jr., et al., March 29, 1951; J. W. Birkenstock to

Learson, April 2, 1951; C. Benton, Jr., to Watson, Jr., et al., April 4, May 2, June 13, October 9, 1951; McDowell to Watson, Jr., November 19, 1951; McDowell to J. G. Phillips, November 28, 1951; McDowell to A. L. Williams, February 18, 1952; and Benton, Jr., to Watson, Jr., February 18, 1952; Watson, Jr., Papers, IBM Corporate Archives. Development costs had exceeded $1 million by November 1951 and an estimated $600,000 more would be required before release for production.

46. IBM *Business Machines*, July 1953, and IBM press release, December 1954.

47. "EDPM Review," April 4, 1955, attached to T. V. Learson to L. H. LaMotte et al., April 13, 1955, LaMotte Papers. A retrospective marketing assessment produced in 1960 indicated that from 1953 through 1955, total rental revenue lagged that expected based on the established 18.5 percent annual growth rate by some $16 million. CMC Minutes, November 16, 1960.

48. The surge more than made up for the sluggish growth of the early 1950s. By 1959, annual rental revenue from data processing was $227 million more than that anticipated based on the targeted 18.5 percent annual growth from 1953 levels. CMC Minutes, November 16, 1960. Eighty-nine percent of the excess came from "the opening of new markets in scientific and engineering applications."

49. IBM *Annual Reports*.

50. T. E. Clemmons to T. J. Watson, Jr., November 7, 1955, Watson, Jr., Papers.

51. Bashe, *IBM's Early Computers*, 171–172, 186.

52. Information on RAMAC and the San Jose facility comes from two internal histories written by David W. Kean: *IBM San Jose—The First Quarter Century* and *IBM San Jose—A Quarter Century of Innovation*. Work on the product, which was not formally announced until September 1956, can be traced in the correspondence of L. H. LaMotte in the IBM Corporate Archives, especially that with F. J. Wesley. The team developing RAMAC briefly detoured from a disk to a drum approach in 1953, when the U.S. Air Force expressed interest in a massive storage system for inventory control, but Wesley helped reorient the project toward the commercial market. See "The IBM 350 RAMAC Disk File," pamphlet prepared by IBM in commemoration of the disk file being designated a national historic landmark by the American Society of Mechanical Engineers, February 27, 1984. The decision to build the plant occurred in June 1955. See W. W. McDowell to L. H. LaMotte, May 2, 1955, and IBM Board Minutes, August 16, 1955. Information on the manufacturing of RAMAC comes from Wight interview with Frizzel, 1985.

53. IBM *Annual Reports* provides information on plant construction. On Owego, see also IBM Board Minutes, September 15, 1955.

54. IBM Board Minutes, September 15, 1955, and IBM *Poughkeepsie News*, May 6, 1960.

55. *United States v. International Business Machines Corporation* 1956 U.S. District LEXIS 3992, 1956 Trade Cas. (CCH) P68, 245 (January 25, 1956). The Department of Justice had filed suit four years earlier, on January 21, 1952. For IBM's response to the suit, see IBM *Annual Reports* and Steven W. Usselman, "Public Policies, Private Platforms: Antitrust and American Computing," in Richard Coopey, ed., *Information Technology Policy: An International History* (New York: Oxford University Press, 2004).

56. IBM *Annual Reports*, and Sobel, *IBM*, 240–241.

57. IBM *Annual Report* (1957).

58. IBM Organization Chart, May 20, 1959, with accompanying remarks by Thomas J. Watson, Jr., and T. Vincent Learson, Organization Papers, IBM Corporate Secretary's Office, Armonk, N.Y. Four years later, General Products still accounted for over two-thirds of domestic revenue and earnings generated from data processing equipment, even though Data Systems had taken responsibility for one of its machines. Data Systems did return 18.6 percent on invested capital, compared to 17.1 percent for General Products. "Financial Highlights, 1963" and "General Company Expense Apportionment," CRM Files, January 29, 1964. That same year, General Products counted 3,875 employees under the heading "engineering." See "GPD Long Range Plan, July 14, 1964, CRM Files, July 16, 1964. Data Systems counted 4,925 employees under this heading. See "DSD Long-Term Plan," CRM Files, July 17, 1964.

59. The letter went out on October 2, 1957, under the signature of director of research and engineering W. W. McDowell. Legal Records, Printed Circuits Capability, PT07825/0013. See also E. M. Piore to T. J. Watson, Jr., July, 31, 1963, Watson, Jr., Papers.

60. Bashe, *IBM's Early Computers*, 444–445. Edward Garvey, program manager for the standard circuit program (known as SMS, or standard modular system), eight months after its creation in October 1957 reported "serious difficulty... in convincing IBM engineers" to use printed circuits of the standard form. E. J. Garvey to W. W. Simmons et al., July 2, 1958, conveying slides from presentation on "SMS Top Priority Meeting," White Plains, N.Y., June 30, 1958, IBM Legal Records, Printed Circuits Capability, PT07829/0043. Garvey later recalled that at the time of announcement, SMS was "to have 22 part numbers and no more." E. J. Garvey interview with P. H. Andreason, July 1, 1971, IBM Legal Papers, Printed Circuits Capability, PT000000/1251. By 1961, the part numbers had proliferated to more than 1,400, and by the following year they exceeded 2,000. "SMS Bulletin," c. 1963, IBM Legal Records, Printed Circuits Capability, PT08525/0095.

61. Lorenz, "History of Printed Wiring" and "History of Automatic Component Assembly"; Edward J. Garvey interview with Steven Usselman, Fort Meyers, Fla., November 5, 1985; and A. H. Johnson, "The Standard Modular System," IBM Technical Report, August 29, 1958, IBM Legal Papers, Printed Circuit Capability, PT07828/0036.

62. Palmer interview with Saphire, 1967, IBM Corporate Archives; Garvey interview with Usselman, 1985; and Garvey interview with Andreasan, 1971.

63. Garvey interview with Usselman, 1985, and E. J. Garvey to L. L. Horn, December 23, 1965, Watson, Jr., Papers.

64. These developments can be traced in the files of the Corporate Management Committee, the Executive Conference Papers, and the Watson, Jr., Papers. A watershed moment came in early January 1962 when the group executive responsible for the System/360 initiative, Vin Learson, reported on the program to Watson. T. V. Learson to T. J. Watson, Jr., and A. L. Williams, January 15, 1962, Watson, Jr., Papers. Though official corporate embrace came several months later, by the start of the year momentum had built behind the new product line using Solid Logic Technology (SLT), transistorized circuits built within IBM. For additional perspective, see Pugh, *Building IBM*.

65. Bashe, *IBM's Early Computers*, 251–254. Watson was later frustrated to learn that the 704 and 705 used different arrangements of core memory. W. W. McDowell to T. J. Watson, Jr., January 19, 1956, Watson, Jr., Papers.

66. Watson, Jr., interview with Usselman and Wight. Several engineers refer to the episode in the Saphire interviews conducted in 1967 and 1968, copies in IBM Corporate Archives. Watson's concern with getting transistors into commercial products dates from at least from early 1954, when management entertained the idea of displaying a one-of-a-kind accounting machine built from transistors in the IBM Hall of Machines in advance of official announcement of the commercial version. F. J. Wesley to L. H. LaMotte, July 20, 1954, with attachments, LaMotte Papers. His interest and frustration mounted throughout 1955, with major top-level meetings on the subject in April and November. On April 1, Watson told his director of engineering that IBM "should proceed with our work on the assumption that we should really get into the transistor business on a long-term basis." W. W. McDowell to A. L. Williams, April 1, 1955, LaMotte Papers. See also W. W. McDowell to L. H. LaMotte, March 31, 1955; Williams to J. W. Schnackel, April 6, 1955; Schnackel to Williams, April 11, 1955; T. V. Learson to LaMotte, April 13, 1955, with attached "EDPM Review" dated April 1, 1955; and LaMotte to Leason, April 13, 1955, LaMotte Papers.

67. On Mervin Kelly, see Watson interview with Usselman and Wight, 1986, and B. N. Slade interview with Steven Usselman, Poughkeepsie, N.Y., November 4, 1985.

68. Lorenz, "History of Printed Wiring" and "History of Automatic Component Assembly," and Bashe, *IBM's Early Computers*, 372–395.

69. J. W. Birkenstock to A. L. Williams et al., February 12, 1954, and T. J. Watson, Jr., to T. V. Learson et al., May 10, 1955, LaMotte Papers; and "Report on the Skytop Conference, December 1955" and related letters, CMC Files and Watson Papers. For similar issues in 1951 and 1952, see Bashe, *IBM's Early Computers*, 143.

70. Bashe, *IBM's Early Computers*, 544–550.

71. By 1963, the direction of IBM Research activities under Piore had become a grave concern of Watson and the management committee. At the CMC meeting of November 18, 1963, Watson grilled Piore, emphasizing "that the Corporation expects Research to provide devices and technologies which can be incorporated into the product development programs of the other divisions. He said further that the Corporation has been disappointed that Research has not provided more such devices and technologies as a result of its operations to date." CMC Minutes, November 18, 1963. The CMC froze the research budget and requested a review of current projects. Reporting back a year later, G. L. Tucker, who directed research under Piore, told the committee "that he will insure that Research does not repeat the past error of overshooting by targeting on exotic developments which are too far removed from IBM's requirements." CMC Minutes, November 13, 1964. Earlier that year, when Piore told Watson he would need additional researchers in order to sketch a plan for future computer projects, Watson wrote Williams: "If the basic responsibility of Research is not to be pre-planning the next family of data processing machines, I think we had better get our thinking straight. It seems to me this is a prime responsibility and if it cannot be done by 1,354 people, I think Piore is out of touch with reality." T. J. Watson, Jr., to A. L. Williams, March 19, 1964, Watson, Jr., Papers.

72. Palmer interview with Saphire, 1967, and Bashe, *IBM's Early Computers*, 256, 462–467.

73. The authorization allocated $800,000 for engineering and $600,000 for tooling, plus an additional $800,000 toward the cost of producing 130,000 usable circuits through the end of 1958. E. A. Wuerthele to A. L. Becker, November 8, 1957, Legal Papers, Printed Circuit Capability, PT07829/0043. As the program took shape over the next six months, tooling costs rose to $2.6 million through 1960. E. J. Garvey to W. W. Simmons et al., July 2, 1958, conveying slides from presentation on "SMS Top Priority Meeting," White Plains, N.Y., June 30, 1958. Legal Papers, Printed Circuit Capability, PT07829/0043. On projected outsourcing, see E. J. Garvey to A. L. Becker, October 15, 1958, Legal Papers, Printed Circuit Capability, PT07829/0043. Subsequent reviews showed that in 1958, purchased material accounted for 77 percent of the cost of an SMS card, with transistors accounting for 70 percent; in 1963, purchased material accounted for 88 percent of total costs, but because of their falling price, transistors accounted for just 45 percent. H. H. Yu to Learson, October 31, 1963, Legal Papers, Endicott Packaging, PT07835/0372.

74. "The IBM-Texas Instruments, Inc. Relationship," Chapter 10 of Components Division Procurement Role in System/360, Legal Papers. This material includes copies of the IBM-TI agreement, December 23, 1957; a memo from P. N. Whittaker to J. W. Birkenstock et al., January 23, 1958 (PT10275), explaining its origins and the philosophy behind it; and the letter officially amending the agreement on May 5, 1960 (PT10276).

75. On the technical dimensions of transistor work at IBM and TI, see Bashe, *IBM's Early Computers*, 390–395, 399–406, and Slade interview with Usselman, 1985. For broader perspective, see Michael Riordan and Lillian Hoddeson, *Crystal Fire: The Birth of the Information Age* (New York: Norton, 1997); Christophe Lecuyer, *Making Silicon Valley: Innovation and the Growth of High Tech, 1939–1970* (Cambridge, Mass.: MIT Press, 2005); and Leslie Berlin, *The Man behind the Microchip: Robert Noyce and the Invention of Silicon Valley* (New York: Oxford University Press, 2005).

76. Slade interview with Usselman, 1985.

77. IBM purchased 940,000 devices from TI in 1958 for $7.9 million; in 1960, it purchased 24.8 million semiconductors from TI for $32.4 million. "The IBM-Texas Instruments, Inc. Relationship." Given prevailing prices at the time and considering IBM's needs, the bulk of these were almost certainly germanium devices. During the production buildup of System/360, IBM continued to purchase large numbers of transistors, and TI remained its largest supplier. From 1964 through 1967, IBM bought just over 300 million transistors from outside suppliers, valued at $133 million. Of these, 68 million (23 percent) were silicon drift devices, valued at $56 million (42 percent of total expenditures); the remainder were germanium alloy transistors. M. J. Devine to J. T. Valetti, February 23, 1972, attaching "Memo to File: IBM vs. Industry Trends Study," Legal Papers, Component Division Procurement Role in System/360, PT10286. IBM purchases from TI during these four years totaled just over $130 million. In 1963, IBM purchased $18.3 million worth of products from TI. L. J. Dalton, "Memo to File: IBM Component Supplier-Texas Instruments, Incorporated," April 4, 1968, Legal Papers, Component Division Procurement Role in System/360, PT10283.

78. As IBM attempted to master silicon technology in the years ahead, it remained in close contact with technical personnel from TI, including many who took up residence at IBM. Garvey interview with Usselman, 1985. Reports to corporate management indicate that some 700 TI personnel were at work in IBM's East Fishkill semiconductor manufacturing facility in early 1966. At least 600 of them had been at East Fishkill for much of the previous year. P. W. Knaplund to T. J. Watson, Jr., October 28, November 19, 1965, January 11, 1966, Watson, Jr., Papers. Subsequent plans called to remove them by mid-1966. J. J. Bricker to Watson, Jr., January 31, 1966; W. J. Pedicord to Watson, Jr., February 11, 1966; and F. T. Cary to Watson, Jr., June 16, 1966; Watson Papers.

79. By mid-1961, these efforts had coalesced around a new standard circuit program known as Solid Logic Technology (SLT). This initiative expressly aimed to provide cost savings over SMS rather than enhanced speed. "Emphasis is placed on high volume activity over which the expense of tooling can be spread," its designers explained. "Further emphasis is placed on developing technologies compatible with the various uses required by IBM's operating divisions." By December 1961, some 255 engineers were at work on the project, with more slated to join. E. Bloch and C. E. Stephens, "Solid Logic Technology Program, 1962–1966 Operating Plan," August 18, 1961, Legal Papers, SLT Program, PT07866/0443. The attached "SLT Program Summary, 1961–1966," dated January 8,

1963, called for total development cost from 1961 to 1966 of $53.9 million, with an additional $41.0 million for manufacturing engineering and research. When Watson and other top management reviewed presentations on the program in early 1962, estimates called for expenditures on development, engineering, and capital equipment of approximately $115 million through 1968, not including construction of a new plant at East Fishkill. Labor and materials would account for another $1 per circuit, with production of 200 million. Bermuda Executive Conference, February 28–March 2, 1962, and "SLT Presentation for Mr. J. W. Haanstra and Others," June 7, 1962, Legal Papers, Endicott Packaging, PT07816/0924. By early 1965, IBM had committed somewhere between $150 million and $200 million toward engineering, development, and capital tooling for SLT through 1968 and had spent another $65 million for construction of new facilities connected with components manufacturing, with production slated for 300 million modules through 1968. M-200 Summary, Recap of Program, February 1965 Legal Records, Endicott Packaging, PT08382/0123; C. E. Frizzel to L. L. Horn, 11/30/65, Legal Records, Endicott Packaging, PT08382/0130; and "SLT Cost Review," revised March 17, 1966, Legal Records, Endicott Packaging, PT08377/0764. On allocations for facilities, see Corporate Review Meeting minutes. By the end of 1967, IBM was producing modules at the rate of 150 million per year or more, and production stayed at a similar level through the decade, after capital and development costs had been written down. M. L. Smalley, Solid Logic Technology Historic Cost Data, Endicott, March 1969, Legal Papers, Endicott Packaging, PT08382/0125.

80. Such concerns occupied a great deal of attention from top management during the early 1960s, and can be followed the CMC files and Watson, Jr., Papers. R. H. Bullen to A. K. Watson, December 8, 1964, presents an especially able summary prepared by the head of IBM's corporate staff throughout this period. CMC Files. See also W. B. McWhirter to T. V. Learson, June 29, 1960, with R. H. Bullen to file, n.d., attached; Learson to A. L. Williams, July 1, 1960; Bullen to file, July 5, 1960; Learson to Bullen, August 3, 1960; and Bullen to T. J. Watson, Jr., March 16, 1961. IBM officially announced formation of a separate Components Division on March 31, 1961. For persistent concern on the part of Watson about measuring the performance of the division against outside suppliers, either by selling its products on the open market or obtaining them from a second source, see T. J. Watson to T. V. Learson, June 14, 1963; A. H. Eschenfelder to Watson, March 3, 1964; Watson to Paul Knapland, February 19, December 3, 1965; and Knaplund to Watson, February 24, 1965; Watson Papers.

81. Watson interview with Usselman and Wight, 1985. For contemporary perspectives emphasizing this line of reasoning and Kelly's role, see Bullen to file, July 5, 1960, and B. L. Havens to E. R. Piore, June 1, 1961, CMC Files.

82. Management effectively committed to this course when they invested in SLT, which needed large volumes in order to generate a satisfactory return on investment. This was clear by early 1962, and the New Product Line to be built from SLT was fairly well articulated at an executive conference held at Bald Peak early the following September. The program came under critical review a year later at

a conference held at Jenny Lake, then marched rapidly toward the dramatic simultaneous announcement of the entire series of machines as System/360 on April 7, 1964. See Executive Conference Papers and Watson, Jr., Papers.

83. On the escalation of System/360 into a worldwide effort across all machines, see T. J. Watson, Jr., to T. V. Learson and A. K. Watson, January 15, 1963, and A. L. Williams to Dr. E. R. Piore, January 22, 1963, Williams Papers, IBM Corporate Archives, Division 10/602, Box 70856.

84. Production records suggest that by 1967, IBM was manufacturing a minimum of 125 million SLT modules, each containing a silicon transistor, and perhaps as many as 160 million. (The lower number is based on recorded production figures; the higher estimate is based on revised program estimates and SLT card production of 32 million, with cards containing on average five transistors.) IBM purchased 17.9 million silicon transistors. Assuming all these purchased transistors made their way onto SLT modules and IBM produced no chips not for use on modules (contrary to its plans, which called for 10 percent greater chip production than module output), IBM produced a minimum of 107 million silicon transistors in 1967. Industry sales of silicon transistors that year totaled 490 million. IBM's internal production thus amounted to 22 percent of the total industry sales. Using less conservative estimates, this figure could exceed 30 percent. This calculation is based on production figures in M. L. Smalley, Solid Logic Technology Historic Cost Data, Endicott, March 1969, Legal Papers, Endicott Packaging, PT08382/0125; C. E. Frizzel to L. L. Horn, November 30, 1965, Legal Records, Endicott Packaging, PT08382/0130; "SLT Cost Review," revised March 17, 1966, Legal Records, Endicott Packaging, PT08377/0764; and purchasing records and market analysis in M. J. Devine to J. T. Valetti, February 23, 1972, attaching "Memo to File: IBM vs. Industry Trends Study." Another large producer was Western Electric, which supplied AT&T and its affiliates that built phone switching systems. This production for internal consumption by firms making products for the commercial market is not typically included in production figures for the semiconductor industry, thus seriously distorting the importance of the military market to the industry. For an interesting discussion of semiconductor manufacturing at Western Electric, where established machinists and assemblers also played a critical role, see Stewart W. Leslie, "Blue Collar Science: Bringing the Transistor to Life in the Lehigh Valley," *Historical Studies in the Physical and Biological Sciences* 32, 1(2001): 71–113.

85. T. A. Wise, "IBM's $5,000,000,000 Gamble," *Fortune* 74 (September 1966): 118–123, 224, 226, 228. The idea may have come to the reporter from the common joke at the time that "IBM Co." stood for "I bet my company." See also T. A. Wise, "The Rocky Road to the Marketplace," *Fortune* 75 (October 1966): 138–143, 199, 201, 205–206, 211–212.

86. Erich Bloch, the IBM engineer who supervised the circuit project, later recalled a difficult moment in 1963 when he informed T. Vincent Learson, the group executive who orchestrated System/360, that IBM had no backup for the

proposed circuit. "Learson," recalled Bloch, "had a difficult time accepting this." Erich Bloch interview with Steven W. Usselman, March 4, 1986, Washington, D.C. The absence of a fallback was the subject of intense retrospective analysis by corporate management throughout 1964, as the components effort drew fire for not pursuing monolithic circuits. See CMC Files and Watson, Jr., Papers. A good summary is D. G. Thoroman to file, attached to Minutes of the Corporate Review Meeting, October 5, 1964, Watson Papers.

87. IBM *Annual Report* (1961). The board certified the appointment on May 16, 1961. On its importance to Wall Street, see Sobel, *IBM*, 210.

88. "DSD Operating Plan, 1961," CMC Minutes, December 16, 1960, and "DSD Long Term Plan," CRM Files, July 17, 1964. The dollar amount spent on product development by DSD in 1963 was $70 million. Total R&D activity across the firm that year, including components ($27 million), advanced system development ($16 million), and central research ($23 million), in addition to the products divisions, totaled roughly $235 million. CRM Papers, May 18, 1964. Another document in the CRM papers from this period, "Major Functional Areas—Costs and Expenses," n.d., pegs development expenses for 1963 at $164 million and credits the Research Division with another $23 million, but may not include the nonproduct divisions.

89. "Divisional Contributions to Gross Income," and "Financial Highlights, 1963," CRM Files, January 29, 1964. Earnings from federal contracts came to just $1.4 million in 1963, compared to $249 million from data processing. Return on investment in data processing exceeded the corporate target of 17 percent; that from federal systems was in the vicinity of 2 percent.

90. Notes from executive conferences at Belleair, April 4–5, 1961, and Bald Creek, September 7, 1962, Executive Conference Papers. Quote from 1961.

91. T. V. Learson to H. M. Sibley, July 12, 1961, and Sibley to Learson, July 17, 1961, CMC Papers, July 11, 1961. "We have never attempted to recover any IBM commercial development expense under military contracts," responded Sibley. "The continuation of the present components commercial development work under a new IBM organization does not justify having the Government pay part of the cost."

92. IBM *Annual Report* (1958) warned that the planned stock dividend of 2.5 percent for early the following year would be the last for the foreseeable future.

93. Foreign income after taxes rose slightly more modestly, from 27 percent of domestic levels in 1958 to 36 percent in 1963.

94. Because IBM began lumping equipment with land and buildings instead of with rental machines beginning in 1965, change in these categories from earlier years must be estimated by making reasonable adjustments based on the declared value of equipment in 1965 (which IBM listed at $761 million without adjusting for depreciation).

95. The cash shortage was caused by an unanticipated surge in work-in-process inventory with the integration into components. In early 1966, IBM estimated

the deficit from this source at a staggering $600 to $700 million. MRC Minutes, May 18, June 2, 1966; T. J. Watson, Jr., to F. T. Cary, August 29, 1966, Watson, Jr., Papers; and Watson and Petre, *Father, Son & Co.*, 358.

96. By the end of 1968, as investment in new machines slowed, IBM had restored the 55 percent ratio.

97. IBM *Annual Report* (1963); presentations to the Corporate Management Committee, January 28, February 26, March 9, 1963, Management Papers; Jenny Lake Executive Conference Notes, September 7, 1963, Executive Conference Papers; and A. L. Williams to T. J. Watson, Jr., February 6, 1963, and W. B. McWhirter to Watson, March 14, 1963. Watson, Jr., Papers.

98. AnnaLee Saxenian, *Regional Advantage: Culture and Competition in Silicon Valley and Route 128* (Cambridge, Mass.: Harvard University Press, 1994) and Kenney, *Understanding Silicon Valley*.

99. Alfred D. Chandler, Jr., *Inventing the Electronic Century: The Epic Story of the Consumer Electronics and Computer Industries* (New York: Free Press, 2001).

100. Usselman, "IBM and Its Imitators."

101. Paul E. Cerruzi, *A History of Modern Computing* (Cambridge, Mass.: MIT Press, 1998).

102. At an executive conference in May 1966, Tom Watson identified the "major lesson" of System/360: "At our size, we can't go 100 percent with anything new again." Executive Conference Papers. Several months later he issued the following to top management: "There is no question that we cannot go through another announcement like 360 where we obsolete virtually our entire installed revenue base at one time and where we commit a very substantial portion of our total production to a new technology. I know that we are continually tightening the controls on the development and introduction of new technologies to maximize the chances of its successful introduction. However, when we have learned from one experience like the 360, it is necessary that we confirm our learning by clearly stating we will never do it again and, accordingly, it should be our policy in the future never to announce a new technology which will require us to devote more than 25% of our production to that technology and equipment dependent upon it during the first year of major production of that technology." Watson to R. H. Bullen, F. T. Cary, W. C. Hume, G. E. Jones, T. V. Learson, A. K. Watson, and A. L. Williams, September 9, 1966, Watson Papers.

103. See Management Files and Watson, Jr., Papers for 1964.

104. W. C. Doud to T. J. Watson, Jr., June 6, 1966, Watson Papers.

105. Usselman, "Public Policies, Private Platforms;" Franklin Fisher et al., *IBM and the U.S. Data Processing Industry* (New York: Praeger, 1983) and *Folded, Spindled, and Mutilated: Economic Analysis and U.S. v. IBM* (Cambridge, Mass.: MIT Press, 1983); and Gerald W. Brock, *The U.S. Computer Industry: A Study in Market Power* (Cambridge, Mass.: Ballinger, 1975).

106. Ross Knox Bassett, *To the Digital Age: Research Labs, Start-Up Companies, and the Rise of MOS Technology* (Baltimore: Johns Hopkins University Press, 2002).

107. This theme is developed more fully in Steven W. Usselman, "Fostering a Capacity for Compromise: Government, Business, and the Paths of Innovation in American Computing," *Annals of the History of Computing* 18, 2 (Summer 1996): 30–39, and "Computer and Communications Technology," in Stanley Kutler, ed., *Encyclopedia of the United States in the Twentieth Century* (New York: Scribner's, 1996), 799–829.

108. On DEC, see Ceruzzi, *A History of Modern Computing* (Cambridge, Mass.: MIT Press, 1998).

109. Usselman, "Public Policies, Private Platforms."

9

Trading Knowledge: An Exploration of Patent Protection and Other Determinants of Market Transactions in Technology and R&D

Ashish Arora, Marco Ceccagnoli, and Wesley M. Cohen

This chapter examines the impact of the strength of patent rights on the growth of technology transactions. Such trade can yield important social welfare benefits. Some of these are familiar to any analysis of the benefits of trade, and attend upon a market-enabled division of labor, which, in the case of technology markets, can emerge across the range of activities comprising invention and its commercialization. These benefits are of several types. First, division of labor may promote specialization and, in turn, the efficiencies from both scale and learning emphasized by Adam Smith. Second, a division of labor can spawn the superior resource allocation that underpins the Ricardian notion of comparative advantage (Arora and Gambardella 1994).[1]

Markets for technology have other benefits that are manifest in dynamic settings. A firm unable or unwilling to commercialize inventions need not leave them on the shelf. Instead, it can derive value by licensing them to others. In a world where commercialization is costly and slow, licensing can also lead to a more rapid diffusion and exploitation of technology.[2] More generally, when the socially optimal pace of exploitation exceeds the innovator's ability, a market for technology can bridge the gap between social and private efficiency.[3]

To realize these various benefits, markets for technology must overcome a number of challenges. These may be distinguished on the basis of whether a new technology for sale is already in hand or has yet to be created—that is, whether the transaction precedes or follows the creation of the technology in question. For innovations that are in hand—those already generated though not yet commercialized—Arrow (1962) and Nelson (1959) highlight market failures that can undercut their sale—

what we call ex post technology contracting. Most notably, and assuming an opportunistic buyer, it is difficult to realize the full value of an innovation without fully disclosing it and thereby quashing the deal. Organization scholars have also highlighted characteristics of knowledge that can also overwhelm such transactions, including the tacit or sticky quality of some knowledge that can impede its transfer across different organizational settings (von Hippel 1990, 1994; Winter 1987; Arora and Gambardella 1994; Kogut and Zander 1992, 1993).

When a firm has a defined need for a new technology but the R&D has yet to be undertaken, there are additional impediments to the purchase of the requisite R&D services—what we call ex ante technology contracting. Uncertainty and information asymmetry between buyers and technology suppliers make the writing of enforceable, complete contracts difficult. Moreover, transaction costs emerge because, to implement an ex ante technology contract, buyers place themselves at risk for either disclosure of proprietary information to rivals via the prospective suppliers or paying monopolistic prices for the new technology. As Mowery (1983) suggests, to effectively contract for the purchase of non-routine R&D services, a prospective buyer will often need to convey to the seller detailed information regarding its needs and capabilities, which can in turn confer a bargaining advantage on the seller. To the extent that such knowledge raises switching costs, it makes the buyer vulnerable to hold-up, and thus discourages the incentive to contract for R&D services to begin with. Perhaps even more daunting, the activities of innovation and its subsequent commercialization are often intertwined, requiring ongoing mutual adjustment between the two (cf. Kline and Rosenberg 1986). This ongoing process of mutual adjustment not only impedes the ability to write complete contracts, but also often requires that the two activities proceed in close proximity with tight and frequent communication links between them, a circumstance perhaps best achieved by integration of the activities within the same firm rather than by market exchange.

As Lamoreaux and Sokoloff (1999) suggest, late in the nineteenth century, there was an extensive trade in patent rights, reflecting the prominent role of independent inventors. As the scale and complexity of the industrial enterprise grew, so did the costs of R&D, along with the advantages of integrating R&D with other functions within the firm (Mowery and Rosenberg 1998). Reflecting a diminished status of the independent inventor, the volume of trade in technology also declined.

Although the dominant mode of organizing innovation in the twentieth century has been the corporate R&D lab that produces and develops inventions for in-house exploitation, available evidence suggests that there has been a rapid growth in a variety of arrangements for the exchange of technologies or technological services over the past two decades. These include R&D joint ventures, licensing agreements, and contract R&D (Anand and Khanna 2000). Estimates for the 1980s suggest that such relationships account for as much as 10 to 15 percent of total civilian R&D in Organization for Economic Cooperation and Development member countries (Arora, Fosfuri, and Gambardella 2001). Yet other estimates of international royalty flows (including royalties for nontechnology-based intellectual property) show a sharp increase in the 1990s (Athreye and Cantwell 2005). Thus, the available evidence points to a renewal of market exchange of technology.

Over the past two decades, patenting, especially in the United States, has also grown rapidly, due in part to the increasing importance of patent-intensive sectors such as pharmaceuticals. However, part of the increase in growth is due to increases in patent propensity in sectors such as semiconductors and electronics (Hicks et al. 2001, Kim and Marschke 2004), in which patents have not traditionally been seen as very important, partly reflecting a rise of firms specializing in chip design (Ziedonis 2003). This rise was arguably facilitated by changes in the legal environment; notably, in 1982, the Court of Appeals for the Federal Circuit was established to make patent law more uniform, which indirectly strengthened it.

The growing incidence of technology trade and the growth in patenting are plausibly related. This chapter focuses on whether variations in patent protection can explain variations in the extent of ex ante and ex post technology contracting that comprise trade in new technology.

As with all other markets, a number of supporting institutions that facilitate effective dissemination of information, standardization, and contracting are vital to the rise and functioning of markets for technology.[4] And patents are one such institution, one that should in principle rectify the key market failure afflicting ex post technology contracting high-lighted by Arrow and Nelson.[5] Indeed, following Coase, economists have argued that defining property rights in innovation should make them easier to exchange. In addition to offering protection from misap-propriation, patents, by virtue of disclosing the key technical details of any patented invention, also serve an additional informational role criti-cal to the functioning of any market (Lamoreaux and Sokoloff 1999, 2002).

Patents, however, vary across industries and firms in the degree to which they provide protection from misappropriation (Scherer et al. 1959; Mansfield 1986; Levin et al. 1987; Cohen, Nelson, and Walsh 2000). They can often be invented around, and themselves disclose the technical details that abet such activity. They can also often be success-fully infringed without substantial penalty (Walsh, Cohen, and Arora 2003) in numerous settings. Arora (1996) suggests that stronger patents can reduce transaction costs in technology licensing contracts. Also, inso-far as stronger patents also enhance the bargaining power of the technol-ogy holder, this encourages firms to offer technologies for licensing or technological capability for hire (Gans, Hsu, and Stern 2002; Arora and Fosfuri 2003). Thus, unused technologies find more willing buyers, and innovators incapable of exploiting their innovations (or unwilling to do so) can appropriate the rents from their innovation by licensing or selling their innovation to others. In many instances, start-up firms in industries such as biotechnology, semiconductors, instruments, and chemicals have used their intellectual property as a means of obtaining financing and corporate partners, both of which are critical for the successful commer-cialization of new knowledge.

Systematic, direct empirical support for the proposition that stronger patents encourage more extensive ex post technology contracting, how-ever plausible, is limited. Using a sample of MIT inventions, Gans, Hsu, and Stern (2002) find that the presence of patents increases the likelihood that an inventor will license to an incumbent rather than enter the prod-uct market by commercializing the invention. Anand and Khanna (2000) find that in the chemicals sector, where patents are believed to be more effective, there are more technology deals, and a larger fraction of these are arm's length, involving exclusive licenses. Also, relative to other sectors, a larger fraction concerns future technologies rather than existing technologies. In contrast, Cassiman and Veugelers (2002) do not find that more effective patents encourage Belgian firms to enter into collaborative R&D arrangements.

Evidence from cross-national data is more mixed. Some studies find a positive association between patents and licensing. Yang and Maskus (2005) report a strong, positive relationship between improved intellectual property right regimes and licensing by U.S. multinational corporations. Analyzing data on international technology licensing contracts of Japanese firms, Nagaoka (2002) finds that weak patent regimes are associated with an increase in the fraction of transfers to an affiliate (such as a subsidiary) rather than to an unaffiliated firm. Smith (2001) finds that U.S. firms are more likely to export or directly manufacture rather than license technology in countries with weak patent regimes. A study using French data on the export of technology services finds that such exports are greater for countries with more effective patent protection, albeit only for higher-income countries (Bascavusoglu and Zuniga 2002).

Other studies cast doubts on the link between patent protection and the extent or form of international technology licensing. Fink (2005) finds a very weak relationship using German data. Puttitanum (2003) reports a higher response of direct investment than licensing to changes in the level of IPR protection. Similarly, Fosfuri (2004) does not find that patent protection significantly affects the extent or composition of technology flow (as joint venture, direct investment, or licensing) in the chemical sector. The mixed nature of the findings is reflected in a recent study by Branstetter, Fisman, and Foley (2004). Using detailed data on the technology royalty payments received by U.S. firms and controlling for country, industry, and firm fixed effects, they find that stronger patent protection does not increase the transfer of technology by U.S. multinationals to unaffiliated parties. However, it does increase the flow of technology to affiliates.

Using a rich data set comprising survey responses from R&D labs in the U.S. manufacturing sector, we explore how technology transactions are conditioned by the effectiveness of patent protection. In this chapter, we specifically consider the influence of firm capabilities on the impact of patent strength on such transactions. Indeed, we suggest that the mixed findings on the impact of patent strength on technology transactions, especially observed in the international studies, may reflect an inattention to the important conditioning role of firm capabilities. Our approach distinguishes between the purchase and sale of technology as well as whether the transaction occurs before (ex ante) or after (ex post) the R&D has been completed. To enrich our analysis of the role of patents, we also consider other appropriation strategies employed by firms, notably secrecy, and control for other firm-specific and technology- and industry-specific factors (e.g., intensity of competition, closeness to science).

Framework

To structure our consideration of the impact of patents on technology transactions, we distinguish between ex ante technology transactions (e.g., contract R&D) and ex post technology transactions (e.g., licensing), and distinguish across parties to the transactions according to whether they are buying or selling. This framework is reflected in table 9.1.

Our data, discussed in more detail below, allow us (within limits) to distinguish transactions between technology sales and purchases. The distinction between ex ante and ex post contracting is important because the impediments to each differ. For example, ex post technology contracting is more likely to be enabled by patents. However, patent rights for prospective inventions that are not yet in hand are likely to be uncertain, and thus likely to be less important for contracting. Moreover, ex ante contracts for prospective technologies are difficult to write and are likely to be highly incomplete. To the degree that such arrangements involve collaborative R&D or exchange of information, both parties have to contend with the possible leakage of proprietary knowledge to

Table 9.1

	Ex ante	Ex post
Buying	Contracting for R&D services	Licensing innovations and know-how from others
Selling	Selling R&D services	Licensing innovations and know-how to others

the other party, and then to others, suggesting that the key to such contracts may be the effectiveness of whatever secrecy strictures that may be put into place.[6] Concerns about know-how and other types of proprietary but nonpatented knowledge leakages are not absent in the transfer for existing technologies, but arguably are less salient.

Model

To structure our empirical analysis of the impact of patent protection on technology transactions, we propose a simple qualitative model that begins with the firm's decision to engage in a technology transaction. We do not observe individual transactions or the decision to seek patent protection for a specific technology, but only the propensity of firm to engage in technology transactions. In principle, both the propensity to patent and the propensity to engage in technology transactions are choices firms make, potentially conditioned by the efficacy of patent protection. In this chapter, we focus on the propensity to engage in technology transactions.

Consider a firm with an R&D project Then consider the decision on whether the project should be conducted in-house or outsourced to a third party (ex ante technology purchase decision). Two sets of factors affect the decision. The first set of factors we call differences in comparative advantage or gains from trade—roughly, which party is more efficient at doing the R&D. The second set of factors we call transaction costs. Transaction costs in turn can be decomposed into contracting costs and transfer costs. It is easier to write contracts for existing technologies that are at least partly protected by patents than for unpatented prospective technologies, implying that contracting costs would be higher for ex ante transactions than for ex post. Contracting costs may also include the cost of locating a suitable partner, and search costs and associated uncertainties are also lower when technologies are in hand. Although patents do not apply to ex ante transactions where technologies are not in hand, they can make contracting more viable for ex post contracting. To the degree that a technology is at least partly patented, any associated contract is easier to write because at least some of the details of the technology are described, and, more important, the rights of the patent holder are established. Moreover, patent protection can encourage sellers to disclose the details of their technologies, diminishing buyers'search costs.

Gains from Trade

We can expect the determinants of gains from trade to differ between ex post and ex ante technology trading. In the case of ex post technology contracting, technology or knowledge that is general purpose, in the sense of finding application in a variety of uses, is more likely to be sold, particularly if its uses are nonrival. The more general purpose the technology, the less likely it is that the technology holder has the ability to exploit all uses (Bresnahan and Trajtenberg 1995). Similar logic suggests that gains from trade will be greater when the technology is held by a small firm, facing larger potential buyers, which can apply the technology to larger and possibly more diverse markets, realizing greater cost-spreading advantages.

Firm size, however, can affect the gains from ex post trade in multiple, and possibly offsetting, ways. Larger firms are likely to possess downstream complementary assets for commercializing innovations and thus are more likely to license technology. But precisely because they can derive more value from an innovation, larger firms are more likely to invest in the R&D itself—notwithstanding the integration of the R&D function with manufacturing, marketing, and other functions, thus obviating the need to acquire external technology (Nelson 1959, Cohen and Klepper 1996).[7] When there are, however, advantages to integrating R&D with other downstream functions because information from downstream activities can inform the R&D process itself, unbundling the R&D activity from the downstream capabilities diminishes R&D performance, thereby obviating the advantages of a market-mediated division of labor. We would also expect considerably less ex ante technology contracting where the bundling of R&D with other capabilities is essential. This discussion implies that larger firms, which possess capabilities necessary for either more effective R&D performance or for realizing the value of innovation, are less likely to want to sell their technology to others. And where firms are less willing to sell technology generally, we would expect patents to play little role in facilitating such transactions.[8]

Transaction Costs

Perhaps the most important source of transactions costs in contract R&D, as highlighted by Mowery and Rosenberg (1998), are the switching costs that can grow over time as the R&D supplier progressively acquires more knowledge specific to the buyer in the course of a long-term relationship, making the buyer vulnerable to supplier opportunism in the face of incomplete contracting. For the purpose of this chapter, we will assume the potential for this source of transaction costs is uniform for a given technology. We also suggest this source of transaction costs is more likely to apply to ex ante transactions, where the prospect for the creation of switching costs is greater. To the extent, however, that licensing relationships are bundled with the acquisition of unpatented knowhow, such switching costs could also apply to long-term ex post technology transactions as well.

From a buyer's perspective, two other factors may affect the costs of technology transactions. First, absorptive capacity is essential to the effective utilization of extramural knowledge, including that which is bought (Cohen and Levinthal 1989). Second, the likelihood of leakage of proprietary information is an important source of transaction costs. The question here is whether buying external knowledge will cause existing proprietary knowledge to leak out to the seller, and perhaps to others. To the extent that we are dealing with ex post technology contracting, leakage is of less concern to the degree that such knowledge is patented and patents are effective. For ex ante technology contracting, however, patents are less relevant because we are talking about research that is not yet completed. Thus, for ex ante contracting, to the extent that leakage can be impeded—that is, where measures to enforce secrecy can be effectively applied—firms are less vulnerable to misappropriation. Similarly, leakage may be less damaging to the extent that lead time confers an effective appropriability advantage on the buyer. Also, the cost of any such leakage would be greater, the more intense the technological rivalry the buyer faces.[9]

The seller faces two related sources of transaction costs. First are the costs of transferring the knowledge. This is likely to be smaller the more articulable is the knowledge. Typically, a stronger grounding in academic science makes knowledge more articulable and less tacit (Winter 1987; Arora and Gambardella 1994; Kogut and Zander 1992, 1993). But the seller also faces the possibility of unintended leakage of information. As with the discussion for the buyer, the costs of such leakages will depend on the effectiveness of secrecy for ex ante technology contracting and patents for ex post contracting. Effective patent protection can also protect a seller against the opportunistic behavior by the buyer, but as suggested above, patent protection should matter more for ex post sales.

Unintended spillovers are more likely in ex ante transactions than in ex post, and their consequences may be perceived as more severe. R&D contracts will typically require richer interactions between the buyer and seller than will ex post contracts. Thus, one implication of the framework developed thus far is that variables such as the effectiveness of secrecy, which condition only the impact of information spillovers without affecting the gains from trade, should have a bigger impact on ex ante trades than ex post.

Interaction between Intellectual Property Rights and Complementary Assets

As noted, intellectual property protection, specifically, patent protection, can lower the transaction costs for technology transactions. Patents may also influence the gains from trade and, more important, do so in ways that depend on the size and complementary capabilities of the firms. This is because patent protection also conditions payoffs from alternatives to licensing: more effective patent protection increases the payoff to commercialization, but this effect is more important for firms that possess commercialization—manufacturing and marketing—capabilities. In other words, patent protection and commercialization capability are strategic complements (cf. Milgrom and Roberts 1990). As patent protection becomes more effective, the opportunity cost of licensing rises for all firms, but it rises less sharply for a firm lacking strong commercialization assets than for a firm capable of effectively commercializing the innovation. As a consequence, the decision to sell technology for a large firm possessing the capabilities required for commercialization is less sensitive to patent protection than a small firm.[10]

The key insight from this discussion is that for the sale of technology, increases in patent strength ought to matter more (in terms of licensing behavior) for small firms and firms lacking complementary assets. For technology purchase, the arguments are, however, less clear-cut. For ex ante purchases of technology, patents could be filed by either party to the transaction depending on the terms of the contract. If filed by the buyer (instead of the R&D performer), the effectiveness of patents, as perceived by the buyer, may also increase the payoff from externally sourced technology. Applying the same argument that we applied to ex post technology transactions, that payoff would, however, depend on whether the buyer possessed the complementary capabilities essential to commercialization. Note that now stronger patent protection should provide greater value to purchasers if they also possess complementary capabilities.

Similar reasoning would apply to the decision to sell a technology on an ex ante basis; effective patent protection would increase the payoff, but with complementary capabilities, the net benefit from such a sale would be less and, furthermore, the greater the complementarity between R&D and complementary assets, the less patent effectiveness would matter. However, it is more likely that ex ante purchases or sales of technology rely heavily on contractual protection and secrecy, in which event, patent strength would have little to do with the decision to engage in an ex ante transaction to begin with.

In the empirical analysis, we try to get at how the impact of patent protection is conditioned or mediated by complementary capabilities by interacting our measure of patent effectiveness with various measures of complementary capabilities. A key empirical problem with ex ante transactions is that either party may own the patent. In other words, the R&D performer may own the patent but agree to provide (ex ante) a license to the purchaser, or more likely, the buyer may file for patent protection itself. Since we do not observe which party files for the patent, it implies that we potentially have significant measurement error in our measure of patent effectiveness for the ex ante transactions. The consequence is, of course, that the estimated coefficient would be biased toward zero, at least in ordinary least squares.

Data

The data used come from the Carnegie Mellon Survey (CMS) on industrial R&D (Cohen, Nelson, and Walsh 2000). The population sampled is that of all R&D labs located in the United States conducting R&D in manufacturing industries as part of a manufacturing firm. The sample was randomly drawn from the eligible labs listed in the *Directory of American Research and Technology* (1995) or belonging to firms listed in Standard and Poor's Compustat, stratified by three-digit SIC industry. R&D lab managers were asked to answer questions with reference to the "focus industry", defined as the principal industry for which the unit was conducting its R&D. Valid responses were received from 1,478 R&D units, with a response rate of 54 percent. The data refer to the period 1991–1993. After trimming for outliers and dropping observations with missing data for the variables of interest, we obtain a final sample that ranges from 695 to 755 observations.[11] Table 9.2 provides descriptive statistics for the variables used in the analysis. Also, although we use the term *firm* for the unit of analysis, the unit of analysis is the business unit within the parent firm operating in the focus industry of the responding R&D lab.

Dependent Variables: Measures of Technology Transactions

Ex Post Technology Sales
The CMS asks respondents to state the percentage of their R&D projects over the last three years that were undertaken with the objective of earning licensing revenues. (Projects could have multiple objectives.) There were five response categories: 0–10 percent, 10–40 percent, 41–60 percent, 61–90 percent, and over 90 percent. We used the midpoints of each response category and treat this as a cardinal variable. (In unreported results, we also estimated an ordered logit specification, which treats the responses as ordinal. The qualitative results are very similar.) Note that this measure may include some contract R&D (ex ante technology sales), if, for example, the firm is doing contract R&D with the provision that if successful, it may earn further licensing revenue.

Table 9.2
Descriptive statistics*

	N	Mean	Standard deviation	Minimum	Maximum
% R&D conducted for licensing revenue (ex post sales)	755	0.06	0.14	0	0.95
% of R&D performed under contract for another firm (ex ante sales)	751	0.05	0.13	0	0.97
% of R&D budget allocated to R&D performed by other firms (ex ante purchase)	707	0.09	0.16	0	1
Secrecy Effectiveness	755	0.53	0.29	0.05	0.95
Patent Effectiveness	755	0.34	0.28	0.05	0.95
Complementary Assets	755	0.62	0.48	0	1
Importance of Science	755	3.28	0.74	1	4
Ph.D. Intensity	741	0.16	0.16	0	0.86
Business Unit Employees (Log)	755	6.71	2.05	2.30	13.01
Number of Technological Rivals	755	4.07	4.98	0	32
Global	755	0.78	0.41	0	1
Foreign	755	0.10	0.30	0	1
Public	755	0.65	0.48	0	1

* See text for units.

Ex Ante Technology Sales

The CMS asks the percentage of the R&D effort that is performed on contract for another firm. The other options were company financed and government contract, and the categories were mutually exclusive and exhaustive.

Ex Ante Technology Purchases

The CMS asks the percentage of the respondent's R&D budget spent on R&D contracts with other firms, institutions, or individuals.

This measure of ex ante technology purchase is subject to a number of limitations. First, we do not measure technology acquisitions by firms without R&D labs, since our sample consists only of R&D performers.

Table 9.3
Technology and R&D trade: Averages, weighted by R&D spending

	N	Mean	Median	Standard error
% R&D conducted for licensing revenue (ex post sale)	755	9.3	5.00	1.00
% of R&D performed under contract for another firm (ex ante sale)	751	2.9	0.00	0.37
% of R&D budget allocated to R&D performed by other firms (ex ante purchase)	707	9.9	5.00	0.50

Moreover, we measure only technology acquisitions made by the R&D lab itself and are thus missing any ex ante technology purchases made elsewhere by the firm.

Ex Post Technology Purchases

Our data unfortunately provided no satisfactory measure of ex post technology purchases (in-licensing). Consequently, we cannot analyze ex post technology purchases.

Extent of Transactions

Before moving to a discussion of the determinants of technology transactions, it is useful to discuss the extent of such transactions. Table 9.3 shows that the weighted average percentage of R&D conducted for the purpose of licensing (where the weight is total R&D spending of the respondent) is 9.3 percent. The weighted average percentage of R&D services sold to other firms is about 3 percent.

The weighted average percentage of firms' R&D that is purchased from other firms is about 9.9 percent. Insofar as our sample overrepresents large manufacturing firms with formal R&D labs, it is likely that R&D services sold to other firms are underestimated. Also, insofar as R&D labs are not the only locus for purchase of R&D services, an underestimation also results. Subject to these caveats, the sample averages suggest that the transactions in technology and R&D in the United States amount to about 10 to 13 percent of R&D. This is in line with other esti-

Table 9.4
Correlations among types of technology transactions

	Ex post sale	Ex ante sale	Ex ante purchase
Ex post sale	1.00	0.09	0.13
Ex ante sale	0.09	1.00	0.37
Ex ante purchase	0.13	0.37	1.00

mates, which range from 12 to 17 percent (Arora, Fosfuri, and Gambardella 2001).

Note also that our estimate of the extent of R&D services sold, 3 percent is substantially smaller than that for purchase, 10.5 percent. Some of this difference may be accounted for by the characteristics of our sample. As noted, our sample consists of R&D labs from U.S. manufacturing firms. This implies that we are likely missing a substantial fraction of firms that may be solely in the business of producing research and invention for sale. Thus, we are likely to underestimate the percentage of technology and R&D sales.[12] At the same time, however, we may be underestimating R&D services purchased because our sample does not include firms that may be buying technology or R&D services but not conducting R&D themselves. Further, we measure only purchases of the R&D unit; purchases of technology (or R&D services) by other parts of the firm are not captured by our measures.[13]

We should also note that our sample, which focuses largely on firms that manufacture, may be better suited to understand how R&D laboratories attached to manufacturing firms (as opposed to research boutiques) participate in the market for technology and how their behavior is conditioned by patent protection and the need to bundle marketing and manufacturing with research.

As table 9.4 shows, respondents in our sample do not necessarily specialize as buyers or sellers. Rather, if they participate in the ex ante market for technology, they tend to participate on both sides of the market. In particular, firms that perform R&D on contract for others are also likely to contract out to others. In general, this would be consistent with some firms having lower transaction costs than others and thus being more willing to participate in such transactions.

There is, of course, considerable interindustry variation in the degree to which firms participate in technology markets, either ex ante or ex post, as shown in table 9A.1 in the appendix. As one might expect, technology transactions are highest in the biopharmaceutical sector. In our empirical specification, we control for industry fixed effects at a more disaggregated level, using eighteen industry dummies.

Explanatory Variables and Controls

Effectiveness of Patent Protection: Patent

The literature does not always distinguish between whether patent rights exist and are well defined and whether patents are in some sense effective. From a Coasian perspective, all that is required is that property rights exist, are enforced, and are well defined (Lamoreaux and Sokoloff 1999). However, patents do in fact vary in the effectiveness of the protection they confer across technologies, industries and firms (Levin et al. 1987; Cohen, Nelson, and Walsh 2000). Given our conjecture that stronger patents (see Arora 1996; Gans, Hsu, and Stern 2002; Arora and Merges 2004) can more effectively overcome the various contracting and opportunism problems attendant upon technology contracts, especially ex post sales, we employ a measure of patent effectiveness as an explanatory variable. The CMS asks respondents to indicate the percentage of their product and process innovations innovations for which patent protection had been effective in protecting their firm's competitive advantage from those innovations during the prior three years. There were five mutually exclusive response categories: (1) less than 10 percent, (2) 10 through 40 percent, (3) 41 through 60 percent, (4) 61 through 90 percent, and (5) greater than 90 percent.[14] We used a weighted average of the product and process scores (using midpoints), with the percentage of R&D effort devoted to product and process innovations as weights, to construct patent effectiveness.

Patent effectiveness may be interpreted in several ways (e.g., Cohen, Nelson, and Walsh 2000). For instance, patents that force rivals to enter into negotiations, or block a rival, may not provide useful protection against opportunism in technology licensing contracts. In a corollary exercise to inform our interpretation of what our patent effectiveness score reflects, we estimated an ordered probit model to analyze the relationship between firms' reasons to patent and the respondents' patent effectiveness scores. We find that the magnitude of the coefficient for conventional motives for patenting such as licensing somewhat exceed those for less conventional reasons, such as using patents to induce rivals to participate in cross-licensing negotiations or for building patent fences (i.e., patenting substitutes) around some core innovation, but the latter reasons for patenting are still significantly related to patent effectiveness. Thus, our measure of effectiveness is likely too broad. This "measurement error" may bias the coefficient estimates toward zero.

The other issue with our patent effectiveness measure is that it may be endogenous. For instance, it may be driven by factors, such as the extent to which knowledge is articulable, that also lower costs of technology transfer. Alternatively, firms may decide to enforce their patents more aggressively, thus making them more effective, in order to license their technologies. In either event, the coefficient estimates may be biased. Arora, Ceccagnoli, and Cohen 2004, investigating the impact of patents on R&D, and Arora and Ceccagnoli 2005, investigating the impact of patents on ex post licensing, find that using the industry average of patent effectiveness for the primary industry of respondent's parent firm as an instrument does not change the basic results, suggesting that the bias, if any, is small.

Effectiveness of Secrecy as means of Protection: Secrecy

Ex ante technology contracting in particular should benefit transactions to the extent that parties to the contract can enforce secrecy to impede misappropriation of either the technology in question or the proprietary know-how and expertise with which it may be combined. To measure the effectiveness of secrecy as an appropriation strategy, respondents were asked to indicate the percentage of their product and process innovations for which secrecy had been effective in protecting their firm's competitive advantage from those innovations during the prior three years. As with our measure of patent effectiveness, there are five mutually exclusive response categories (less than 10 percent, 10 to 40 percent, 41 to 60 percent, 61 to 90 percent, and over 90 percent).[15] To construct our final measure of the effectiveness of secrecy, we used a weighted average of the product and process scores (using midpoints), with the percentage of R&D effort devoted to product and process innovations as weights.

Integration between Innovation and Complementary Capabilities: Complementary Assets

To the extent that a firm's R&D performance depends on close interaction between R&D and the firm's marketing and manufacturing activities, we would expect the firm to be less engaged in the market for technology. Put differently, the difference between the value a firm derives by commercializing its own innovation and the value it derives by licensing or selling the innovation is conditioned by the degree to which R&D and commercialization assets, such as marketing and manufacturing, are complementary.

In general, measuring the degree of complementarity is difficult. Previous studies have used measures of manufacturing or sales force (Tripsas 1997, Nerkar and Roberts 2004) or financial assets as measures of complementary capabilities (Helfat 1997). The CMS provides a measure for the frequency of face-to-face interaction between personnel from R&D and production, measured with a four-point Likert scale.[16] We constructed a binary variable, which takes value 1 if R&D and manufacturing personnel interact daily. We construct an analagous variable for the interaction between R&D and marketing personnel interaction. We construct a binary variable, complementary asset, which takes value 1 if either of the binary variables (for manufacturing or marketing) is positive and zero otherwise. This is our measure of the integration between R&D and complementary assets.

Our measure does not simply reflect a firm's ownership of complementary manufacturing or marketing capabilities, because most firms in our sample have manufacturing capability. It also reflects the extent of interaction between the R&D and complementary marketing or manufacturing functions—the essence of the notion of co-specialized complementary assets (Teece 1986).[17]

There are other possible, and broader, measures of specialized complementary assets, such as the importance of complementary manufacturing and marketing assets in appropriating the profits from innovations

(Cohen, Nelson, and Walsh 2000; Shane 2001; Gans, Hsu, and Stern 2002). In unreported regressions, we find that their use yields qualitatively similar—albeit less precise—results as those presented here.

It is plausible that firms that intend to license may choose more modular organizations, with fewer interactions between R&D and manufacturing or marketing. However, for manufacturing, Arora and Ceccagnoli (2005) probe the robustness of our results to such endogeneity in the context of ex-post licensing and find that the bias, if any, is small.

Business Unit Size: Size

Business unit size, measured by the natural logarithm of the number of business unit employees, is included as a determinant of the value of an innovation.[18] Business unit size is likely to decrease the value from technology sales (due to the firm's opportunity cost of not commercializing the innovation itself) and may increase the value from technology purchases due to cost spreading (Cohen and Klepper 1996).

Interaction between Patent Effectiveness and Complementary Assets

We explore how the impact of patent effectiveness is conditioned by the degree of integration between the R&D function and complementary assets by including the interaction term of Patent and Complementary Assets. Insofar as size may proxy for other unmeasured complementary assets, we also include an interaction term between Size and patent effectiveness.

Number of Technological Rivals: Technological Rivals

The CMU survey provides measures for the total number of technological rivals, as categorical variables in the following ranges: 0, 1–2, 3–5, 6–10, 11–20, or more than 20 competitors.[19] These responses were recoded to category midpoints. These variables vary across respondents within industries because they represent each respondent's assessment of his or her focus industry conditions, often reflecting a particular niche or market segment. We conjecture that as the number of rivals increases, the degree to which spillovers dissipate rents will increase, and the less likely that any one firm is likely to sell its technology. To the extent that this variable also reflects the number of possible buyers and suppliers of new technology, it may increase the likelihood of a transaction. Accordingly, our prior on the effect of the number of technological rivals is ambiguous.

Nature of Knowledge: Science

We also attempt to control for the degree to which knowledge may be codified, which increases the potential for market transactions. As a proxy for this notion, we use the CMS survey question that provides a measure of the degree to which knowledge is science based. In particular, we used a measure of the importance of research findings from the field of science or engineering that contributed the most to the firm's R&D.[20]

Absorptive Capacity: Absorptive Capacity

As has been well established in the literature (Cohen and Levinthal 1989), a firm's absorptive capacity conditions its ability to benefit from external technology. Since absorptive capacity includes the ability to identify and assess external technology alternatives, we use the percentage of R&D employees who have a Ph.D. or M.D. as our measure. This is also related to the extent to which knowledge is science based. In unreported specifications, we also use a second measure: the percentage of the R&D personnel's time devoted to monitoring and gathering information on new scientific and technical developments, as well as R&D intensity as an additional control. The results are similar to those reported here.

Other Controls

We include binary variables indicating whether the firm owning the lab is global (sells products in Japan or Europe), foreign (the respondent R&D lab is located in the United States but the parent firm is located abroad), or public, to reflect possible differences in market access, cost of capital, and cost of contract enforcement.

Conditioning on Industry Characteristics

Note that our data are at the level of the business unit, not that of the transaction. Thus, for any chosen measure of market-mediated knowledge flow across firms, we observe the characteristics of the buyer or of the seller, but not both. In order to control for industry effects, there are two possible responses. The first is to assume that both buyers and sellers belong to the same industry. If so, the characteristics of the other side of the transaction can be captured by a common industry dummy or by industry level measures of the relevant variables. This assumption has the virtue of simplicity but is not accurate because firms commonly buy or sell technology to firms from other industries. For instance, Arora, Fosfuri, and Gambardella (2001) find that although firms in SIC 28 (Chemicals and Pharmaceuticals) tend to sell technology mostly to firms in the same industry, they also tend to buy technology from other industries, such as software, computers, and electronics. The second option is to develop firm-specific measures of the "other side," which our data do not allow. Thus, we will employ industry dummies assuming that buyers and sellers of technology are in the same industry. We use nineteen industry dummy variables, constructed as binary variables using the SIC code assigned to the focus industry of each respondent, where focus industry was defined as the principal industry for which the unit was conducting its R&D.[21]

Empirical Specification and Results

Our theory is derived from thinking about individual transactions. Our measures, however, are grouped at the level of the firm: ex post sales is measured in five percentage-based cardinal categories, and the two ex ante dependent variables are measured as percentages. It is natural, therefore, to have an empirical specification that treats each individual transaction as probabilistic. In other words, suppose $y_{ij} = 1$ is the event that firm i licenses its jth innovation (for ex post sales) and $y_{ij} = 0$ otherwise.[22] Then our theoretical discussion involves specifying $\Pr(y_{ij} = 1 \mid X_i)$, where X_i represents firm and environmental characteristics, such as size, complementary assets, and patent effectiveness, and developing conjectures about how this probability responds to changes in X_i. To bring this theory to data, we note that we have measures equivalent to the percentage of licensed projects, whose expected value is simply $\Pr(y_{ij} = 1 \mid X_i)$. In other words, if Y_i is the share of licensing for firm

i, then $Y_i = \Pr(y_{ij} = 1 \mid X_i) + \varepsilon_i$, where ε_i is sampling error, and hence mean zero and independent across firms.

For convenience, we use a linear random utility specification and assume that the stochastic terms are independently and identically distributed across innovations, with type 1 extreme value distributions. This implies that $\Pr(y_{ij} = 1 \mid X_i) = \exp(X_i)/(1 + \exp(X_i))$ (cf. McFadden 1973), so that the share of licensing for firm i, Y_i, is $Y_i = \exp(X_i)/(1 + \exp(X_i)) + \varepsilon_i$.

We estimated this specification using nonlinear least squares for each of the three dependent variables. We also estimated ordinary least squares (OLS) specifications for all three.[23] Since absorptive capacity applies only to technology buyers, we include it only in the ex ante purchase equation. Although patent effectiveness ought to matter principally for sales, we include it in the purchase equation because, as noted, a buyer may have the right to file for patent protection. The other explanatory variables and controls are common across all three estimated equations.

This estimation procedure implies that the three decisions are structurally independent of each other. The alternative of modeling the mutual dependence would require us to use instrument variables. However, it is conceptually and practically difficult to identify factors that affect only one decision variable but not others, with the exception of absorptive capacity, which arguably affects only purchases. Given the exploratory nature of our analysis here, we decided in favor of simplicity.

Table 9.5 reports OLS estimates. Table 9.6 shows the corresponding estimates using the nonlinear (logistic) specification. For each dependent variable we report a specification without interactions between patent effectiveness and size or complementary assets (specification I) and one with interactions (specification II). Note that for each variable, the logistic provides a better fit, especially for ex ante and ex post sales. This implies that that in some cases, the standard errors of the estimates are smaller. Since the results are similar, we focus on table 9.6.

We do not report the individual industry effects, which are included as controls in all specifications. However, these effects together have significant explanatory power. This is true even if one were to use the industry averages of the right-hand-side variables, in addition to the respondent-level ones. This confirms that there are industry-level factors that condition the extent to which technology and R&D services are traded, consistent with Anand and Khanna (2000).

Turning to the results for the specification without interactions (I), we see that the effectiveness of secrecy is positively associated with ex ante sales, as expected. It has a statistically insignificant impact on ex post sales and ex ante purchases. Although we did not have strong priors, the lack of significance is encouraging in the sense that it suggests that the result for ex ante sales is not a statistical artifact. Contracts for ex ante transactions have greater potential for unintended knowledge spillovers, and effective secrecy is an important safeguard for sellers. Patents, as we argued, are less important for ex ante transactions since patent protection is unavailable at the time of writing the contract, but can help structure efficient licensing contracts through provisions assigning future rights. Indeed, patent effectiveness has a small and statistically insignificant coefficient for ex ante sales and purchases. Patent effectiveness is positive and significant in ex post sales, as expected.

Knowledge that is more closely related to science (as measured by importance of science) facilitates licensing, perhaps by reducing transfer costs and also because such knowledge may be more general purpose, thereby implying higher gains from trade. However, these variables have statistically insignificant coefficients in ex ante sales and purchases. Absorptive capacity is expected to increase ex ante purchases, and indeed, the measures of absorptive capacity, namely, the percentage of Ph.D. and M.D. holders among R&D employees (i.e., Ph.D. Intensity), are positively associated with ex ante purchases.

Finally, when R&D is integrated with other complementary functions (reflected in the Complementary Assets measure), the knowledge developed by a firm would be customized to its complementary capabilities, making it harder to transfer to others. Conversely, it would also make such capabilities less suited for exploiting externally generated knowledge as well. This is exactly what we find for ex post sales in specification I in table 9.6. We find similar results for ex ante sales as well as ex ante purchases, although the coefficient is not statistically significant.

Table 9.5
Determinants of ex post and ex ante technology transactions: OLS results

	Ex post sales		Ex ante sales		Ex ante purchase	
	I	II	I	II	I	II
Intercept	0.026	−0.046	0.059a	0.032	0.009	−0.027
	(0.036)	(0.043)	(0.034)	(0.040)	(0.041)	(0.048)
Secrecy Effectiveness	0.011	0.014	0.048**	0.049**	−0.003	0.001
	(0.018)	(0.018)	(0.017)	(0.017)	(0.020)	(0.020)
Patent Effectiveness	0.079**	0.287**	0.011	0.093	0.017	0.133a
	(0.020)	(0.073)	(0.018)	(0.068)	(0.022)	(0.082)
Complementary Assets	−0.023*	0.025	−0.010	−0.001	−0.015	−0.033a
	(0.011)	(0.017)	(0.010)	(0.016)	(0.012)	(0.019)
Business Unit Employees (Log)	−0.004	0.002	−0.004	−0.001	0.000	0.007
	(0.003)	(0.004)	(0.003)	(0.004)	(0.003)	(0.005)
Patent Effectiveness × Complementary Assets		−0.145**		−0.028		0.049
		(0.038)		(0.036)		(0.043)
Patent Effectiveness × Business Unit Employees (Log)		−0.018a		−0.010		−0.022*
		(0.009)		(0.009)		(0.011)
Ph.D. Intensity					0.102*	0.109**
					(0.038)	(0.038)
Importance of Science	0.015*	0.014*	−0.003	−0.003	0.013a	0.013a
	(0.007)	(0.007)	(0.007)	(0.007)	(0.008)	(0.008)

Number of Technological Rivals	0.0003	0.0002	5E-05	−1E-05	0.002	0.002
	(0.001)	(0.001)	(0.001)	(0.001)	(0.001)	(0.001)
Global	0.005	0.012	−0.030*	−0.028*	−0.024^a	−0.024^a
	(0.013)	(0.013)	(0.012)	(0.012)	(0.014)	(0.015)
Foreign	−0.006	−0.005	0.001	0.000	0.029	0.026
	(0.020)	(0.020)	(0.019)	(0.019)	(0.023)	(0.022)
Public	0.006	0.007	−0.021^a	−0.021^a	−0.007	−0.009
	(0.013)	(0.013)	(0.012)	(0.012)	(0.015)	(0.015)
N	755	755	751	751	695	695
Adjusted R^2	0.05	0.07	0.08	0.08	0.07	0.07

Note: Standard errors in parentheses. **, *, ^{a}: Significantly different from zero at the 0.01, 0.05, and 0.10 confidence levels. A full set of eighteen industry dummies is included in all specifications

Table 9.6
Determinants of ex post and ex ante technology transactions: Nonlinear OLS results

	Ex post sales		Ex ante sales		Ex ante purchase	
	I	II	I	II	I	II
Intercept	-3.498**	-4.665**	-4.126**	-4.396**	-3.706**	-4.168
	(0.747)	(0.821)	(0.948)	(1.048)	(0.465)	(0.556)
Secrecy Effectiveness	0.246	0.272	1.240**	1.240**	-0.052	-0.012
	(0.315)	(0.308)	(0.350)	(0.350)	(0.241)	(0.244)
Patent Effectiveness	1.248**	3.663**	0.199	0.969	0.206	1.525
	(0.312)	(1.263)	(0.396)	(1.416)	(0.234)	(1.010)
Complementary Assets	-0.404*	0.359	-0.249	-0.075	-0.177	-0.420[a]
	(0.194)	(0.315)	(0.212)	(0.378)	(0.139)	(0.231)
Business Unit Employees (Log)	-0.071	0.023	-0.094	-0.064	-0.005	0.093
	(0.057)	(0.076)	(0.077)	(0.096)	(0.040)	(0.063)
Patent Effectiveness × Complementary Assets		-1.805**		-0.486		0.641
		(0.626)		(0.776)		(0.493)
Patent Effectiveness × Business Unit Employees (Log)		-0.223		-0.081		-0.268[a]
		(0.159)		(0.230)		(0.144)
Ph.D. Intensity					1.151*	1.250*
					(0.418)	(0.444)
Importance of Science	0.275*	0.277*	-0.086	-0.098	0.181[a]	0.171
	(0.143)	(0.140)	(0.154)	(0.156)	(0.109)	(0.110)
Number of Technological Rivals	0.005	0.004	-0.003	-0.002	0.020	0.018
	(0.019)	(0.018)	(0.023)	(0.023)	(0.014)	(0.013)

Global	0.145	0.257	-0.603*	-0.573*	-0.289[a]	-0.282[a]
	(0.211)	(0.215)	(0.242)	(0.248)	(0.171)	(0.170)
Foreign	-0.050	-0.092	0.070	0.056	0.319	0.318
	(0.334)	(0.339)	(0.435)	(0.438)	(0.274)	(0.269)
Public	0.136	0.111	-0.469[a]	-0.473	-0.082	-0.102
	(0.258)	(0.251)	(0.270)	(0.266)	(0.179)	(0.179)
N	755	755	751	751	695	695
Adjusted R^2	0.08	0.10	0.12	0.12	0.07	0.09

Note: Standard errors in parentheses. **, *, [a]: Significantly diferent from zero at the 0.01, 0.05, and 0.10 confidence levels. A full set of eighteen industry dummies is included in all specifcations.

We would also have expected larger firms to be less likely to sell technology because such firms are perhaps better placed to commercialize technology. For instance, Cohen and Klepper (1996) argue that insofar as the returns to commercialization are proportional to sales, larger firms have less to gain by selling technology, and perhaps more to gain by buying it. However, we find that the coefficient of size is negative in all specifications, though never statistically significant. This suggests that although the lack of statistical significance argues for caution, larger organizations are in general less likely to engage in the market for technology, be it sales or purchases.

Table 9.6, specification II, shows the results for the specification where we interact Patent Effectiveness with Size and with the degree of integration between R&D and Complementary Capabilities. For the interacted variables, we see that the interactions are negative and significant in ex post sales, as expected. We would expect similar results for ex ante sales, except that we expect patent protection per se to matter less when the knowledge to be protected is yet to be produced. And indeed, we see that the coefficients are statistically insignificant. The results are similarly mixed in the ex ante purchase equation, where we expect the interaction to have a positive effect. We do indeed find that the coefficient is positive but the statistical significance is low. However, somewhat unexpectedly, we find that the interaction with size is negative. As discussed earlier, this may reflect either the lower salience of patents in structuring ex ante contracts, or the fact that the ex ante contracts may assign patents to either the buyer or the seller, resulting in measurement error.

Overall, our results indicate that technology suppliers rely on secrecy (and by extension, contractual provisions) to facilitate transactions for ex ante contracts, but secrecy appears to matter less for buyers. Patent protection matters principally for ex post sales of technology. However, the interaction specifications imply that stronger patents induce more licensing only for smaller firms and for firms where R&D has less frequent interaction with manufacturing and with marketing and sales. In other words, when the innovator is well positioned to commercialize the innovation, stronger patents matter much less for licensing. Once stated, the result is sensible, even obvious. However, much of the literature that has sought to empirically explore the impact of patent protection on technology licensing has not adequately accounted for it.

Conclusion

Although corporate R&D labs have many successful innovations to their credit, basic economic theory leads one to suspect that integrating R&D inside large commercial operations may not always be the most efficient form of organizing innovation. Large organizations can be bureaucratic, with restricted information flows and long decision lags. Inventors (and teams of inventors) can rarely be provided the sorts of incentives that a smaller organization can provide. Separating invention from other types of economic activities and putting it in specialized organizations may provide great social benefits in the form of higher rates of innovation under some circumstances. And yet, organizations specializing in inventive activity face a number of challenges, including financial constraints, finding buyers for their inventions, and protecting their inventions and their knowledge from misappropriation.

This chapter is a first step in understanding this market and the institutions and factors that facilitate its working. We use a rich data set to investigate the determinants of technology transactions, spanning technology licensing and the purchase and sale of R&D services. The literature has pointed to a number of factors that might be important. In particular, it suggests that even when there are gains from trade in transactions in technology, such trades may not take place if the participants cannot adequately safeguard themselves against misappropriation of their proprietary technology and against opportunistic behavior by potential trading partners. Our results suggest that formal patent protection can provide such safeguards, particularly for technology licensing. They also suggest that secrecy, the ability to prevent unintended disclosure of proprietary information, can provide a similar protection for R&D contracts. Finally, our results confirm the notion that where tight links between R&D capabilities and other functions within the firm are essential to R&D performance, and specifically to the protection of profits from R&D, technology sales tend to be less extensive and patent protection is less effective in inducing licensing.

Technology markets, like all other markets, aggregate individual actions. However, individual actions in turn are responsive to perceived and expected market outcomes. Thus, to the degree that historical forces spawned an increase in firm size, and capital market imperfections combined with increasing R&D costs forced an integration of R&D with manufacturing and other complementary functions, our results suggest that these would have long-term effects. Later, even if R&D costs ceased to be decisive, organizations where R&D and manufacturing interacted closely would be less likely to participate in markets for technology and would also be less sensitive to changes in patent protection. In turn, this would have reduced the economic space available for specialized technology suppliers, further reinforcing the growing position of in-house R&D labs. However, when there is a significant technical advance strongly rooted in formal knowledge, such as biotechnology or semiconductor design, and patent protection is effective, space for specialized technology suppliers may open up once again, as appears to have been the case in the past two decades. As technologies become more complex, even established incumbents may appreciate the benefits of a division of labor associated with active technology markets, even in industries where patent protection is not that strong.

Our empirical results, subject to a number of limitations and qualifications discussed in the text, are only suggestive. A more careful modeling of the decisions to sell and buy technology and the factors that condition such decisions is required for definitive answers. However, our results, insofar as they point in the right direction, suggest that the more modular organization structures in the nineteenth century were conducive to trade in technology and that along with rising R&D costs, other changes that increased integration across different functions contributed to the decline of markets for technology in the early twentieth century and, with it, the specialized technology suppliers. Looking ahead, improvements in communication technology, the growing benefits from a division of innovative labor associated with greater technological complexity, the increasing formalization of knowledge bases, and the greater reliability of patent protection are likely to sustain the trend toward growing markets for technology, and with them, of technology specialists, in the twenty-first century.

Appendix

Table 9A.1
Descriptive statistics, within industry groups*

Industry group	Oil-chemical	Biopharmaceutical	Computer-Electronics	Transportation	Instruments
Ex Post Sales	0.05 (Mean) (N = 159 SD = 0.13)	0.17 (40; 0.24)	0.06 (124; 0.15)	0.06 (65; 0.14)	0.06 (105; 0.14)
Ex Ante Sales	0.01 (158; 0.03)	0.12 (40; 0.24)	0.07 (123; 0.16)	0.06 (64; 0.11)	0.08 (104; 0.19)
Ex Ante Purchase	0.05 (142; 0.12)	0.18 (38; 0.25)	0.10 (120; 0.19)	0.11 (63; 0.16)	0.12 (98; 0.17)
Secrecy Effectiveness	0.59 (159; 0.27)	0.64 (40; 0.26)	0.46 (124; 0.32)	0.55 (65; 0.26)	0.50 (105; 0.32)
Patent Effectiveness	0.36 (159; 0.29)	0.51 (40; 0.35)	0.30 (124; 0.27)	0.38 (65; 0.28)	0.37 (105; 0.28)
Complementary Assets	0.65 (159; 0.48)	0.60 (40; 0.50)	0.55 (124; 0.50)	0.52 (65; 0.50)	0.67 (105; 0.47)
Importance of Science	3.35 (159; 0.76)	3.63 (40; 0.59)	3.30 (124; 0.70)	3.23 (65; 0.72)	3.38 (105; 0.73)
Ph.D. Intensity	0.22 (153; 0.17)	0.27 (39; 0.15)	0.12 (122; 0.17)	0.13 (63; 0.17)	0.12 (104; 0.13)
Business Unit Employees (Log)	6.70 (159; 1.84)	5.80 (40; 1.96)	6.32 (124; 2.27)	7.49 (65; 2.21)	5.97 (105; 1.84)

* See text for units.

Table 9A.1
(continued)

Industry group	Oil-chemical	Biopharma-ceutical	Computer-Electronics	Transportation	Instruments
Number of Technological Rivals	4.62	7.10	3.88	3.06	3.71
	(159; 4.96)	(40; 9.00)	(124; 4.72)	(65; 2.10)	(105; 3.91)
Global	0.82	0.68	0.73	0.77	0.80
	(159; 0.38)	(40; 0.47)	(124; 0.45)	(65; 0.42)	(105; 0.40)
Foreign	0.20	0.13	0.12	0.08	0.03
	(159; 0.40)	(40; 0.33)	(124; 0.33)	(65; 0.27)	(105; 0.17)
Public	0.57	0.68	0.68	0.63	0.78
	(159; 0.50)	(40; 0.47)	(124; 0.47)	(65; 0.49)	(105; 0.42)

Notes

1. Indeed, economic theory suggests that organizing innovation through an employment relationship—which is the way that most innovation is structured in our economy—may not always be the most efficient form of organizing innovation. For instance, if the likelihood of successful invention depends on the efforts of the inventor and if such efforts cannot be monitored adequately, a salaried inventor is less likely to succeed than one with a bigger stake in outcome. Aghion and Tirole (1994) provide a model where financial constraints can lead to residual rights to the innovation being taken away from the inventor, leading to suboptimal outcomes.

2. This has been the motive for licensing by a number of large, well-established firms, such as Exxon, Union Carbide, Du Pont, Dow Chemicals, Boeing, Procter and Gamble, Honeywell, and many others. See Arora, Fosfuri, and Gambardella (2001) for more details on licensing of existing technologies.

3. A similar logic underlies Nelson's (1959) insightful analysis of incentives for investment in basic research. Absent a market for technology, only a large, diversified firm would be reasonably sure of benefiting from its basic research.

4. Arora, Fosfuri, and Gambardella (2001) survey this literature in greater detail.

5. Intellectual property rights and especially patents have, however, been thought of primarily in terms of providing incentives for innovation rather than facilitating the creation of markets for technology.

6. Arora and Merges (2004) provide a simple stylized model of such an ex ante contract with information leakages. They show that stronger patent protection for the "supplier" enhances the efficiency of arm's-length contracts. By contrast, weak patent protection favors in-house research. However, the effects also depend in subtle ways on the nature of the information spillovers.

7. In the end, this probably boils down to the different degrees of "lumpiness" in downstream assets and R&D capability.

8. Larger firms may also possess the legal and financial resources that may plausibly affect contracting costs; we unfortunately do not possess measures of such resources beyond size itself.

9. Ceccagnoli (2005) analyzes the implications of technological and market competition for R&D with spillovers (non-market-mediated flows of technology).

10. The formal arguments for the case of ex post licensing are developed more fully in Arora and Ceccagnoli (2005).

11. This also reflects the exclusion of business units with fewer than ten employees, R&D units reporting more than fifty patent applications per million dollars of R&D, and those reporting an R&D budget of less than, or equal to, $100,000. Notice, however, that including very small units does not affect our results but is consistent with Arora, Ceccagnoli, and Cohen (2003) and Cohen, Nelson, and Walsh (2000).

12. On the other hand, we are using midpoints of the categories for percentage of R&D for licensing, which may overestimate the lowest category.

13. It is also instructive to compare these figures to NSF reported figures. NSF reports company-financed R&D contracted to outside organizations by R&D performing companies. From 1995 to 1997, the period for which this calculation can be made, the share of company-financed R&D contracted to outside organizations averaged around 5 percent. Moreover, this share appears stable across different firm size categories. Medium-sized firms (500–10,000 employees) do appear to contract out less than small (fewer than 500 employees) or large (greater than 10,000 employees) firms. Whereas medium-sized firms contract out a little more than 3.5 percent of company-financed R&D, this figure is slightly above 5 percent for the other firm size categories. The NSF does not report on technology licensing, so we cannot compare to our sample averages for licensing. Some of the difference between the sample average for contract R&D services and the NSF reported figures is due to differences in coverage and definition. NSF figures may include R&D contracted to universities and individuals, which are excluded from our measure. Further, our sample typically excludes research boutiques and others likely to specialize as technology suppliers. This may also partially explain why our estimate for R&D services purchased is over 10 percent, which is significantly higher than the NSF estimate of 5 percent. There are, however, other possibilities as well. One is that NSF figures relate only to company-financed R&D outsourced, whereas our figures include all sources of financing. Even so, since the bulk of R&D performed is company financed (82 to 84 percent for the sample period), this difference is unlikely to be quantitatively very significant. Another possible explanation is different coverage of industry sectors or firm size categories between the NSF and our sample. It is plausible that such outsourcing is more prevalent in sectors such biotechnology and pharmaceuticals, which may be overrepresented in our sample. Another possible reason is that if some of the contracting expenses are in kind (as in shared equipment or material) and the respondents in our sample included these in-kind payments, it might account for part of the difference since NSF estimates exclude such payments.

14. The CMS distinguishes between product and process innovations in a number of cases, including patent effectiveness. The results reported here are for product innovations alone.

15. The CMS distinguishes between product and process innovations in a number of cases, including the effectiveness of secrecy. The results reported here are for product innovations alone.

16. Respondents were asked: "How frequently do your R&D personnel talk face-to-face with personnel from the 'Production,' 'Marketing or Sales,' and 'Other R&D units' functions?"

17. Of course, some industry-specific studies have the advantage of being able to use finer measures of specialized complementary assets. Tripsas (1997) uses font libraries in her study of the typesetting industry, Thomke and Keummerle (2002) use therapeutical area-specific chemical libraries in pharmaceuticals, and Penner-

Hahn and Shaver (2005) use the stock of fermentation patents as a measure of complementary manufacturing capabilities for Japanese pharmaceutical firms.

18. The category of business unit employees is reported by R&D managers from the CMU survey. We also experimented using total firm employees, obtained from sources such as Compustat, Dun and Bradstreet, Moody's, and Ward's, and obtained similar results.

19. Technological rivals are defined in the CMS questionnaire as the number of U.S. competitors capable of introducing competing innovations soon enough such that they can effectively diminish the respondent's profits from an innovation in the lab's focus industry.

20. Respondents were asked to indicate which field contributed the most to their R&D activity and to rate its importance on a scale from 1 to 4 (Not Important, Slightly Important, Moderately Important, or Very Important).

21. The industry dummies are constructed using the SIC code assigned to the focus industry of each respondent: Food and Tobacco (SIC 20, 21), Industrial Chemicals (SIC 281–82, 286), Drugs (SIC 283 excluding Biotech), Biotech (various in 283, 384), Other Chemicals (SIC 284–85, 287–89), Petroleum (SIC 13, 29), Rubber (SIC 30), Metals (SIC 33–34), Computers (SIC 357), Machinery (SIC 35, exc. 357), Communication Equipment (SIC 366), Electronic Components (SIC 367 excl. 3674), Semiconductors (SIC 3674), Other Electrical Equipment (361–65, 369), Transportation (SIC 37 excl. 372, 376), Aircraft and Missiles (SIC 372, 376), Instruments (SIC 38 excluding 384), Medical Instruments (SIC 384), and Other Manufacturing (SIC 22–27, 31–32, 39).

22. For ex ante sales, this is the event that the firm licenses its *j*th R&D project. One can treat ex ante purchases analogously.

23. For the OLS specification, we also estimated the three equations jointly (not reported here) to allow for correlations in the residuals. Though the residuals are correlated, there is little change in the results. We also estimated an ordered probit for ex post sales, which treats the five licensing categories as purely ordinal. Once again, the results are qualitatively similar to those reported here.

References

Aghion, P., and J. Tirole. 1994. "The Management of Innovation." *Quarterly Journal of Economics* 109, 1185–1209.

Anand, B. N., and T. Khanna. 2000. "The Structure of Licensing Contracts." *Journal of Industrial Economics* 48, 103–135.

Arora, A. 1996. "Contracting for Tacit Knowledge: The Provision of Technical Services in Technology Licensing Contracts." *Journal of Development Economics* 50, 233–256.

Arora, A., and M. Ceccagnoli. 2005. "Patent Protection, Complementary Assets, and Firms' Incentives for Technology Licensing." *Management Science*.

Arora, A., M. Ceccagnoli, and W. M. Cohen. 2003. "R&D and the Patent Premium." Working paper 9431, NBER, Cambridge, Mass.

Arora, A., and A. Fosfuri. 2003. "Licensing the Market for Technology." *Journal of Economic Behavior and Organization* 52, 277–295.

Arora, A., A. Fosfuri, and A. Gambardella. 2001. *Markets for Technology: Economics of Innovation and Corporate Strategy.* Cambridge, Mass.: MIT Press.

Arora, A., and A. Gambardella. 1994. "The Changing Technology of Technical Change: General and Abstract Knowledge and the Division of Innovative Labor." *Research Policy* 23, 523–532.

Arora, A., and R. Merges. 2004. "Specialized Supply Firms, Property Rights and Firm Boundaries." *Industrial and Corporate Change* 13, 451–475.

Arrow, K. 1962. "Economic Welfare and the Allocation of Resources for Invention." In Richard R. Nelson, ed., *The Rate and Direction of Inventive Activity.* Princeton, N.J.: Princeton University Press.

Athreye, S., and J. Cantwell. 2005. "Creating Competition? Globalisation and the Emergence of New Technology Producers." Open University Economics Discussion paper no. 52, Milton Keynes, England.

Bascavusoglu, E., and M. P. Zuniga. 2002. "Foreign Patent Rights, Technology and Disembodied Knowledge Transfer Cross Borders: An Empirical Application." Unpublished manuscript, University of Paris I, France.

Branstetter, L. G., R. Fisman, and C. F. Foley. 2004. "Do Stronger Intellectual Property Rights Increase International Technology Transfer? Empirical Evidence from U.S. Firm-Level Panel Data." Unpublished manuscript, Columbia Business School.

Bresnahan, T., and M. Trajtenberg. 1995. "General-Purpose Technologies— Engines of Growth?" *Journal of Econometrics* 65, 83–108.

Cassiman, B., and R. Veugelers. 2002. "R&D Cooperation and Spillovers: Some Empirical Evidence from Belgium." *American Economic Review* 92, 1169–1184.

Ceccagnoli, M. 2005. "Firm Heterogeneity, Imitation, and the Incentives for Cost Reducing R&D Effort." *Journal of Industrial Economics* 53, 83–100.

Cohen, W. M., and S. Klepper. 1996. "A Reprise of Size and R&D." *Economic Journal* 106, 925–951.

Cohen, W. M., and D. A. Levinthal. 1989. "Innovation and Learning: The Two Faces of R&D." *Economic Journal* 99, 569–596.

Cohen, W. M., R. R. Nelson, and J. P. Walsh. 2000. "Protecting Their Intellectual Assets: Appropriability Conditions and Why U.S. Manufacturing Firms Patent or Not." Working paper no. 7552, NBER, Cambridge, Mass.

Directory of American Research and Technology. 30th ed. 1995. New Providence: Bowker.

Fink, C. 2005. "Intellectual Property Rights and U.S. and German International Transactions in Manufacturing Industries." In C. Fink and K. E. Maskus, eds.,

Intellectual Property and Development: Lessons from Recent Economic Research. New York: World Bank and Oxford University Press.

Fosfuri, A. 2004. "Determinants of International Activity: Evidence from the Chemical Processing Industry." *Research Policy* 33, 1599–1614.

Gans, J., D. Hsu, and S. Stern, 2002. "When Does Start-Up Innovation Spur the Gale of Creative Destruction?" *RAND Journal of Economics* 33, 571–586.

Grindley, P., and D. Teece. 1997. "Managing Intellectual Capital: Licensing and Cross Licensing in Semiconductors and Electronics." *California Management Review* 39, 8–41.

Helfat, C. E. 1997. "Know-how and Asset Complementarity and Dynamic Capability Accumulation: The Case of R&D." *Strategic Management Journal* 18, 339–360.

Hicks, D., T. Breitzman, D. Olivastro, and K. Hamilton. 2001. "The Changing Composition of Innovative Activity in the US—A Portrait Based on Patent Analysis." *Research Policy* 30, 681–704.

Kim, J., and G. Marschke. 2004. "Accounting for the Recent Surge in U.S. Patenting: Changes in R&D Expenditures, Patent Yields, and the High-tech Sector." *Economics of Innovation and New Technology* 13, 543–558.

Kline, S. J., and N. Rosenberg. 1986. "An overview of innovation." In R. Landau and N. Rosenberg (eds.), *The Positive Sum Strategy*. Washington, D.C.: the National Academies Press.

Kogut, B., and U. Zander. 1992. "Knowledge of the Firm, Combinative Capabilities, and the Replication of Technology." *Organization Science* 3, 383–397.

Kogut, B., and U. Zander. 1993. "Knowledge of the Firm and Evolutionary Theory of the Multinational Corporation." *Journal of International Business Studies* 24, 625–645.

Lamoreaux, N. R., and Kenneth L. Sokoloff. 1999. "Inventors, Firms, and the Market for Technology in the Late Nineteenth and Early Twentieth Centuries." In N. R. Lamoreaux, D. M. Raff, and P. Temin, eds., *Learning by Doing in Firms, Markets, and Nations*. Chicago: University of Chicago Press.

Lamoreaux, N., and K. Sokoloff. 2002. "Intermediaries in the U.S. Market for Technology, 1870–1920." In S. Engerman, P. Hoffman, J. Rosenthal, and K. Sokoloff, eds., *Finance, Intermediaries, and Economic Development*. New York: Cambridge University Press.

Levin, R. C., A. K. Klevorick, R. R. Nelson, and S. G. Winter. 1987. "Appropriating the Returns from Industrial R&D." *Brookings Papers on Economic Activity*, 783–820.

Mansfield, E. 1986. "Patents and Innovation: An Empirical Study." *Management Science* 32, 173–181.

McFadden, D. 1973. "Conditional Logit Analysis of Qualitative Choice Behavior." In P. Zarembka, ed., *Frontier in Econometrics*. New York: Academic Press.

Merges, R. 1998. "Property Rights, Transactions, and the Value of Intangible Assets." Unpublished manuscript, Boalt School of Law, University of California, Berkeley.

Milgrom, P., and J. Roberts. 1990. "The Economics of Modern Manufacturing: Technology, Strategy, and Organization." *American Economic Review* 80, 511–529.

Mowery, D. 1983. "The Relationship between Contractual and Intrafirm Forms of Industrial Research in American Manufacturing, 1900–1940." *Explorations in Economic History*, 351–374.

Mowery, D., and N. Rosenberg. 1998. *Paths of Innovation: Technical Change in twentieth Century America*. Cambridge: Cambridge University Press.

Nagaoka, S. 2002. "Impact of Intellectual Property Rights on International Licensing: Evidence from Licensing Contracts of Japanese Industry." Unpublished manuscript, Hitotsubashi University, Japan.

Nelson, R. R. 1959. "The Simple Economics of Basic Scientific Research." *Journal of Political Economy* 67, 297–306.

Nerkar, A., and P. W. Roberts. 2004. "Technological and Product-market Experience and the Success of New Product Introductions in the Pharmaceutical Industry." *Strategic Management Journal* 25, 779–799.

Penner-Hahn, J., and M. Shaver. 2005. "Does International Research and Development Increase Patent Output? An Analysis of Japanese Pharmaceutical Firms." *Strategic Management Journal* 26, 121–140.

Puttitanun, T. 2003. "Intellectual Property Rights and Multinational Firms' Modes of Entry." Unpublished manuscript, San Diego State University.

Scherer, F. M., et al. 1959. *Patents and the Corporation*, 2nd ed., Boston: Privately published.

Shane, S. 2001. "Technology Regimes and New Firm Formation." *Management Science* 47, 1173–1190.

Smith, P. J. 2001. "How Do Foreign Patent Rights Affect U.S. Exports, Affiliate Sales, and Licenses?" *Journal of International Economics* 55, 411–439.

Teece, D. J. 1977. "Technology Transfer by Multinational Firms: The Resource Cost of Transferring Technological Know-How." *Economic Journal* 87, 242–261.

Teece, D. J. 1986. "Profiting from Technological Innovation: Implications for Integration, Collaboration, Licensing and Public Policy." *Research Policy* 15, 285–305.

Thomke, S., and W. Kuemmerle. 2002. "Asset Accumulation, Interdependence and Technological Change: Evidence from Pharmaceutical Drug Discovery." *Strategic Management Journal* 23, 619–635.

Tripsas, M. 1997. "Unraveling the Process of Creative Destruction: Complementary Assets and Incumbent Survival in the Typesetter Industry." *Strategic Management Journal* 18, 119–142.

von Hippel, E. 1990. "Task Partitioning: An Innovation Process Variable." *Research Policy* 19, 407–418.

von Hippel, E. 1994. "Sticky Information and the Locus of Problem Solving: Implications for Innovation." *Management Science* 40, 429–439.

Walsh, J. P., W. M. Cohen, and A. Arora. 2003. "Effects of Research Tool Patents and Licensing on Biomedical Innovation." In W. M. Cohen and S. A. Merrill, eds., *Patents in the Knowledge-Based Economy*. Washington, D.C.: The National Academies Press.

Winter, S. G. 1987. "Knowledge and Competence as Strategic Assets." In D. J. Teece, ed., *The Competitive Challenge*. Balling Publishing Company.

Yang, G., and K. Maskus. 2005. "Intellectual Property Rights and Licensing: An Econometric Investigation." In C. Fink and K. E. Maskus, *Intellectual Property and Development: Lessons from Recent Economic Research*. New York: World Bank and Oxford University Press.

Ziedonis, R. H. 2003. "Patent Litigation in the U.S. Semiconductor Industry." In W. M. Cohen and S. A. Merrill, eds., *Patents in the Knowledge-Based Economy*, Washington, D.C.: The National Academies Press.

10

The Governance of New Firms: A Functional Perspective

Josh Lerner

Entrepreneurs frequently view outside investors with suspicion. The control rights and protective provisions that they demand are often seen as onerous by company founders, the amount of equity they demand for a given capital infusion excessive. In the United States, such misgivings are seen in the prerogatives that entrepreneurs use to describe such investors, such as "vulture capitalists" and "dumb money." In Europe and Asia, where there has been a shorter tradition of professional investors' funding young firms, such suspicions are, if anything, deeper.

In many cases, however, these resentments and suspicions are not well founded. Entrepreneurs, optimistic about the prospects of their businesses, often do not stop to consider the risks that their outside investors face. In many cases, the provisions employed by outside investors are a necessary response to the substantial problems posed by limited information and intangible assets.

This chapter explores the frequently contentious interactions between entrepreneurs and their financiers. To explore this challenging terrain, it takes a "functional approach" (see Merton 1995). Rather than focusing on any one nation or institution, this chapter highlights the four general problems seen in these settings, as well as the six classes of solutions that financiers use to address them. While the perspective is of necessity more macro than others in this book, I illustrate these points when appropriate with references to the other chapters.

At the same time, the chapter emphasizes that outside investors frequently do not have magic bullets to resolve these problems. In particular, I highlight three circumstances that can lead to outside investors' being unable to fund entrepreneurial firms appropriately: the presence of inexperienced investors, difficulties in reaching the optimal contract, and pathologies identified in the behavioral finance literature. Stated in this manner, these problems may sound quite abstract. To make these limitations more concrete, I illustrate them with examples from two very different settings: alliances between biotechnology and pharmaceutical firms in the United States and stock purchase contracts by Asian venture capital funds.[1]

The Challenge of Financing Young Firms

Young growth-oriented firms, particularly in high-technology industries, frequently require substantial capital to develop and deploy their ideas. The entrepreneurs who run these firms rarely have the capital to see their ideas to fruition and must rely on outside financiers. Meanwhile, those who control capital—for instance, pension fund trustees and university overseers—are unlikely to have the time or expertise to invest directly in young or restructuring firms.

The Critical Four Factors

A variety of factors limit access to capital for some of the most potentially profitable and exciting firms. These difficulties can be sorted into four critical factors: uncertainty, asymmetric information, the nature of firm assets, and the conditions in the relevant financial and product markets. At any one point in time, these four factors determine the financing choices that a firm faces. As a firm evolves over time, however, these factors can change in rapid and unanticipated ways. In each case, the firm's ability to change dynamically is a key source of competitive advantage, but also a major problem to those who provide the financing.

The first of these four problems, uncertainty, is a measure of the array of potential outcomes for a company or project. The wider the dispersion of potential outcomes, the greater the uncertainty is. By their very nature, new companies are associated with significant levels of uncertainty. Uncertainty surrounds whether the research program or new product will succeed. The response of the firm's rivals may also be uncertain. High uncertainty means that investors and entrepreneurs cannot confidently predict what the company will look like in the future.

Uncertainty affects the willingness of investors to contribute capital, the desire of suppliers to extend credit, and the decisions of firms' managers. If managers are averse to taking risks, it may be difficult to induce them to make the right decisions. Conversely, if entrepreneurs are over-optimistic, then investors want to curtail various actions. Uncertainty also affects the timing of investment. Should an investor contribute all the capital at the beginning or stage the investment through time? Investors need to know how information-gathering activities can address these concerns and when they should be undertaken.

The second factor, asymmetric information, is distinct from uncertainty. Because of his day-to-day involvement with the firm, an entrepreneur knows more about his company's prospects than investors, suppliers, or strategic partners. Various problems develop in settings where asymmetric information is prevalent. For instance, the entrepreneur may take detrimental actions that investors cannot observe: perhaps undertaking a riskier strategy than initially suggested or not working as hard as the investor expects. The entrepreneur might also invest in projects that build up his reputation at the investors' expense.

Asymmetric information can also lead to selection problems. The entrepreneur may exploit the fact that he knows more about the project or his abilities than investors do. Investors may find it difficult to distinguish between competent entrepreneurs and incompetent ones. Without the ability to screen out unacceptable projects and entrepreneurs, investors are unable to make efficient and appropriate decision choices regarding where to invest.

The third factor affecting a firm's corporate and financial strategy is the nature of its assets. Firms that have tangible assets, such as machines, buildings, land, or physical inventory, may find financing easier to obtain or may be able to obtain more favorable terms. The ability to abscond with the firm's source of value is more difficult when it relies on physical assets. When the most important assets are intangible, raising outside financing from traditional sources may be more challenging.

Even among intangible assets, certain types of assets are particularly hard to protect. For instance, in the biotechnology industry, firms have tended to rely on patent protection to protect assets. But in many other industries, key assets are protected by trade secrets or informal know-how. These firms have found attracting investors or entering into licensing agreements to be particularly difficult.[2]

Market conditions also play a key role in determining the difficulty of financing firms. Both the capital and product markets may be subject to substantial variations, as, for instance, Lamoreaux, Levenstein, and Sokoloff's account of financing in Cleveland in chapter 1 in this volume highlights. The supply of capital from public investors and the price at which this capital is available may vary dramatically. These changes may be a response to regulatory edicts or shifts in investors' perceptions of future profitability, as Neal and Davis suggest in chapter 3 and O'Sullivan in chapter 4. Similarly, the nature of product markets may vary dramatically, whether due to shifts in the intensity of competition with rivals or in the nature of the customers. If there is exceedingly intense competition or a great deal of uncertainty about the size of the potential market, firms may find it very difficult to raise capital from traditional sources.

The Consequences of These Problems

Described in this manner, these problems may appear to be quite abstract. But they have very real implications for entrepreneurs and executives seeking to commercialize early-stage technologies.

Jensen and Meckling (1976) demonstrate that conflicts between managers and investors ("agency problems") can affect the willingness of both debt and equity holders to provide capital. If the firm raises equity from outside investors, the manager has an incentive to engage in wasteful expenditures (e.g., lavish offices) because he may benefit disproportionately from these but does not bear their entire cost. Similarly, if the firm raises debt, the manager may increase risk to undesirable levels. Because providers of capital recognize these problems, outside investors demand a higher rate of return than would be the case if the funds were internally generated.

Even if the manager is motivated to maximize shareholder value, informational asymmetries may make raising external capital more expensive or even preclude it entirely. Myers and Majluf (1984) and Greenwald, Stiglitz, and Weiss (1984) demonstrate that equity offerings of firms may be associated with a "lemons" problem (first identified by Akerlof 1970). If the manager is better informed about the investment opportunities of the firm and acts in the interest of current shareholders, then managers issue new shares only when the company's stock is overvalued. Indeed, numerous studies have documented that stock prices decline on the announcement of equity issues, largely because of the negative signal sent to the market.

These information problems have also been shown to exist in debt markets. Stiglitz and Weiss (1981) show that if banks find it difficult to discriminate among companies, raising interest rates can have perverse selection effects. In particular, the high interest rates discourage all but the highest-risk borrowers, so the quality of the loan pool declines markedly. To address this problem, banks may restrict the amount of lending rather than increasing interest rates. The difficulties entrepreneurs face in obtaining formal debt financing are highlighted in many of the chapters in this book.

More generally, the inability to verify outcomes makes it difficult to write contracts that are contingent on particular events. This inability makes external financing costly. Many of the models of ownership (Grossman and Hart 1986 and Hart and Moore 1990) and financing choice (Hart and Moore 1998) depend on the inability of investors to verify that certain actions have been taken or certain outcomes have occurred. While actions or outcomes might be observable, meaning that investors know what the entrepreneur did, they are assumed not to be verifiable: investors could not convince a court of the action or outcome. Start-up firms are likely to face exactly these types of problems, making external financing costly or difficult to obtain.

Responses by Intermediaries

If the information asymmetries could be eliminated, financing constraints would disappear. Financial economists argue that specialized intermediaries, such as venture capital organizations or corporations with related lines of business, can address these problems. By intensively scrutinizing firms before providing capital and then monitoring them afterward, intermediaries can alleviate some of the information gaps and reduce capital constraints. It is the nonmonetary aspects employed in these settings that are critical to success.

These intermediaries have a variety of mechanisms at their disposal to address these changing factors. Careful crafting of financial contracts and firm strategies can alleviate many potential roadblocks. These responses can be divided into two broad classes.

The first set of roles relates to the manner in which the firms are financed. To begin with, from whom a firm acquires capital is not always obvious. Each source—private equity investors, corporations, and the public markets—may be appropriate for a firm at different points in its life. Furthermore, because the firm may be very different in the future, the appropriate source of financing may change.

To illustrate this role, consider the role of U.S. venture capitalists. These groups rarely seek to provide all the financing that an entrepreneurial firm seeks. Rather, they will complement their own funds with those of other investors. The initial financing may be syndicated with other venture groups, typically of equal sophistication. In this way, the groups can get a "second opinion" on the firm and have a peer to share the oversight of the firm with. Later financing rounds typically involve less established (or less sophisticated) venture groups. These groups are willing to contribute capital at a higher valuation, in exchange for the chance to invest in a firm whose prospects are more secure and which places less demands on them for oversight.

Second, the form of financing plays a critical role in reducing potential conflicts. Financing provided in these cases can be simple debt or equity, or involve hybrid securities like convertible preferred equity or convertible debt. These financial structures can potentially screen out overconfident or underqualified entrepreneurs. The structure and timing of financing can also reduce the impact of uncertainty on future returns.

A third element is the division of the profits between the entrepreneurs and the investors. The most obvious aspect is the pricing of the investment: for a given cash infusion, how much of the company does the intermediary receive? Compensation contracts can be written that align the incentives of managers and investors. Incentive compensation can be in the form of cash, stock, or options. Performance can be tied to several measures and compared to various benchmarks. Carefully designed incentive schemes can avert destructive behavior.

Strategic alliances, as discussed by Graham in chapter 6 in this book, provide a good example of this crafting of incentives. The payments in these alliances typically are of several distinct types. For instance, it is not uncommon for an agreement to include an initial up-front payment, a purchase of equity (which the financing firm may be able to force the R&D firm to repurchase if the alliance is unfruitful) or warrants, commitments to contract for R&D on specific topics, milestone payments contingent on the achievement of technological and marketing objectives or the renewal of the agreement, and a royalty on the eventual sales generated by the product. A rich array of theoretical works (Gallini and Wright 1990, Kamien and Tauman 1986) suggests that payments in alliances may have profound consequences on the success of these arrangements.

The second set of activities of venture capital and corporate investors relates to the strategic control of the firm. Monitoring is a critical role. Both parties must ensure that proper actions are taken and that appropriate progress is being made. Critical control mechanisms—such as active and qualified boards of directors, the right to approve important decisions, and the ability to fire and recruit key managers—need to be effectively allocated in any relationship between an entrepreneur and investors.

These investors can also encourage firms to alter the nature of their assets and thus obtain greater financial flexibility. Patents, trademarks, and copyrights are all mechanisms to protect firm assets. Understanding the advantages and limitations of various forms of intellectual property protection, and coordinating financial and intellectual property strategies are essential to ensuring a young firm's growth. Investors can also shape firms' assets by encouraging certain strategic decisions, such as the creation of a set of locked-in users who rely on the firm's products.

Evaluation is the final, and perhaps most critical, element of the relationship between entrepreneurs and private equity investors. The ultimate control mechanism exercised by the private equity investors and corporations is to refuse to provide more financing to a firm. In many cases, the investor can, through direct or indirect actions, even block the firm's ability to raise capital from other sources.

A natural question is what special attributes allow intermediaries like venture capitalists and corporations to perform these roles, while other potential intermediaries (such as banks) appear generally ineffective in these settings. While it is easy to see why individuals may not have the expertise to address these types of agency problems, it might be thought that bank credit officers could undertake this type of oversight. Yet even in countries with exceedingly well-developed banking systems, such as Germany and Japan, policymakers today are seeking to encourage the development of alternative mechanisms to ensure more adequate financing for risky entrepreneurial firms.

The limitations of banks stem from several of their key institutional features. First, because regulations in the United States limit banks' ability to hold shares, they cannot freely use equity to fund projects. Taking an equity position in the firm allows an intermediary to proportionately share in the upside, guaranteeing that the investor benefits if the firm does well. Second, banks may not have the necessary skills to evaluate projects with few tangible assets and significant uncertainty. In addition, banks in competitive markets may not be able to finance high-risk projects because they are unable to charge borrowers rates that are high enough to compensate for the firm's riskiness. Finally, private equity funds' high-powered compensation schemes give these investors incentives to monitor firms more closely, because their individual compensation is closely linked to the funds' returns. Banks (and corporations) that undertake such investments without such high-powered incentives have found it difficult to retain personnel once the investors have developed a performance record that enables them to raise a fund of their own.

When Can the System Break Down?

The discussion so far has highlighted the ways in which venture capitalists or corporate alliances can successfully address agency problems in the firms in their portfolios. But at the same time, there appear to be situations or time periods when the optimal level of financing is not provided or the firms are not governed properly. In this section, we review three apparent reasons that the most desirable arrangements may not be achieved.

Inexperienced Parties

Historical research into financial contracting suggests that parties often do not anticipate all the problems that will emerge in a given transaction. Instead, financial contracts have tended to evolve over time, addressing new problems as they are discovered. For instance, the protections that bondholders enjoy have increased over time to address new forms of opportunistic behavior by corporate managers (Lehn and Poulsen 1992, Tufano 1997). Similarly, financial innovations have frequently been repeatedly refined, as the limitations of the initial products become apparent (van Horne 1985, Miller 1986).

The same evolutionary process may be at work in the financing of young firms. For instance, pioneering venture capitalists in a given country or industry may not anticipate all the opportunistic ways in which managers can act (or vice versa). Over time, the agreements governing the relationships between these parties are likely to evolve to reflect these concerns.

Contracting Difficulties

An alternative possibility is that the two parties are sophisticated, but due to the early stage of the project cannot write a contract that covers every possible contingency. As noted above, numerous economic models consider incomplete contracting between a principal and an agent.

In these works, a typical assumption is that it is impossible for the two parties to write a verifiable contract that could be enforced in a court of law and specifies the effort and final output of the two parties. This is because there are many possible contingencies, all of which cannot be anticipated at the time the contract is drafted. Due to this nonverifiability problem, these models argue that it is optimal for ownership of the project to be assigned to the party with the greatest ability to affect the outcome at the margin. This party, who will retain the right to make the decisions that cannot be specified in the contract, should also receive any surplus that results from the project. Because of this incentive, the party will make the decisions that maximize—or come close to maximizing—the returns from the project.

Aghion and Tirole (1994) adapt this general model to the specific setting of financing young firms. In particular, they consider an R&D alliance between an established corporation and a small, research-intensive firm without financial resources of its own, which cannot borrow any funds and has no ability to commercialize the innovation itself. Because the small firm cannot raise financing directly from the public markets, it turns for financing to a customer—a firm that may intend to use the product itself or to resell it to others but cannot make the discovery independently. (In refinements of the model that will not be discussed here, the authors allow the young firm to instead choose to finance the project through a third party, such as a venture capitalist, and to commercialize the project itself.) The success of the research project is an increasing function, though at a decelerating rate, of both the effort provided by the young firm and the resources provided by the customer.

Developing a contract between the two parties is challenging. While the ownership of the product can be specified in an enforceable contract and the resources provided by the customer may be so specified, uncertainty precludes writing a contract for the delivery of a specific innovation. Similarly, an enforceable contract cannot be written that specifies the level of effort that the young firm will provide.

Aghion and Tirole consider two polar cases: when the young firm has the ex ante bargaining power and when the customer has this power. When the young firm has the bargaining power, the ownership of the research output will be efficiently allocated. If the marginal impact of the young firm's effort on the innovative output is greater than the marginal impact of the customer's investment, then the young firm will receive the property rights. If not, the young firm will transfer ownership to the customer in exchange for a cash payment. This result is similar to that of Grossman and Hart (1986).

When the customer has the bargaining power, a different pattern emerges. If it is optimal for the customer to own the project, it will retain the project. If, however, it would maximize innovation for the property rights to be transferred to the young firm, the ideal outcome will not be achieved. In particular, the customer will be willing to transfer ownership, but the cash-constrained young firm will not have enough resources to compensate the customer. As a result, an inefficient allocation of the property rights occurs, with the customer retaining the rights to the invention.

While this is just one model, we could anticipate that the fundamental problems delineated above could deter the ideal governance of an entrepreneurial firm, no matter how sophisticated the parties.

Behavioral Explanations

A third class of explanations may be referred to as behavioral in nature. In the past decade, financial economists have increasingly appreciated that rational explanations may not exist for many phenomena, from the discounts to net asset value at which most closed-end funds trade to the poor performance of firms after offering equity to the public. As an alternative, they have offered behavioral explanations, hypothesizing that investors may be prone to systematic overoptimism or to shifts in sentiment. (For an overview, see the discussion in Shleifer 2000.)

A similar effect may characterize the financing of young firms. The venture capital industry of many nations appears to have gone through periods when there were dramatic shifts in the supply of capital available from institutional and individual investors. These shifts in the fundraising environment appear to have had profound effects on the investments made in new firms. That this is not just a phenomenon of the past few years is clear from many of the chapters in this book.

While there are not any theoretical discussions of this phenomenon of which I am aware, discussions of such patterns have appeared in the trade press since at least the 1960s. The first extended discussion, however, was in Sahlman and Stevenson (1986). The authors chronicle the exploits of venture capitalists in the disk drive industry during the early 1980s. Sahlman and Stevenson assert that a type of market myopia affected venture capital investing in the industry. During the late 1970s and early 1980s, nineteen disk drive companies received venture capital financing. Two-thirds of these investments came between 1982 and 1984, the period of rapid expansion of the venture industry. Many of these companies also went public during this period. While industry growth was rapid during this period of time (sales increased from $27 million in 1978 to $1.3 billion in 1983), Sahlman and Stevenson question whether the scale of investment was rational given any reasonable expectations of industry growth and future economic trends.[3] Similar stories are often told concerning investments in software, biotechnology, and the Internet. The phrase "too much money chasing too few deals" is a common refrain in the venture capital market during periods of rapid growth.

Gompers and Lerner (2000) systematically examine one facet of these claims through a data set of over 4,000 venture investments between 1987 and 1995 developed by the consulting firm VentureOne. They construct a price index that controls for various firm attributes that might affect firm valuation, including firm age, stage of development, and industry, as well as the inflow of funds into the venture capital industry. In addition, they control for public market valuations through indexes of public market values for firms in the same industries and average book-to-market and earnings-to-price ratios.

The results support contentions that a strong relation exists between the valuation of venture capital investments and capital inflows. While other variables also have significant explanatory power—for instance, the marginal impact of a doubling in public market values was between a 15 and 35 percent increase in the valuation of private equity transactions—the inflows variable is significantly positive. A doubling of inflows into venture funds leads to between a 7 and 21 percent increase in valuation levels.

The results are particularly strong for specific types of funds and funds in particular regions. Because funds have become larger in real dollar terms, with more capital per partner, many venture capital organizations have invested larger amounts of money in each portfolio company. Firms have attempted to do this by moving to finance later-stage companies that can accept larger blocks of financing. Similarly, because the majority of money is raised in California and Massachusetts, competition for deals in these regions should be particularly intense, and venture capital inflows may have a more dramatic effect on prices in those regions. The results support these contentions. The effect of venture capital inflows is significantly more dramatic on later-stage investments and investments in California and Massachusetts.

They also examine whether increases in venture capital inflows and valuations simultaneously reflect improvements in the environment for young firms. If shifts in the supply of venture capital are contemporaneous with changes in the demand for capital, their inferences may be biased. Success rates—whether measured through the completion of an initial public offering or an acquisition at an attractive price—did not differ significantly between investments made during the early 1990s, a period of relatively low inflows and valuations, and those of the boom years of the late 1980s. The results seem to indicate that the price increases reflect increasing competition for investment.

On a more speculative level, it may be possible that the tremendous concentration of the firms backed by venture capitalists is also problematic in terms of social welfare. Several models argue that institutional investors frequently engage in "herding": making investments that are too similar to one another. These models suggest that a variety of factors—for instance, when performance is assessed on a relative, not an absolute, basis—can lead to investors' obtaining poor performance by making too similar investments. (Much of the theoretical literature is reviewed in Devenow and Welch 1996.) As a result, social welfare may suffer because value-creating investments in less popular technological areas may have been ignored.

Illustrations of These Phenomena

In this final section, I present two case studies that illustrate these phenomena. These examples, drawn from recent research projects, illustrate how the problems of financing and governing young firms can be overcome, as well as the barriers that remain. In this way, I hope to make the somewhat abstract-sounding typology of problems in the previous section more tangible.

Case Study 1: Biotechnology Alliances in the United States

Biotechnology research has numerous features that resemble the setting depicted in the theoretical literature on incomplete contracts.[4] Biotechnology projects, particularly early-stage efforts, are highly complex and uncertain, making it very difficult to specify the features of the product to be developed. As one biotechnology executive relates: "Redefining the work when the unexpected happens, as it invariably will, [is essential]. Research is by its very nature an iterative process, requiring constant reassessment depending on its findings. If there is a low risk of

Table 10.1
Characteristics of the sample

Variable	Mean	Median	SD	Minimum	Maximum
Basic characteristics					
Date of alliance	July 1991	Mary 1992	3.1 years	January 1980	December 1995
Minimum length of alliance (years)	3.91	3.00	3.14	0.92	31.00
Stage of product at time of alliance					
Discovery/lead molecule	0.63			0	1
Preclinical development	0.14			0	1
Undergoing regulatory review	0.22			0	1
Approved for sale	0.01			0	1
Focus of alliance					
Human therapeutics	0.92			0	1
Human diagnostics	0.04			0	1
Agricultural or industrial applications	0.04			0	1
Condition of biotech equity markets					
Total public equity raised in prior quarter	402.86	184.04	467.31	0.00	1,699.87
Total public equity raised in prior year	1,600.42	1,150.67	1,323.35	0.00	4,832.43
Biotech index at end of prior quarter	1.67	1.61	0.46	0.91	2.75
Financial position of financing firm					
Revenues in prior year	6,420.52	4,210.80	8,103.25	0.09	48,959.37
R&D expenditures in prior year	562.36	398.46	536.35	2.68	2,075.79
Net income in prior year	562.82	353.33	614.91	−457.44	2,231.98
Cash and equivalents at end of prior year	913.92	538.30	1,038.30	0.70	4,938.42

Total assets at end of prior year	6,902.54	4,564.25	7,448.66	5.24	35,253.06
Shareholders' equity at end of prior year	3,449.73	2,216.58	3,669.47	0.22	17,504.68
Financial position of R&D firm					
Revenues in prior year	9.92	1.29	40.31	0	494.57
R&D expenditures in prior year	11.17	5.57	21.32	0	229.11
Net income in prior year	−11.13	−4.99	30.75	−284.06	47.69
Cash and equivalents at end of prior year	25.89	7.36	59.49	0	554.24
Total assets at end of prior year	48.24	18.04	119.95	0.49	1,325.02
Shareholders' equity at end of prior year	33.44	14.26	88.62	−17.08	1,021.88
Patent holdings of R&D firm					
Total patent awards	5.32	0	14.44	0	114
Patent awards related to alliance	1.14	0	7.80	0	102

Note: The sample consists of 200 technology alliances initiated between biotechnology and pharmaceutical companies or between biotechnology firms in the 1980–1995 period. The table summarizes the financial market conditions around the time of the alliance and the characteristics of the firms in the alliance. The date variable is expressed as a decimal (e.g., July 1, 1995, is coded as 1995.5). The stage of product measures are all dummy variables. The public equity raised and financial position variables are expressed in millions of 1995 dollars. The biotechnology index reflects inflation-adjusted public equity values and is normalized to 1.0 on January 1, 1978.

Table 10.2
Percentage of alliances allocating control rights to the firm financing the R&D activity

Control Right	1980–1987	1988–1990	1991–1992	1993–1995	Total sample
Key aspects of alliance management					
1. Right to manage clinical trials	64%	62%	46%	62%	57%
2. Right to undertake process development	4	3	9	11	8
3. Right to manufacture final product	50	66	66	64	63
4. Right to market universally	89	53	69	63	67
5. Right to market product alone	96	91	82	68	80
Determination of alliance scope					
6. Right to expand alliance	7	9	7	15	10
7. Right to extend alliance	32	25	21	18	22
8. Right to terminate alliance without cause	46	50	33	18	32
9. Right to terminate particular projects	11	12	12	11	12
10. Right to sublicense	18	25	31	23	26
11. Right to license after expiration/termination	39	41	54	41	45
12. Right to shelve projects	96	94	99	86	93
Control of intellectual property					
13. Ownership of patents	18	6	7	10	10
14. At least partial patent ownership	71	56	73	78	72
15. Control of patent litigation	29	25	22	25	25
16. Right to know-how transfer	54	28	43	51	45
17. Ownership of core technology	11	0	9	5	6
18. Right to delay publications	14	22	33	51	35
19. Right to suppress publications	32	9	16	19	18
Governance structures					
20. Control of top project management body	7	12	3	5	6
21. Seat on R&D firm's board	14	34	15	23	21
22. Equity in R&D firm	32	56	45	62	51
23. Right to participate in R&D firm's financings	18	34	21	15	20
24. Right to register R&D firm's stock	18	25	36	33	30

Table 10.2
(continued)

Control Right	1980–1987	1988–1990	1991–1992	1993–1995	Total sample
25. Ability to make public equity purchases	89	81	81	66	76
Mean number of control rights in each agreement	9.6	9.2	9.3	9.2	9.3
Number of observations	28	32	67	73	200

Note: The sample consists of 200 technology alliances initiated between biotechnology and pharmaceutical companies or between biotechnology firms in the 1980–1995 period. The table divides the tabulations into four chronological periods. The mean number of control rights is the average number of control rights (out of the possible twenty-five) included in alliances in each subperiod.

unexpected findings requiring program reassessment, then it is probably not much of a research program" (Sherbloom 1991, pp. 220–221). Similarly, the complexity and unpredictability of the research present challenges in drafting an enforceable agreement that specifies the contributions of the young firm. In particular, firms that contract to perform R&D in alliances frequently have ongoing research projects of their own, in addition to the contracted efforts. In case of a dispute, it may be very difficult for the financing firm to prove that the new firm has employed alliance resources to advance projects that are not part of the alliance.[5]

When we analyze the way in which alliances with new biotechnology firms are governed, we see evidence quite consistent with incomplete contracting theory. To illustrate this, I present an analysis of the allocation of control rights in a sample of 200 alliances between two biotechnology firms or between a biotechnology and pharmaceutical firm. In particular, I identify the twenty-five rights that the firm providing the bulk of the financing (which is typically much larger than the smaller firm) demands in at least 5 percent of these agreements. Table 10.1 highlights the discrepancy in size between the firms that provide the financing and those undertaking the bulk of the research. Table 10.2 summarizes the key control rights.

Table 10.3 presents the critical regression analyses. I seek to explain the number of control rights, using as explanatory variables the number

Table 10.3
Regression analysis of the control rights allocated to the funding party in biotechnology alliances

Independent variables	Using OLS Specification		Using Ordered Logit Specification	
A: Dependent variable is Number of Control Rights Out of 25 Rights				
R&D firm's patent awards related to alliance	0.08 [2.16]	0.08 [2.27]	0.06 [2.05]	0.04 [1.43]
Total public equity raised in prior quarter	0.001 [0.00]		−0.01 [0.03]	
Biotech index at end of prior quarter		−0.31 [0.80]		−0.24 [0.82]
R&D firm's shareholders' equity at end of prior year	−11.44 [3.47]		−11.31 [3.79]	
R&D firm's total assets at end of prior year		−8.16 [3.53]		−6.70 [3.25]
Constant	9.59 [38.89]	10.12 [14.84]		
F-statistic	4.21	4.55		
χ^2-statistic			15.50	14.47
p-value	0.01	0.00	0.00	0.00
Adjusted R^2	0.05	0.06		
Log likelihood			−401.47	−393.47
Number of observations	180	176	180	176
B: Using Interaction Terms				
R&D firm's patent awards related to alliance	0.19 [3.45]	0.20 [3.49]	0.13 [3.09]	0.13 [3.18]
Total public equity raised in prior quarter	0.03 [0.07]		0.02 [0.08]	
Biotech index at end of prior quarter		−0.31 [0.79]		−0.23 [0.78]
R&D firm's shareholders' equity at end of prior year	−7.18 [1.97]		−6.21 [1.82]	
R&D firm's total assets at end of prior year		−5.38 [2.14]		−3.93 [2.03]
Patent Awards × Shareholders' Equity	−0.17 [2.57]		−0.16 [1.92]	
Patent Awards × Total Assets		−0.13 [2.64]		−0.13 [2.10]

Constant	9.41 [37.11]	9.93 [14.72]		
F-statistic	4.91	5.28		
χ^2-statistic			22.24	23.25
p-value	0.00	0.00	0.00	0.00
Adjusted R^2	0.08	0.09		
Log likelihood			−398.10	−389.08
Number of observations	180	176	180	176

Note: The sample consists of 200 technology alliances initiated between biotechnology and pharmaceutical companies or between biotechnology firms in the 1980–1995 period. The dependent variable is the number of control rights included in each alliance out of the twenty-five rights appearing in between 5 and 95 percent of the alliances. The public equity raised and the financial position variables are in billions of 1995 dollars. The biotechnology index reflects inflation-adjusted public equity values and is normalized to 1.0 on January 1, 1978. The dummy variable for early-stage alliances is coded as 1.0 for projects in the discovery through preclinical research phase. The second panel includes interaction terms between the patent counts and the financial variables. The constant terms are not reported for the ordered logit regressions. Absolute *t*-statistics in brackets.

of patent awards to the R&D firm in fields related to the alliance, the total amount of equity collectively raised by biotechnology firms from the public markets in the previous quarter, an index of biotechnology equity valuations, and two measures of the financial resources of the R&D firm (shareholders' equity and total assets).

In each case, the coefficients on the measures of the financial condition of the firm are significantly negative, at least at the 95 percent confidence level. When the R&D firm is in a stronger financial position, it retains more of the control rights in the alliance. The coefficients suggest that these considerations have a significant economic impact. For example, in the first regression in panel A, a one-standard-deviation increase in the R&D firm's shareholders' equity at the mean of the independent variables leads to an 11 percent drop in the predicted number of control rights assigned to the financing firm, declining from a predicted 9.3 to 8.3.

The results concerning the maturity of the project are also in the predicted direction. As reported in panel A, the coefficient on the number of patents is positive. R&D firms entering into alliances with strong patent positions (where we may anticipate that much of the initial research is already completed, the relative contribution of the R&D firm to the alliance will be more modest, and the writing of an enforceable contract covering outcomes and effort will be easier) are assigned fewer control rights. At the mean of the independent variables in the first regression in panel A, a one-standard-deviation increase in the number of related patent awards to the R&D firm at the time of the alliance leads to an increase from 9.3 to 9.9 rights assigned to the financing firm.

In panel B, we add an interaction between the number of related patent awards and the shareholders' equity or the total assets of the R&D firm. The interaction term is significantly negative. The R&D firm's additional patents are associated with more control rights being assigned to the financing firm, but only when the R&D firm has few financial resources. If the R&D firm has more financial resources, a stronger patent position leads to the R&D firm's ceding fewer control rights. This finding may be considered the strongest evidence for the presence of financial effects on the allocation of control rights. While there may be alternative explanations for why patents should lead to more or fewer control rights being assigned to the financing firm, they would be hard-pressed to explain this interaction pattern.

Thus, the results suggest a two-sided view of corporate alliances as a way to fund and govern young firms. On the one hand, the governance of these alliances on the whole is consistent with maximizing the pursuit of innovation. On the other hand, the difficult environment sometimes precludes structuring the ideal agreement.

Case Study 2: Venture Capital Contracts in Asia

A second illustration is the evolution of contracts between venture capitalists and entrepreneurs in Asia. These agreements appear to have included increasing provisions that defend the rights of the venture capitalist, consistent with the suggestion that investors have become more sophisticated and learned about the kinds of problems that can emerge in these settings.

Anecdotal accounts suggest that these agreements in the 1980s and early 1990s were quite simplistic. In many cases, the venture groups appear to have purchased common stock, which gave them no special control rights. This was particularly problematic when the firm purchased a minority stake. Even when preferred stock was purchased, accounts suggest that these agreements were frequently short and contained few protections for the investors. In fact, in many cases, no board seats were granted to the investors. (For discussions of these patterns, see, for instance, Vonk 1988, Chia and Wong 1989, Green 1991, Whyte 1996, and frequent discussions in the *Asian Venture Capital Journal*.)

Observers attributed this failure to secure control rights to two factors. First, many Asian entrepreneurs were unfamiliar with private equity and thus resisted ceding control vigorously. (The venture capital groups were so young that they had not had time to create a reputation for fair dealing, which might have overcome these objections.) Second, many of the organizations themselves had grown out of merchant banks, which typically invested in relatively mature companies, whether already-public concerns or private companies that were due to shortly complete an initial public offering. While the companies in which the venture groups invested were frequently riskier and less transparent, the fund managers

426   Josh Lerner

Table 10.4
Summary statistics of Asian and U.S. venture capital transactions

	Asian sample	U.S. sample
Number of observations in sample	19	200
Year of financing		
Mean	1998	NA
Median	1999	1997
Size of financing		
Mean	74.4	7.1
Median	44.1	4.8
Share of equity purchased		
Mean	40.2%	47.6%
Median	45.0%	47.9%

Note: Four leading private equity groups provide the Asian sample; the U.S. sample is from Kaplan and Stromberg (2000). The financing size variable is in millions of 2000 dollars. The equity ownership in each case is the minimum amount (before any contingencies for nonperformance).
NA = This information cannot be determined.

did not always realize that these differences necessitated demanding much stronger control rights.

These agreements appear to have become increasingly sophisticated over recent years. This is illustrated in a comparison of nineteen recent Asian venture capital stock purchase agreements (from China, Malaysia, the Philippines, South Korea, and Taiwan) with the sample of U.S. venture capital transactions assembled by Kaplan and Stromberg (2000). As in that article, I approached a number of leading private equity groups and asked them to submit documentation of representative transactions that they had undertaken in recent years. I asked them to include a typical mixture in terms of stage, industry, and ultimate success (insofar as it could be determined).

The summary statistics concerning these agreements is in table 10.4. One difference is the substantially larger size of the Asian transactions. This reflects, in part, the differences in the definition of venture capital in the two regions. In the United States, the term *venture capital* typically is not used to describe consolidation, restructuring, and certain other

Table 10.5
Governance provisions on venture capital transactions

	Asian sample	U.S. sample
A. *Board-related provisions*		
Percentage with VC board seat	84%	95%
Of those with board seats		
Number of seats granted		
Mean	2.7	2.5
Median	3.0	NA
Venture capital share of board of directors		
Mean	39.5%	41.4%
Median		40.0%
B. *Presence of other contractual provisions*		
Mandatory redemption provision	37%	84%
Antidilution provision	42%	95%
Approval required for:		
New equity issues by firm	37%	66%
Large expenditures by firm	53%	56%
Transfer of managers' equity holdings	37%	62%

Note: The Asian sample consists of nineteen transactions by four private equity groups; the U.S. sample, 200 transactions by fourteen groups (from Kaplan and Stomberg 2000). The board representation corresponds to that after immediately after the financing, not that if the firm fails to meet its financial or operating targets. The last three entries in Panel B for the United States are taken from the sample of fifty agreements discussed in Gompers (1998).
NA = This information cannot be determined.

later-stage investments. Instead, they are typically described as private equity transactions. Outside the United States, this distinction is frequently not made: the terms *venture capital* and *private equity* are used interchangeably.

Table 10.5 compares the provisions regarding the governance of these agreements. The most striking pattern is the absence of differences, particularly in regard to the most critical aspect of control: service on the board (see the discussion in Lerner 1995.) As panel A reports, the presence of venture capitalists on the boards of Asian firms—on both an absolute and percentage basis—does not appear to differ significantly from their U.S. counterparts at the end of this period.

Panel B reports on other control-related provisions.[6] Here some differences remain: in particular, the U.S. firms are substantially more likely to successfully demand a variety of provisions that reinforce their holdings. But even so, the contrast to the absolute lack of these protections that discussions suggest characterized the Asian venture capital industry in the 1980s is still quite striking.

In short, the experience of Asian venture capital investing investments appears to be illustrative of the discussion of evolution above. Initially, anecdotal accounts suggest that they did seek adequate control provisions. Over time, however, the need for such provisions has been appreciated, as the analyzed transactions highlight.

More generally, Antoinette Schoar and I (2005) are examining the evolution of private equity transactions in developing countries across the world. This research shows that the choice of security employed appears to be driven by the setting of the firm. Investments are far less likely to employ common stock, and somewhat more likely to employ convertible securities, in common law and high-GDP nations. In nations where property rights are less established, private equity groups emphasize equity holdings. They are likely to rely on majority ownership of the firm's equity in order to ensure control.

Conclusions

This chapter has sought to take a systematic look at the process through which young firms are financed. Contrary to the often overly optimistic view of entrepreneurs, young firms often pose substantial risks to potential financiers. To address these risks, those who invest in these firms, such as venture capitalists and corporations, have developed substantial tool kits to address these problems. This chapter systematically catalogues both the problems and the solutions.

At the same time, professional investors are not universally successful. In particular, several common problems emerge to limit the effectiveness of these investors. Through discussion of economic theory and two empirical case studies, I have sought to illustrate both the strengths and the limitations of the investors who dare to operate in this challenging arena.

Notes

This chapter was prepared for the Social Science Research Council's Project on Financing Major Innovations. I thank several Asian private equity groups for contributing private placement data, Philip Bilden of HarbourVest Partners for providing introductions to these groups, and Mark Edwards of Recombinant Capital for the biotechnology alliance data. Ken Sokoloff provided helpful comments. Trang Tran provided research assistance. I thank Harvard Business School's Division of Research for its financial support.

1. The next two sections are based in part on Gompers and Lerner (1999, 2001) and Lerner (2000).

2. For instance, in almost all nations, trade secrets offer exceedingly narrow intellectual property protection, protecting only against misappropriation. This term is defined as "the acquisition of a trade secret by a person who knows or has reason to know that the trade secret was acquired by improper means" (Milgrim 1993). Thus, a firm cannot sue a rival who discovers its trade secret independently or through reverse engineering (the disassembly of a device to discover how it works). This is unlike patent protection, which allows the awardee to prosecute others who infringe, regardless of the source of the infringers' ideas. Pooley (1989) notes that very few "naked" trade secret licenses are observed, suggesting that the information covered only through this very narrow property right is difficult to transfer in an arm's-length exchange.

3. Lerner (1997) suggests, however, that these firms may have displayed behavior consistent with strategic models of technology races in the economics literature. Because firms had the option to exit the competition to develop a new disk drive, it may have indeed been rational for venture capitalists to fund a substantial number of disk drive manufacturers.

4. This analysis is discussed in more detail in Lerner and Merges (1998) and Lerner, Shane, and Tsai (2003).

5. At the same time, biotechnology alliances present a more complex picture than many incomplete contract models. Typically, these models assume a one-time contracting process between the two parties. In reality, pairs of firms undertake repeated sets of alliances on different topics. These prior interactions may allow firms to develop reputational capital and at least partially address some of the contracting problems.

6. Kaplan and Stromberg (2000) do not report on a number of control provisions of interest. Consequently, I use the tabulations in Gompers (1998) for the final three rows of panel B. Gompers analyzes a smaller sample of fifty contracts, but considers a broader array of measures.

References

Aghion, Phillipe, and Jean Tirole. 1994. "On the Management of Innovation." *Quarterly Journal of Economics* 109, 1185–1207.

Akerlof, George A. 1970. "The Market for 'Lemons': Qualitative Uncertainty and the Market Mechanism." *Quarterly Journal of Economics* 84, 488–500.

Chia, Robert Kay Guan, and Kwei Cheong Wong. 1989. *Venture Capital in the Asia Pacific Region.* Singapore: Toppen.

Devenow, Andrea, and Ivo Welch. 1996. "Rational Herding in Financial Economics." *European Economic Review* 40, 603–615.

Gallini, Nancy, and Brian D. Wright. 1990. "Technology Transfer under Asymmetric Information." *Rand Journal of Economics* 21, 237–252.

Gompers, Paul A. 1998. "Ownership and Control in Entrepreneurial Firms: An Examination of Convertible Securities in Venture Capital Investments." Unpublished working paper, Harvard University.

Gompers, Paul A., and Josh Lerner. 1999. *The Venture Capital Cycle.* Cambridge, Mass.: MIT Press.

Gompers, Paul A., and Josh Lerner. 2000. "Money Chasing Deals? The Impact of Fund Inflows on Private Equity Valuations." *Journal of Financial Economics* 55, 239–279.

Gompers, Paul A., and Josh Lerner. 2001. *The Money of Invention.* Boston: Harvard Business School Press.

Green, Milford B. 1991. *Venture Capital: International Comparisons.* New York: Routledge.

Greenwald, Bruce C., Joseph E. Stiglitz, and Andrew Weiss. 1984. "Information Imperfections in the Capital Market and Macroeconomic Fluctuations." *American Economic Review Papers and Proceedings* 74, 194–199.

Grossman, Sanford J., and Oliver D. Hart. 1986. "The Costs and Benefits of Ownership: A Theory of Lateral and Vertical Integration." *Journal of Political Economy* 94, 691–719.

Hart, Oliver D., and John Moore. 1990. "Property Rights and the Nature of the Firm." *Journal of Political Economy* 98, 1119–1158.

Hart, Oliver D., and John Moore. 1998. "Default and Renegotiation: A Dynamic Model of Debt." *Quarterly Journal of Economics* 113, 1–41.

Jensen, Michael C., and William H. Meckling. 1976. "Theory of the Firm: Managerial Behavior, Agency Costs, and Ownership Structure." *Journal of Financial Economics* 3, 305–360.

Kamien, Morton, and Yair Tauman. 1986. "Fees versus Royalties and the Private Value of a Patent." *Quarterly Journal of Economics* 101, 471–493.

Kaplan, Steven N., and Per Stromberg. 2000. "Financial Contracting Theory Meets the Real World: An Empirical Analysis of Venture Capital Contracts." Discussion paper 2421, Centre for Economic Policy Research.

Lehn, Kenneth, and Annette B. Poulson. 1992. "Contractual Resolution of Bondholder-Stockholder Conflicts in Leveraged Buyouts." *Journal of Law and Economics* 34, 645–673.

Lerner, Josh. 1995. "Venture Capitalists and the Oversight of Private Firms." *Journal of Finance* 50, 301–318.

Lerner, Josh. 1997. "An Empirical Examination of a Technology Race." *Rand Journal of Economics* 28, 228–247.

Lerner, Josh. 2000. *Venture Capital and Private Equity: A Casebook*. New York: Wiley.

Lerner, Josh, and Robert P. Merges. 1998. "The Control of Technology Alliances: An Empirical Analysis of the Biotechnology Industry." *Journal of Industrial Economics* 46, 125–156.

Lerner, Josh, and Antoinette Schoar. 2005. "Does Legal Enforcement Affect Financial Transactions: The Contractual Channel in Private Equity." *Quarterly Journal of Economics* 120, 223–246.

Lerner, Josh, Hilary Shane, and Alexander Tsai. 2003. "Do Equity Financing Cycles Matter? Evidence from Biotechnology Alliances." *Journal of Financial Economics* 67, 411–446.

Merton, Robert C. 1995. "A Functional Perspective of Financial Intermediation." *Financial Management* 24 (Summer), 23–41.

Milgrim, Roger M. 1993. *Milgrim on Trade Secrets*. New York: Matthew Bender.

Miller, Merton H. 1986. "Financial Innovation: The Last Twenty Years and the Next." *Journal of Financial and Quantitative Analysis* 21, 459–471.

Myers, Stewart C., and Nicholas S. Majluf. 1984. "Corporate Financing and Investment Decisions When Firms Have Information That Investors Do Not Have." *Journal of Financial Economics* 13, 187–221.

Pooley, James. 1989. *Trade Secrets: A Guide to Protecting Proprietary Business Information*. New York: American Management Association.

Sahlman, William A., and Howard Stevenson. 1986. "Capital Market Myopia." *Journal of Business Venturing* 1, 7–30.

Sherbloom, James P. 1991. "Ours, Theirs, or Both? Strategic Planning and Deal Making." In R. Dana Ono, ed., *The Business of Biotechnology: From the Bench to the Street* (213–224). Stoneham, Mass.: Butterworth-Heineman.

Shleifer, Andrei. 2000. *Inefficient Markets: An Introduction to Behavioral Finance*. New York: Oxford University Press.

Stiglitz, Joseph E., and Andrew Weiss. 1981. "Credit Rationing in Markets with Incomplete Information." *American Economic Review* 71, 393–409.

Tufano, Peter. 1997. "Business Failure, Judicial Intervention, and Financial Innovation: Restructuring U.S. Railroads in the Nineteenth Century." *Business History Review* 71, 1–40.

Van Horne, James C. 1985. "Of Financial Innovations and Excesses." *Journal of Finance* 40, 620–631.

Vonk, Gerro. 1988. *Venture Capital Beyond Boundaries*. Utrecht, Netherlands: Driestar Publishing.

Whyte, Robert. 1996. "Special Study on Asian Private Equity Investing." Unpublished consulting report, Robert Whyte Associates.

11
Real Effects of Knowledge Capital on Going Public and Market Valuation

Michael R. Darby and Lynne G. Zucker

Major technological innovations are extremely important for the success of a high-technology firm, particularly innovations that other firms cannot easily emulate, either because of broad patent protection or because of natural excludability, often the result of important tacit knowledge underlying the technology that is unavailable to potential emulators. These constitute the main parts of a firm's knowledge capital. Firms exploiting such legal or natural excludability enjoy a higher probability of going public and a higher rate of return until other firms are able to catch up or overtake the firm's growing technology portfolio with similar or more advanced technologies.

We argue that the strong, positive effects of access to knowledge capital on firm performance alter the financial prospects of the firm, making it easier to find capital and to obtain it in larger amounts. In this chapter, we show that this is true of early funding by venture capitalists, of funding through an initial public offering (IPO), and later market returns. Since firms with collaborating relationships with star scientists are more likely to go public, previous studies that have excluded firms that do not go public seriously underestimate the total effects of stars on financial success of firms. Here, we investigate the full process, from birth to IPO to later financial market performance, of a substantial portion of those dedicated biotechnology firms that ever go public through 1992. We end observations at that time in order to have patent citation data for the following five years to allow reliable estimation.

Our approach adds indicators for knowledge capital, generally viewed as intangible capital in financial analyses of firms. What is new in this chapter is that we now quantify the effects of knowledge capital on the bottom line of firms' market performance, effectively pricing the value of several forms of knowledge capital. Early in commercialization, we have found significant effects of knowledge capital on the birth of new biotechnology and nanotechnology firms and on the number of products in development in biotech, but it can be argued that the acid test is a financial market one, a test that becomes possible only at later stages of a new high-technology industry's development.

After we present the theoretical approach, we detail the data used, present the empirical results on IPOs, follow the firms that go public to examine their subsequent stock price performance, and end by offering our conclusions.

Theoretical Approach

Our theoretical approach to determinants of financial market success rests on two fundamental conceptualizations. First, metamorphic progress can create new industries and transform old and is often fueled by discontinuities in knowledge derived from basic science breakthroughs (Darby and Zucker 2003). Second, knowledge capital, considered intangible in commercial accounting frameworks, has both tangible and measurable consequences on firm founding and performance. We argue that these effects of knowledge capital are driven by natural excludability based on tacit knowledge that often results from breakthrough, discontinuous, scientific discoveries.

Most firms achieve perfective progress, incrementally improving commodities or productivity. But technological progress is concentrated in a few firms achieving metamorphic progress: forming or transforming industries with technological breakthroughs. Literally hundreds, if not thousands, of firms entered each time to exploit such breakthroughs as the internal combustion engine, the integrated circuit, biotechnology, and, most recently, nanotechnology. Biotechnology is characteristic of many industries born based on scientific or technological breakthroughs that make it possible to do what was before impossible or impossibly uneconomical.[1] Unless congruent with incumbents' science and technology base, metamorphic progress promotes entry.

Scientific breakthroughs embodied in discovering scientists, protected by natural excludability, and transferred by learning-by-doing-with at the bench generally drive metamorphic progress. Embodied knowledge is rivalrous and leads to entry and industry dominance by star-scientist-linked firms. It is often complemented and reinforced by formal intellectual property rights, particularly patents. We use *knowledge capital* to refer to the distinctive collection of know-how and know-why that gives a particular high-technology firm its value.

As the industries mature and rapidly grow, a relatively small number of firms grow rapidly, while most stagnate, shrink, or exit through merger or failure. The entrant biotech firm must win or at least place in repeated rounds of beauty contests if it is to attract sufficient funding from angel investors, venture capitalists, and ultimately public investors to fund the long years of research and clinical trials that typically lie between brilliant idea and marketed, revenue-producing product. The firms that will be most successful in passing through rounds of financing and simultaneously achieving sufficient R&D success to be an attractive contestant will be those with the most valuable knowledge capital.

For private firms, we do not have the rich data sets available for public firms, but our previous work has identified some key variables that serve as useful indicators of a high-quality knowledge capital and of interim progress toward ultimate profitability. The single most powerful indicator of ultimate success is the active working involvement of star scientists who, although small in number, are responsible for a large fraction of the most important discoveries. When these scientists become involved in commercializing their discoveries, they frequently have the insight and scientific taste to identify the sweet spot where scientific possibility and economic payoff are combined. We have found that articles authored by star scientists either with or as firm employees is powerful evidence of that involvement with strong predictive value for the success of the firm's R&D program. Other such indicators of knowledge capital will be detailed in the next section.

We also expect that the firm's history of receiving venture capital investments will have a positive impact on the probability of the firm's going public. First, venture capitalists are investors themselves, and their funding connotes their expectation that they will be able to exit after a few years through a successful public offering. Their investment and due diligence also provide a behavioral signal on which uninformed investors can cascade (Banerjee 1992; Bikhchandani, Hirshleifer, and Welch 1992; Welch 1992).

As is well known, there are waves, or windows, in which financing of particular technologies is in vogue and IPOs are particularly favorable for firms that are ready to go in terms of a technology at least reasonably proven in the laboratory. We thus expect firms more likely to go public in periods following rapid increases in biotech stock prices.

IPOs for high-technology firms are things of remarkable beauty. Investors provide large amounts of money to firms with few employees and relatively few assets beyond some lab equipment and the brains of the personnel represented by accumulated losses that have proven— more or less—the plausibility of the firm's nascent technology succeeding in the market. Investors in these firms seek to diversify their portfolios in the hope that two or three winners will much more than compensate for the inevitable forty or fifty losers. Not surprisingly, a significant minority of IPOs involve firms that are more or less fraudulent. We hypothesize that investors will foresee this possibility and provide more funding to firms in which star scientists are involved in bench-level science for the firm (indicated for us by joint publications) than merely lending their name to a scientific advisory board.

Aghion and Tirole (1994) provide a more general analysis of why investors would place a higher value on firms in which the research principals are deeply involved. Deeds, Decarolis, and Coombs (1997) give a signaling interpretation to their finding that proceeds increase with the number of citations to publications authored by the full-time executives and employees of the firm. Stephan and Everhart (1998) confirm for two years of IPOs the value of highly cited scientists associated with the firm. We have emphasized the importance of having the best scientists to the real productivity of the firm's research and development efforts (Zucker, Darby, and Armstrong 1998; Darby, Liu, and Zucker 2004).

Without necessity to choose among these motivations here, we hypothesize that the amount of IPO proceeds raised if and when the firm goes public will be greater the deeper the knowledge capital of the firm is.

We also expect that the firm's history of receiving venture capital investments and hot market conditions indicated by prior high biotech returns will have positive impacts on the amount the firm is able to raise for very much the same reason that each factor increases the probability that the firm goes public.

One other factor that should affect the probability of making an IPO is whether the firm has been organized from the beginning with making an IPO as an important intermediate goal or instead aims at remaining a moderately sized, privately owned business or perhaps achieving proof of concept and being purchased by a large firm. Romanelli (1991) reviews the largely supportive evidence for the organizational imprinting hypothesis: firms tend to reflect the environments in which they were born and embody the intentions of their founders. A homier version derived from our fieldwork is that a number of firms were founded in frank imitation of the IPO success of a nearby colleague, possibly a personal rival. Our study adds further empirical evidence in support of organizational imprinting.

Data

This chapter is unique in beginning with a universe of nonpublic firms and analyzing the determinants of the probability a firm will go public in a given year and, if it does, the amount of capital that it raises in the IPO. Deeds, Decarolis, and Coombs (1997) and Stephan and Everhart (1998) have previously examined the amount raised in biotech IPOs for those firms that go public during their time frame; we are able to add to and improve on their analyses in several ways. First, because of our rich data resources from the larger biotechnology project, we examine all IPOs occurring over a much longer period than could either of those research teams. Second, because we have the universe of private firms, we are able to apply the Tobit technique to properly estimate the funding amounts conditional on the probability that a firm goes public.

Finally, and most important, we have much more information on the scientists involved in firms and are able to identify publications with star scientists that are joint with the firms. These joint publications identify the labor effort that has been moved over to the firm by top scientists—

those who have made breakthrough discoveries in genetic sequencing.[2] The importance of this labor effort has been documented in a series of substantive analyses, predicting a wide variety of performance measures from patents and products to employment growth. In our most recent analyses, we examine publications joint with firm scientists by all scientists in the relevant science areas and by the stars (Zucker, Darby, and Armstrong 2002). We find that stars' joint publications provide a significant increment above the significant positive effect of all scientists who are copublishing with firm scientists on a variety of performance measures, especially strong and consistent in panel analyses.

Table 11.1 reports descriptive statistics for the variables used in our empirical analyses. Part A of that table pertains to the variables used to explain the duration of time from birth to IPO for a private biotechnology firm, as well as the amount raised in the IPO. Section B of the table pertains to the variables used in the analysis of stock price behavior of those biotech firms that went public between 1976 and 1992. The structure of this section corresponds to the table.

Data for IPO Analysis

Estimating the probability that private firms go public is inherently data constrained since by definition, they are under no compulsion to make public disclosures in the systematic and detailed ways that public firms must. However, a surprising amount of information can be gained by coding the data contained in commercial directories that serve to attract customers and by matching to these data information on recipients of venture capital, assignees of patents, and affiliations of authors of scientific papers covering areas closely related to the firms' technologies. Since 1988 we have been building a matched, linked, and cleaned database for examining the interaction among firms, universities, research institutes, and scientists involved in biotechnology.[3]

The use of modern biotechnology mushroomed from nonexistent in 1975 to over 700 firms by 1990 in the United States alone. Of those firms, some 512 were new entrants (new biotech firms), most of which were attempting to apply the new bioscience breakthroughs to create commercially valuable products and to go public, ensuring the survival of the firm and wealthy founders and early investors. By 1992, 162 of

Table 11.1
Descriptive statistics

Variable	Mean	SD	Minimum	Maximum	N
A. IPO analysis					
Knowledge capital					
Star-Firm Articles to Date	0.010	0.23	0	9	3,675
Patents Granted with Application to Date	0.027	0.25	0	8	3,675
Use of rDNA Technology by Firm	0.42	0.49	0	1	3,675
Local Top-Quality Universities	1.23	1.13	0	3	3,675
Number of SBIR Grants to Date	0.031	0.21	0	5	3,675
Firm strategy indicator					
IPOs/Private NBFs in Year and Region Firm Born	0.064	0.056	0	0.17	3,675
Market indicators					
Biotech Returns Two Years Prior	0.21	0.35	−1.65	1.03	3,675
Biotech Returns One Year Prior	0.37	0.62	−1.65	2.05	3,675
Capital raised					
IPO Proceeds Raised by Firm	2.18	6.88	0	7,997	3,675
Firm Has Received Venture Capital	0.22	0.41	0	1	3,675
Rounds of Venture Capital Received	0.63	1.55	0	13	3,675
B. Stock price analysis					
Knowledge capital					
R&D Stock—expenses cumulated subject to 20 percent depreciation	19.73	48.80	0.0033	623.114	717
Star-Firm Articles to Date/R&D Stock	0.084	0.46	0	9.00	717
Citations to Patents Granted with Application to Date/R&D Stock	0.72	2.68	0	29.51	717
Claims in Patents Granted with Application to Date/R&D Stock	6.79	20.40	0	301.85	717
Use of rDNA Technology by Firm	0.60	0.49	0	1	717
Local Top-Quality Universities	1.49	1.12	0	3	717

Table 11.1
(continued)

Variable	Mean	SD	Mini-mum	Maxi-mum	N
Age variables					
Age—number of years the firm has been in biotech	6.87	3.13	0	16	717
Age of the Biotech Industry (year—1975)	12.06	2.93	1	16	717
Financial variables					
Market Value of Firm Equity	155.05	508.30	0.29	7,334	717
Book Value of Firm's Total Physical Assets	49.14	104.44	2.00	979.56	717
Debt Level of Firm	6.58	18.94	0.04	292.19	717
Maturity of Debt (McCauley Duration)	1.29	0.43	1.00	4.32	717
IPO Lead Underwriter Reputation Rank	6.07	3.23	1	9	717
Short-Term Interest Rate (6-month T-bill rate)	0.067	0.20	0.036	0.14	717

Note: Dollar values are given in millions of 1984 U.S. dollars.

these firms had traded publicly, including 156 (30.5 percent of the 512 entrants) that made an IPO.[4] On average firms were in the at-risk population for 7.18 years before exiting by IPO or otherwise or right-censoring after 1992, making for a total of 3,675 firm-year observations.

We identified which firms went public, the date of IPOs, and the proceeds raised using *Bioscan* (1989–1997), its precursor Cetus Corp. (1988), the *IPO Reporter*, and the Securities Data Company (1998a) *Global New Issues* online electronic database. *Moody's* manual and purchased copies of the IPO prospectus were used to resolve a few instances of conflict among these sources. In sum, 512 firms were present for one or more years in our at-risk population for going public, and 156 of those firms exited the at-risk population via an IPO.

Measures of the Firm's Knowledge Capital
Table 11.1.A lists five measures or indicators of a firm's knowledge capital for which we were able to obtain data. The first of these, Star-Firm Articles to Date, is a count of articles written by star scientists as or with

the firm's employees, counting each such article once for each star author on it. Zucker and Darby (1996) defined a star as one of the 327 most productive biotech researchers in the world defined as those who had discovered over forty genetic sequences or published twenty or more articles reporting genetic sequence discoveries by April 1990 (GenBank 1990). We hand-collected all the stars' sequence discovery articles and identified all the authors and their affiliations on each article. Zucker, Darby, and Armstrong (1998) showed that a cumulative count of the "tied" articles written by stars either as employees or with employees of the firms is an accurate indicator of the extent of their involvement in the actual research of the firm and hence an excellent predictor of the firm's eventual success. We use the current Zucker-Darby biotechnology database, which includes over 2,000 more articles by these stars that have been collected based on genetic sequence discoveries reported in Gen-Bank (1994), and so are able to compute and cumulate the number of star-firm articles for each of the firms through 1992.

Patents granted are often used as an indicator of R&D productivity, although it is often remarked that it may be a better indicator of past R&D expenditures than production of valuable innovation (Griliches 1990). Since we do not have a history of R&D expenditure data available for the private firms in our at-risk population, this collinearity may be a blessing despite implying some difficulty in interpretation. For this chapter, we obtained the patent data for each biotech firm from the U.S. Department of Commerce (1993) CD-ROM *Patent Technology Set: Genetic Engineering*, produced by the Office of Electronic Information Products and Services of the U.S. Patent and Trademark Office. The Patent Technology Sets include patents issued through January 26, 1993, and included in the above classifications as of June 29, 1993. We counted the number of patents granted to the firm using year of application date through 1991. We used year of application because the work underlying the patent had by then been done and would be available for disclosure (and normally would be disclosed) to the financial markets in the event of an IPO. We also calculated the total patents applied for by each firm from 1976 to 1991 that were later granted.

A simple indicator of the level of technological sophistication of a biotech firm in the 1980s was whether it was actively using the recombinant DNA (rDNA) technology, also referred to as genetic engineering. We coded (1 yes, 0 no) whether the firms reported using the recombinant DNA technology (rDNA) in *BioScan*, the leading commercial directory of biotech-producing firms, or were licensed by Stanford and the University of California to do so. This rDNA technology is a marker of a high-science firm.

The presence of nearby top-quality universities appears to be a valuable knowledge resource to a firm, although it is not clear whether this represents the presence of stars unidentified by our empirical definition, geographically localized knowledge spillovers, or the pecuniary externalities inherent in subsidized training of labor and geographic agglomerations generally. There are twenty U.S. universities with biotech-relevant departments receiving the highest ratings (rated above 4 on a scale of 5) on overall scholarly quality as reported in the 1982 National Academy of Sciences' reputational survey of doctoral programs (Jones, Lindzey, and Coggeshall 1982).[5] Biotech-relevant departments included biochemistry, cellular and molecular biology, and microbiology. Local top-quality universities is a count of the number of universities in the same region with one or more such department. Same region refers to the functional economic areas (also referred to as BEA areas) into which the United States is divided by the Bureau of Economic Analysis in U.S. Department of Commerce (1992).

While the first four measures are previously proven scientific indicators, we also try in some regressions a more novel indicator: count of the Number of Small Business Innovation Research (SBIR) Grants to Date received by the firm.[6] The number of SBIR grants received might be expected to reduce the probability of remaining private either because the resources permit creation of more intellectual property or serve as a government certification that the firm is doing good science (the halo effect) or for both reasons.

Firm Strategy Indicator

In order to test the organizational imprinting hypothesis (Romanelli 1991), we distinguish those that are founded in hopes of becoming a major player by proceeding through rounds of private financing to the IPO and beyond and those that in fact are intended by the proprietor either to remain indefinitely a small or medium-sized firm with a narrow scope and private ownership or to be sold to a major firm after proof of concept. The only variable we have found that distinguishes at all effectively between firm types is IPOs/Private NBFs (New Biotech Firm) in Year and Region Firm Born, the fraction of private biotech firms going public in the year, and the region of the firm's founding. Our fieldwork suggests that a number of scientists founded eventually very successful firms in order to imitate, if not outdo, the commercial success of the suddenly rich rival down the university corridor.

Market Indicators

We define a biotech equity index and use its returns to measure the overall performance of biotech stocks. It is generally observed that IPOs in an industry increase after a large run-up in the valuation of the industry. We considered two different portfolios constructed to calculate monthly weighted returns from January 1975 to December 1995. First, our basic biotech equity index is defined based on a portfolio consisting of all publicly traded biotech-using firms (including incumbent adopters) in our larger project biotech firm data set. Second, we defined an alternative portfolio consisting of all publicly traded firms in the biotech-relevant Standard Industrial Classifications 2830 through 2836, inclusive. We retrieved return data from CRSP and calculated the monthly weighted returns for each portfolio. Based on these monthly returns, we calculated annualized returns for each portfolio. In this chapter, we use the annualized weighted returns of the first portfolio since the second portfolio included numerous nonbiotech-using firms.

Capital Raised

For each of the 156 IPOs observed, we used the same sources already cited to obtain the proceeds raised. IPO values were converted (deflated) to 1983 U.S. dollars by using the Consumer Price Index for All Urban Consumers as reported in the *Economic Report of the President* (February 1995). In work not reported here, we also examined use of the cumulative number of biotech IPOs and the cumulative value of biotech IPO proceeds as control variables.

Firm Has Received Venture Capital (1 yes, 0 no) and Rounds of Venture Capital Received (cumulative count through current year) are based on data available in the *VentureXpert* online electronic database (maintained by the Securities Data Company 1998b). The database contains detailed information on the date, stage, and amount for each round of funding for each firm. We drop "Bridge," "Bridge Loan," "Open Market Purchase," "Other Spec Situation," "Secondary Purchase," and "Turnaround" stage funding. We drop observations where there are missing values on date or on the amount of venture capital investment. Venture capitalists sometimes see SBIR and similar government programs as competitors, so including venture capital and SBIR funding in the same regression permits us to assess whether these funding sources are equally valued in the financial markets.

Data for Stock Price Analysis

Of the 156 firms reported as going public, we can find complete firm data and subsequent trading records for only 129 firms in the COMPUSTAT database. For those 129 firms, we retrieved such data as number of outstanding shares, closing stock price, total assets, debts with different maturities, and R&D expenditures. These firms are an attractive natural experiment for understanding the IPO process and IPO proceeds. To follow later market returns, we use subsequent trading records for these same 129 firms in Computstat, which was our source for basic data on the number of outstanding shares, closing stock price, total assets, debts with different maturities, and R&D expenditures for the year or as of the end of the year as the variable is a flow or a stock. Straightforward variables are defined in table 11.1B without further elaboration here. The period of analysis for each firm is from the year of founding through 1992. The year 1992 was chosen because it was the last date for which we have full information on the star-firm articles variable and because we needed to allow another five years to find out which pending patents are granted and to accumulate patent citation data.[7]

Measures of the Firm's Knowledge Capital

Once the firm goes public, we have data on R&D expenditures beginning with two years prior to the IPO. The R&D stock is a cumulative sum of these expenditures deflated by the CPI to 1984 dollars and depreciated at a conventional compounding rate of 20 percent per year Griliches (1990). Since R&D expenditures are available in Computstat only back two years before the firm went public, we experimented with extrapolating these expenditures backward to the firms' founding date. This effort proved difficult to implement in practice and did not significantly affect the estimates and so was dropped for reasons of simplicity.[8] We also tried total assets and the number of employees to control for firm size. The empirical results are qualitatively the same whichever variable is used to scale the knowledge capital measures.

It is conventional to depreciate the knowledge indicated by R&D output measures at a rate like the 20 percent that we use for patents. It is not so clear that procedure makes sense for input measures like star-firm articles. Consider, for example, two firms otherwise identical except that one has one star-firm article each of the five years since its birth, while the other has five star-firm articles, all in the current year. It could be argued that if the star-firm articles were productive inputs, the first of these firms has more knowledge capital than the second. This is an argument for cumulating star-firm articles with accrued interest instead of depreciation. We found that the results were qualitatively the same whether we used a 20 percent depreciation rate, a 20 percent appreciation rate, or just cumulated star-firm articles over time, as done in Zucker, Darby, and Armstrong (1998). We used the last measure in the results reported here since it had been previously validated in other studies. Accordingly, we divide Star-Firm Articles to Date by the R&D Stock to obtain a measure of the depth of the firm's star-firm articles relative to its R&D investment. Patents Granted with Application to Date is used in two weighted forms to measure patent quality. Citations to Patents Granted with Application to Date/R&D Stock weights each patent by the number of citations received by over the next five years, to obtain a measure of patent quality, and then again scaled by the R&D Stock to obtain an indication of the success of the firm's R&D efforts relative to their costs. Future citations are the standard measure of patent quality in the literature, but are inherently unknowable by financial market participants for recently granted patents. Claims in Patents Granted with Application to Date/R&D Stock is an alternative measure that is knowable by market participants at the time the patent is granted.

Use of rDNA Technology by Firm and Local Top-Quality Universities have the same meaning as in the IPO analysis. The firm's Age (meaning the number of years it has been in biotech) is implicit in the IPO duration analysis and explicit in the stock price analysis.

Financial Variables

Market Value of Firm Equity, Book Value of Firm's Total Physical Assets, Debt Level of Firm, Maturity of Debt (McCauley Duration), and Short-Term Interest Rate (six-month T-bill rate) are all straightforward variables from Compustat. We are indebted to an anonymous referee for the suggestion that we include IPO Lead Underwriter Reputation Rank, the reputation ranking of the lead underwriter that brings the firm public, in our set of instrumental variables. This variable is taken from Carter, Dark, and Singh (1998) and follows a 1-to-9 scale, with 9 representing the most reputable underwriter. Perhaps surprisingly, there is considerable variation in this variable, with some firms brought public by less reputable underwriters. We thought that potential discovery of fraud would present a second source of stock price jumps not covered by our model and experimented with dropping firms not brought public by a reputable underwriter. The results were not significantly different from those reported, so that line of investigation was dropped.

Empirical Results

We first report estimates on the probability per period that a private bio-tech firm will go public and assess the cumulative impact of the firm's knowledge capital on this probability. We then report estimates of Tobit regressions explaining the amount raised given that the firm goes public. The effects of the firm's knowledge capital are even stronger for proceeds than for duration to IPO. Finally, we examine what appears to be a reciprocal causal relationship between the firm's knowledge capital and its receipts of venture capital funding.

Duration to IPO Results

In table 11.2 we report estimates of standard Weibull survival models explaining the duration from the founding of the firm to the firm's IPO (if any). In the (unfortunate) terminology used for these models, positive coefficients indicate greater probabilities of surviving as a private firm, so negative coefficients indicate a greater probability of exiting—that is, making an IPO.

In column a of table 11.2 we present our simplest model in which the duration to IPO is predicted by six variables: three indicators of the firm's knowledge capital, the percentage of biotech firms going public the year the firm was founded, and two indicators of whether there is a hot biotech market. The three variables describing the firm's knowledge capital are the number of articles written to date by star scientists as or with a firm employee, the number of eventually granted patents applied for to date, and whether the firm uses rDNA technology. Despite some multicollinearity, each of these variables is significant at the 5 percent or better level, and the chi-squared test at the bottom of the table indicates that we can reject the joint hypothesis that all the knowledge-capital coefficients in this model (or any of the models in table 11.2) are zero at better than the 0.001 level of confidence. The percentage of biotech firms going public in the year the firm was founded is also very significant, indicating that a high IPO rate leads to imitative entry of firms pursuing the same strategy. Finally, whether the biotech market has experienced high returns in the previous year also significantly reduces the probability that a private firm will remain private. Thus, these results all closely correspond to our basic hypotheses about what drives the process of high-tech firms going public.

In columns b and c, we consider the issue of whether in fact the knowledge capital is important only because it attracts venture capital support or whether it works separately in terms of pure productivity of producing new, valuable intellectual property. After some experimentation, we found that the best available indicators of venture capital support were simply the dummy variable for ever having such support or the count of the number of rounds of support received. All the indicators are highly correlated, and we suspect that omissions in the SDC survey coverage introduce significant measurement error into the alternative venture-capital-funds-received variables that we tried. We introduce the dummy and cumulative rounds separately because multicollinearity becomes severe when both are in the same equation. In any case, we get

Table 11.2
Estimates of Weibull survival model of duration from founding to initial public offering

	(a)	(b)	(c)	(d)	(e)
Constant	3.760***	4.111***	4.093***	3.963***	4.170***
	(0.235)	(0.315)	(0.349)	(0.280)	(0.376)
Knowledge capital					
Star-Firm Articles to Date	−0.207*	−0.161	−0.235	−0.195*	−0.236
	(0.086)	(0.098)	(0.187)	(0.093)	(0.206)
Patents Granted with Application	−0.266*	−0.167	−0.268	−0.208†	−0.225
to Date	(0.117)	(0.162)	(0.172)	(0.119)	(0.176)
Use of rDNA Technology by Firm	−0.519***	−0.271†	−0.379*	−0.494**	−0.365*
	(0.153)	(0.145)	(0.169)	(0.153)	(0.170)
Local Top-Quality Universities	—	—	—	−0.143*	−0.052
				(0.066)	(0.072)
Number of SBIR Grants to Date	—	—	—	−0.302†	−0.328
				(0.171)	(0.221)
Firm strategy indicator					
IPOs/Private NBFs in Year and	−4.347***	−4.050***	−4.304**	−4.392***	−4.248**
Region Firm Born	(1.192)	(1.210)	(1.365)	(1.204)	(1.376)
Market indicators					
Biotech Returns One Year Prior	−0.339**	−0.376**	−0.393**	−0.337**	−0.396**
	(0.121)	(0.132)	(0.146)	(0.122)	(0.149)
Biotech Returns Two Years Prior	−0.248	−0.240	−0.325	−0.258	−0.348
	(0.192)	(0.203)	(0.232)	(0.196)	(0.237)

Capital raised					
Firm Has Received Venture Capital	—	-1.205*** (0.247)	—	—	—
Rounds of Venture Capital Received	—	—	-0.209*** (0.056)	—	-0.204*** (0.056)
Sigma	0.774*** (0.089)	0.779*** (0.397)	0.883*** (0.135)	0.778*** (0.091)	0.886*** (0.137)
Log likelihood	-607.6	-566.5	-579.5	-602.9	-577.9
Chi-squared test [d.f.]	39.56 [3]***	15.56 [3]***	24.48 [3]***	49.98 [5]***	27.68 [5]***

Note: Standard errors in parentheses. $N = 3,675$.
Prob($|t| \geq x$) or Prob($|\Pi^2| \geq x$): * < 0.05, ** < 0.01, *** < 0.001.
The chi-squared test is for the null hypothesis that the coefficients of the firm science-base variables all $= 0$. The degrees of freedom for the statistic are given in square brackets.

similar results: The venture capital variables are indeed highly enough correlated with the knowledge capital variables to make it difficult to measure their separate effects robustly, but—based on the chi-squared test—we see that both knowledge-capital variables as a group and venture capital support increase the likelihood of going public. The results for the strategy and market indicators are qualitatively the same for all the models reported in table 11.2.

In columns d and e, we report the results from experimentation with other knowledge capital indicators that we have used in our other work on biotechnology. As noted above, the number of SBIR grants received might be expected to reduce the probability of remaining private either because the resources permit creation of more intellectual property or serve as a government certification that the firm is doing good science (the halo effect) or for both reasons. Similarly, the number of nearby universities with top-ranked biotech-relevant doctoral programs serves as an indicator of geographically localized knowledge spillovers or favorable labor cost conditions.[9] Again, the knowledge-capital coefficients are all negative, as expected. However, although as a group they are highly significant in explaining the probability of going public, the individual coefficients are not robustly statistically significant except for whether the firm does genetic engineering (uses rDNA technology).

One way to interpret the survival models is to simulate them using different assumptions for the values of the determining variables. This is done in table 11.3 for model e of table 11.2. Table 11.3 reports the predicted number of years from firm founding that it would take for a cohort of firms with given characteristics to reach various percentages of firms having gone public.[10] The base case assumes for each year of the simulation that the surviving firms have the mean values for firms their age of each of the variables in model e. Case c sets all the knowledge capital variables at 0 instead of their mean values and reruns the simulation. The difference between cases a and c indicates the estimated reduction in duration to IPO due to the combined effects of all the firms' knowledge-capital variables. Case b is like case a except it sets at 0 all the knowledge capital variables other than the number of Articles to Date by Stars with Firm and the indicator for Use of rDNA Technology. Comparing all three cases, we see that overall, the knowledge-capital variables reduced

Table 11.3
Percentage of private biotech firms going public, by age and science base

Percentage of public firms	Base case all variables = Sample means (a)	Case with patents, top universities, and SBIR all = 0 (b)	Case with all science-base variables = 0 (c)
5	1.43	1.77	2.25
10	2.49	3.10	3.95
20	4.47	5.56	7.07
30	6.44	8.00	10.18
40	8.51	10.58	13.46
50	10.79	13.41	17.07
60	13.41	16.67	21.21
70	16.58	20.61	25.93
80	20.78	25.83	32.86
90	27.45	34.12	43.42

the time to IPO by a little over a third, and a little more than half of this reduction is attributable to the number of articles the firms have authored with stars and to the use of genetic engineering. Figure 11.1 plots the simulated effects in table 11.3, illustrating graphically the economically significant effect of the firm's knowledge capital, holding constant the market-condition variables and the firm's strategy and receipts of venture capital.

Taken as a whole, table 11.2 implies that market conditions and the firm's knowledge capital, strategy, and receipts of venture capital all play significant roles in determining the probability per year that a particular private biotech firm will go public. We cannot robustly characterize which knowledge-capital indicators are the most important, but the use of genetic engineering was always significant in determining which firms were able to go IPO when the market conditions were favorable.

Results on Proceeds from the IPO

Going public is much like the consumer buying a car; there is not only the question of whether it is done at all in any given period, but also, if it is done, how much money is involved. Accordingly, we turn to estimating the funds raised in those IPOs that do occur using the Tobit technique.

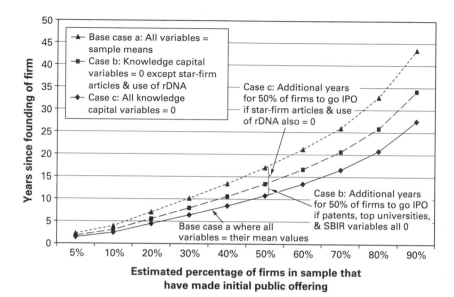

Figure 11.1
Percentage of private biotech firms going public, by age and knowledge capital

The results are reported in table 11.4. Although the reported coefficient estimates do not have the easy interpretation of OLS regressions, significant positive coefficients do mean that the variable results in a predicted increase in IPO proceeds if an IPO occurs.

Models a, b, and c estimate the Tobit regression with the core knowledge capital variables (articles by stars with or as firm employees, patents granted applied for by this date, and use of rDNA technology), two-year lagged biotech returns as our market indicator, and no, dummy, or round count venture capital variables.[11] Unlike the survival models, we obtain distinct, highly significant, positive coefficients for each of these variables in explaining how much money the firm can raise. The market indicator and venture capital variables also have robustly significant, positive coefficients in table 11.4.

In models d and e, we add the SBIR and local top-quality universities variables, but only the latter has significant, positive coefficients. It appears that investors in IPOs are much more impressed by investments by other private investors than by government bureaucrats allocating SBIR funds.

Table 11.4
Estimates of Tobit regressions for IPO proceeds

	(a)	(b)	(c)	(d)	(e)
Constant	-23.026*** (1.034)	-25.017*** (1.017)	-24.909*** (0.349)	-27.610*** (1.220)	-28.044*** (1.215)
Knowledge capital					
Star-Firm Articles to Date	6.806*** (1.440)	5.765*** (1.235)	6.549*** (1.382)	6.001*** (1.399)	5.951*** (1.356)
Patents Granted with Application to Date	10.970*** (1.348)	6.718*** (1.156)	8.548*** (1.295)	9.945*** (1.317)	8.158*** (1.279)
Use of rDNA Technology by Firm	10.570*** (0.915)	6.105*** (0.817)	8.506*** (0.893)	10.232*** (0.895)	8.428*** (0.880)
Local Top-Quality Universities	—	—	—	3.917*** (0.383)	2.937*** (0.377)
Number of SBIR Grants to Date	—	—	—	0.393 (1.838)	-1.036 (1.853)
Market indicator					
Biotech Returns Two Years Prior	6.390*** (1.232)	6.649*** (1.109)	7.890*** (1.213)	6.376*** (1.208)	7.682*** (1.196)
Capital raised					
Firm Has Received Venture Capital	—	20.711*** (0.932)	—	—	—
Rounds of Venture Capital Received	—	—	3.612*** (0.244)	—	3.235*** (0.241)
Sigma	18.996*** (0.566)	16.247*** (0.468)	18.214*** (0.537)	18.410*** (0.545)	17.832*** (0.524)
Log likelihood	-4,274.0	-3,972.1	-4,152.0	-4,219.3	-4,121.2

Note: Standard errors in parentheses. $N = 3,675$. $\text{Prob}(|t| \geq x)$: * < 0.05, ** < 0.01, *** < 0.001.

Table 11.5 computes the partial derivatives of expected IPO proceeds with respect to each of the variables in the table 11.4 models, assuming in each case that all the variables are at their sample-mean values. Each article written by a star as or with a firm employee increases IPO proceeds by from $0.9 to $1.3 million, depending on model specification. The effect on the value of the firm will typically be a significant multiple of these derivative amounts since only a fraction of the equity value is sold at IPO. Each patent is worth between $1 and $2 million, as is the use of rDNA technology. A nearby top-quality university adds from $0.5 to $0.7 million to IPO proceeds. Receiving venture capital is worth about $3.3 million in IPO proceeds as a yes or no matter or about $0.6 million per round of venture capital received. Higher prior returns also increase current IPO proceeds. We believe that the about $1 million impact on IPO proceeds of an article written by a star as or with a firm employee is a significant underestimate of the actual impact of star involvement, since that is likely to have increased the number of patents obtained by the firm and, at least for the earlier firms, enabled adoption of rDNA technology.

As with the probability of going public, the firm's knowledge capital, market conditions, and receipts of venture capital all play distinct and significant roles in determining the amount of money raised by the firms that indeed do go public. In this case, we can robustly characterize the impact of four distinct indicators of the knowledge capital, with only SBIR grants failing to make a significant positive contribution. Also as with the probability of going public, the use of genetic engineering and prior high-biotech returns always significantly increase expected IPO proceeds.

Granger Causality Analysis of the Firm's Knowledge Capital and Venture Capital Funding

Table 11.6 reports the results of Granger causality tests to see whether venture capital funding enables the firm to engage in more joint research with star scientists or whether more joint research with star scientists enables the firm to get venture capital funding. As is often the case with these experiments, the results are somewhat ambiguous, but we believe they are most consistent with the view that the variables are mutu-

Table 11.5
Marginal effects of explanatory variables on IPO proceeds

	(a)	(b)	(c)	(d)	(e)
Knowledge capital					
Star-Firm Articles to Date	1.2727***	0.9127***	1.1379***	1.0822***	1.0127***
	(0.2725)	(0.1991)	(10.2436)	(0.2553)	(0.2340)
Patents Granted with Application to Date	2.0515***	1.0634***	1.4853***	1.7934***	1.3882***
	(0.2577)	(0.1874)	(0.2298)	(0.2433)	(0.2224)
Use of rDNA Technology by Firm	1.9766***	0.9664***	1.4779***	1.8452***	1.4344***
	(0.1674)	(0.1288)	(0.1530)	(0.1583)	(0.1478)
Local Top-Quality Universities	—	—	—	0.7064***	0.4998***
				(0.0681)	(0.0639)
Number of SBIR Grants to Date	—	—	—	0.0709	−0.1764
				(0.3314)	(0.3154)
Market indicator					
Biotech Returns Two Years Prior	1.1950***	1.0526***	1.3709***	1.1498***	1.3073***
	(0.2292)	(0.1747)	(0.2091)	(0.2169)	(0.2020)
Capital raised					
Firm Has Received Venture Capital	—	3.2787***	—	—	—
		(0.1590)			
Rounds of Venture Capital Received	—	—	0.6275***	—	0.5506***
			(0.0428)		(0.0416)

Note: Standard errors in parentheses. $N = 3{,}675$. $\text{Prob}(|t| \geq x)$: * < 0.05, ** < 0.01, *** < 0.001.

Table 11.6
Causality analysis between star-firm articles and venture capital rounds (Poisson regressions)

Dependent Variable	Star-Firm Articles to Two Years Prior[a] (a)	Star-Firm Articles to Two Years Prior[a] (b)	Rounds of Venture Capital Received (c)	Rounds of Venture Capital Received (d)
Constant	-4.501***	-4.791***	-2.190***	-2.197***
	(0.173)	(0.202)	(0.055)	(0.055)
Star-Firm Articles to Three Years Prior[a]	2.345***	2.244***	—	0.636*
	(0.219)	(0.218)		(0.251)
Rounds of Venture Capital Received as of One Year Prior	—	0.972***	1.054***	1.054***
		(0.194)	(0.050)	(0.050)
Log likelihood	-214.4	-205.9	-1255.6	-1253.4
Chi-squared tests				
H_0: Venture capital variable does not cause star article: $\chi^2(1) = 17.0$***				
H_0: Star Article does not cause Venture Capital: $\chi^2(1) = 4.4$*				

Notes: Standard errors in parentheses. $N = 2,989$. Significance level for coefficients and chi-squared tests: * < 0.05; ** < 0.01; *** < 0.001.

[a] We assume it takes two years for an article to be finished and published (i.e., if an article is published in year t, we assume its authors began the work with the firm on the project in year $t - 2$).

ally reinforcing in a virtuous circle in which more of each increases the other.

The particular difficulty in using the Granger methodology here is that we do not have a clear dating of when the work that culminates in an article published on a given date was actually done. The firm would be sure that potential venture capital funders were aware of star involvement in the firm while it was actually in process as opposed to completed, submitted, revised, and eventually published. Table 11.6 reports Granger causality tests assuming that the work was done precisely two years before the article was published. In this case, we can reject both the hypothesis that star articles do not cause venture capital funding and the hypothesis that venture capital funding does not cause star articles, implying reciprocal causality. If one believes that the R&D work plus publication must have significantly exceeded two years, then the hypothesis that venture capital funding does not cause star articles becomes tenable. If, on the other hand, one believes that the R&D work plus publication takes only one year, then the hypothesis that star articles do not cause venture capital funding becomes tenable.

We see reciprocal causality as the most reasonable interpretation of the results. On the other hand, based on extensive interviews with the scientists involved, we find it much more plausible that the work plus publication process significantly exceeds two years than that it is significantly less. Thus, there is stronger evidence for star articles to venture funding than for the reverse.

Analysis of Stock Price Behavior of Firms That Go Public

This section analyzes the stock price behavior for those 129 new biotech firms that went public between 1976 and 1992 for which data are available in Compustat. A much fuller analysis is contained in Darby, Liu, and Zucker (2004). This section summarizes and extends that analysis. As we have seen, the firms with deep ties to stars go public faster due to the direct effect and any indirect effects by patents granted and technology used. Those advantages continue after they are public, but we also find evidence that high-science companies follow a different high-stakes strategy pursuing high-risk, high-payoff R&D programs—typical of top scientists' personal research programs (Merton 1968). It is rational both for these firms to follow this strategy and for other firms not to do so because the probability of success is higher for firms with the highest level of knowledge capital.

Darby, Liu, and Zucker (2004) base their estimates of stock price behavior on a five-equation option pricing model. (The interested reader is referred there for the details.) The relevant part for our discussion is that biotech stock prices appear to evolve according to a standard Brownian motion interrupted by occasional stock price jumps. Such jumps occur, for example, when it is revealed that a risky R&D project has evidenced success or failure. We model the frequency of these jumps as dependent on either one or two of the knowledge stock variables. Table 11.7 reports the parameters for that equation estimated by the generalized method of moments (GMM). The Poisson parameter λ (expected frequency of stock price jumps per year) is estimated in the first three columns as $\lambda = \lambda_0 + \lambda_1 X_1$ where X_1 is alternatively Star-Firm Articles Published to Date/R&D Stock, Citations to Patents Granted with Application to Date/R&D Stock, or Claims in Patents Granted with Application to Date/R&D Stock.[12] We see that in all three cases, the knowledge stock variable is positive and highly significant, but only the Claims in Patents Granted with Application to Date/R&D Stock passes the goodness-of-fit test at conventional levels.

Since Star-Firm Articles Published to Date/R&D Stock also passed the same goodness-of-fit test in a truncated sample with observations omitted for any year in which the firm reported assets was less than $2 million, we were encouraged to try a computationally challenging two-variable version of the Poisson parameter equation:

$$\lambda = \lambda_0 + \lambda_1 X_1 + \lambda_2 X_2,$$

where X_1 is alternatively Star-Firm Articles Published to Date/R&D Stock or Citations to Patents Granted with Application to Date/R&D Stock and X_2 is always Claims in Patents Granted with Application to Date/R&D Stock. These estimates are reported in columns d and e of table 11.7, and we see in both cases that the knowledge stock variables are all positive and highly significant and the estimation passes the goodness-of-fit test at conventional levels.

Table 11.7
GMM estimation of Poisson parameter $\lambda = \lambda_0 + \lambda_1 X_1$ or $\lambda = \lambda_0 + \lambda_1 X_1 + \lambda_2 X_2$: Expected frequency of stock-price jumps

	(a)	(b)	(c)	(d)	(e)
Constant [λ_0]	0.1952*** (0.0002)	0.3078*** (0.0010)	0.0837*** (0.0000)	-0.1109*** (0.0005)	0.0784*** (0.0000)
Star-Firm Articles to Date/R&D Stock	0.3471*** (.0004)	—	—	0.8969*** (0.0041)	—
Citations to Patents Granted with Application to Date/R&D Stock	—	0.1056*** (0.0004)	—	—	0.1688*** (0.0001)
Claims in Patents Granted with Application to Date/R&D Stock	—	—	0.2851*** (0.0001)	0.0626*** (0.0000)	0.2377*** (0.0001)
Goodness-of-fit statistic [$\chi^2(f)$]	18.01*	17.08*	12.74	12.26	10.48
Degrees of freedom = f	7	7	7	6	6
Newey and West test [$\chi^2(2)$] of both Knowledge capital coefficients = 0	—	—	—	22.13***	27.86***

Notes: Standard errors in parentheses. $N = 727$. Significance level for coefficients and chi-squared tests: * < 0.05; ** < 0.01; *** < 0.001. The estimations use the following fourteen regressors to form moment conditions: Debt Level of the Firm, Maturity of Debt (McCauley Duration), Book Value of Firm's Total Physical Assets, R&D Stock, Use of rDNA Technology by Firm, Local Top-Quality Universities, IPO Lead Underwriter Reputation Rank, Age, Age of the Biotech Industry (year—1975), constant; and four current-year versions of knowledge-stock measures (Star-Firm Articles Published in Year/R&D Stock, Citations to Patents Granted with Application in Year/R&D Stock, Patents Granted with Application in Year/R&D Stock, and Claims in Patents Granted with Application in Year/R&D Stock).
Source: Darby, Liu, and Zucker (2004, 361–362).

The obvious next step was to include all three knowledge stock variables as determinants of the Poisson parameter, but this was computationally impossible. As a result, we have two statistically acceptable ways to describe the jump process: model d using star-firm articles and patent claims to date or model e using patent citations and patent claims to date. Model d is based on an input measure and an output measure of knowledge capital. Model e is based on two output measures of knowledge capital.

Given uncertainty and risk in the research process, we would expect model e to be more accurate than model d, but model d has the advantage of requiring less information not known to the market at the time the stock is being priced. The number of star-firm articles published to date is knowable in current time as market participants price the firm. The number of claims granted in patents applied for to date is knowable in current time for any patents that have been granted and can be estimated for pending patent applications based on the claims made in the application and applying probabilities to the granting of each claim. The number of citations received over the first five years after the patent is granted to patents granted that have been applied for to date is unknowable in current time for recently granted and for those applied for which may or may not be granted. Of course, the last variable is the econometrician's proxy for the unobserved value of the patent, and stock market participants may be able to form their own estimates faster than it takes for citations to accrue. Nonetheless, model d using star-firm articles and patent claims to date may more closely approximate the state of knowledge in the stock market on a given day than model e.

Table 11.8 illustrates the implications as simulated using the options-pricing model of different amounts of knowledge capital. Panel A refers to simulations based on model d, and panel B refers to simulations based on model e. The two models give very similar estimates of $110 and $109 million, respectively, for the market value of a firm with knowledge capital measures equal to 0 but all other variables at their mean value. As we move down each panel with increasing intensity of measured knowledge capital, the impact on market value of each assumed increase in knowledge capital is much larger for the all-output-measures model e than for the input-and-output-measures model d: at the mean plus two

Table 11.8
Effects of knowledge capital on the firm's estimated market valuation

Value of Star-Firm Articles/R&D Stock (assumed for comparison)	Value of Claims in Patents granted with Application to Date/R&D Stock (assumed for comparison)	Market value derived from the option pricing model (US$ millions)	Marginal increase in value (relative to 0 knowledge capital case) (US$ millions)
A. *Estimated effects of alternative assumed values of knowledge capital based on model d from table 11.7*			
0	0	110.41	—
0.09 = mean	6.65 = mean	112.30	1.89
0.63 = mean + 1 SD	27.05 = mean + 1 SD	118.02	7.61
1.17 = mean + 2 SD	47.45 = mean + 2 SD	123.66	13.25
B. *Estimated effects of alternative assumed values of knowledge capital based on model e from table 11.7*			
0	0	109.01	—
2.17 = mean	6.65 = mean	119.98	10.97
11.18 = mean + 1 SD	27.05 = mean + 1 SD	148.54	39.53
20.29 = mean + 2 SD	47.45 = mean + 2 SD	179.79	70.78

Notes: In this table, we calculate how changes in the firm-specific knowledge capital measures (Claims in Patents Granted with Application to Date/R&D Stock and either Star-Firm Articles to Date/R&D Stock or Citations to Patents Granted with Application to Date/R&D Stock) affect the biotech firm's estimated market values. Here, we assume a hypothesized firm that has the mean values for all the other variables as we vary the assumed value of the firm's knowledge capital. Panel A presents the results of using the parameters estimated from model d in table 11.7, which uses Star-Firm Articles to Date/R&D Stock and Claims in Patents Granted with Application to Date/R&D Stock as the knowledge-capital measures. Panel B presents the results of using parameters estimated from model e, which uses Citations to Patents Granted with Application to Date/R&D Stock and Claims in Patents Granted with Application to Date/R&D Stock as the knowledge-capital measures. The mean value of the R&D stock is $19.73 million (1984 dollars), so the mean value of Star-Firm Articles to Date is 1.66, the mean value of Citations to Patents Granted with Application to Date is 14.21, and the mean value of Claims in Patents Granted with Application to Date is 133.97.
Source: Darby, Liu, and Zucker (2004, 365–366).

standard deviations on the knowledge capital variables, model e prices the firm at $180 million compared to $124 million for model d. This confirms that while it is useful to know inputs when that is the only information available to an investor, it is even better to know more about how successful the firm's research program has been.

The approach used in Darby, Liu, and Zucker (2004) provides a possible way of pricing a firm's knowledge capital. We can value the importance of a star scientist or patent by considering the contribution they may make to the firm's market value. Traditional accounting procedures fail to capture the real value of the firm's intangibles, especially knowledge capital. The methodology described here, however, values a firm's intangible assets by taking into account their contribution to the firm's market value.

Conclusions

This chapter starts from a universe of private biotech firms and estimates survival models to explain the probability each year that a particular firm will go public. We find that more knowledge capital, venture capital funding, and a hot market all increase the probability that the firm will go public. Our evidence on duration to IPO also supported the organizational imprinting hypothesis: firms founded in years with exceptionally high IPO rates are themselves significantly more likely to eventually go public.

We also find that more knowledge capital, venture capital funding, and a prior hot market all increase the expected amount of money that the firm will raise if it does go public. In the case of proceeds raised, we can estimate strongly significant and distinct positive contributions for four of our five more knowledge capital indicators—the only exception being the number of SBIR grants received. Each article written by a star scientist as or with a firm employee increases IPO proceeds by around $1 million and each patent by $1 to $2 million. Each nearby top-quality university or round of venture capital adds about $0.6 million to IPO proceeds. Thus, star articles are worth nearly as much as a patent in terms of firm value and about twice as much as a round of venture capital financing. As to the chicken-and-egg question, the evidence is most consistent with mutual (i.e., two-way) star-firm articles and venture capital causation, but if we had to choose one-way causation, our reading would be that star involvement in a firm is more likely to bring venture capital financing than is venture capital financing to attract star involvement.

Darby, Liu, and Zucker's (2004) complementary study uses mostly the same variables to explain the stock price for the firms after they go public. Knowledge capital again plays a key role in valuing the firm, reflecting underlying differences in research strategics pursued by those firms with very high endowments of knowledge capital. Measuring knowledge capital by star-firm articles and patent claims, the estimated value of a firm with knowledge capital endowment two standard deviations above the mean is 12 percent higher than that of an otherwise similar firm with no measured knowledge capital. If we know more about the value of the patents produced and can measure knowledge capital by future citations to the firm's patents and patent claims, the estimated value of a firm with knowledge capital endowment two standard deviations above the mean is 65 percent higher than that of an otherwise similar firm with no measured knowledge capital.

We conclude that a high level of knowledge capital is a continuing source of both research and financial success for the new biotechnology firms. Underwriters and initial investors correctly foresaw that these companies had the best chance of ongoing success and placed their bets accordingly. While a few of these firms have become full pharmaceutical firms, most of them have specialized in drug discovery, leaving regulatory, manufacturing, and marketing tasks to others. Of course, a few of these early firms had an entirely different vision or insight on how to apply biotechnology and have prospered—or failed—in other industries altogether. The available data on the financing and knowledge capital of private biotech firms are sufficiently rich to permit meaningful estimation of duration and proceeds models for IPOs and stock prices for the firms after they go public. It remains for future research to apply the same methods to other high-tech industries.

Notes

This research has been supported by grants from the National Science Foundation (SES 0304727 and SES 9012925), the University of California Systemwide

Biotechnology Research and Education Program, the University of California President's Initiative for Industry-University Cooperative Research, and the Alfred P. Sloan Foundation through the NBER Research Program on Industrial Technology and Productivity. Ivo I. Welch provided comments and advice far above and beyond collegial standards, for which we are truly grateful. Qiao Liu was primarily responsible for integrating the new variables into the Zucker-Darby high-technology relational database, conducting the analysis of IPOs, and leading the analysis of market value. Marc Junkunc, Stephanie Hwang, and Andrew Jing did substantial work in developing the new variables. This chapter is a part of the NBER's research program in productivity. Any opinions expressed are our own and not those of the National Bureau of Economic Research.

1. On nanotechnology, see Darby and Zucker (2006). In addition to biotech and nanotech, semiconductors, lasers, and parts of the multimedia/Internet industry are characterized by relatively recent metamorphic change.

2. Factors that predict stars working with firms are identified in Zucker, Darby, and Torero (2002).

3. The basic data sets developed for this project are described in Zucker, Darby, and Brewer (1994, 1998). Details of how we defined star ties to firms through coauthorship on scientific articles—either by affiliation to the firm or by linkage via co-authorship with firm employees—can be found in Zucker, Darby, and Armstrong (1998). These data have been or will be archived on completion of the project in the Data Archives of the Center for International Science, Technology and Cultural Policy (CISTCP) at the UCLA School of Public Affairs.

4. The other six firms became publicly traded through merger with a public company or spin-off.

5. Reputational ratings were based on responses from approximately 15 percent of the faculty in the fields studied.

6. The Small Business Innovation Research (SBIR) sets aside for small business contracts and grants a portion of all federal research funding programs, and a number of the private biotech firms obtained research funding from this source.

7. In principle, we could have extended the ties data set to the present since we are interested only in articles authored by stars as or with employees of these comparatively few publicly traded firms. However, 1997 was the last full year of patent grants and citations available when we started this research, and the need to wait to see which patents are granted or are important is an inherent weakness in the use of patent measures as an indicator of knowledge capital.

8. The likely reason for the unimportance of how we handle R&D expenditures more than two years prior to going public is that combining rapid growth rates with 20 percent depreciation rates implies that three or more year-distant expenditures have a very low weight. For example, if R&D expenditures are growing at 20 percent, a conservative figure for these firms, the three-year prior weight is only 0.262.

9. Top-quality universities are defined as those with a quality ranking of 4.0 or above on a scale of 1 to 5 for one or more of the biotech-relevant doctoral programs in biochemistry, cellular and molecular biology, and microbiology in the 1981 National Research Council survey (Jones, Lindzey, and Coggeshall 1982). Twenty U.S. universities were so defined as top quality. *Nearby* is defined as in the same BEA area as the firm. Zucker, Darby, and Brewer (1998) found that top-quality universities had a significant positive effect on entry of biotech firms over and above that of the number of local actively publishing star scientists.

10. This table would look like a conventional survival table if we subtract the percentage public from 100 percent to get the percentage "surviving" as private firms. This table presents the predicted number of years from founding for various percentages of a cohort of entering private firms to have gone public based on model e in table 12.2. Model e is simulated for three different cohorts. In the case a, the firms have the mean values each year of all the firms of a given age in our sample. Case b is the same as case a except that the values are set to 0 for Patents Granted with Applications to Date ("Patents"), Local Top-Quality Universities ("Top Universities"), and Number of SBIR Grants to Date ("SBIR"). The simulated increase in time to IPO moving from case a to case b indicates the effect of these variables at their mean values versus 0. Case c is the same as case a except that the values of all the science variables are set to 0. The simulated increase in time to IPO moving from case b to case c indicates the effect of the number of Star-Firm Articles to Date and the Use of rDNA Technology by Firm at their mean values versus 0.

11. The models were unstable when one-year and two-year lagged biotech returns were included simultaneously and those with two-year returns generally had a higher log likelihood. We did not include the strategy variable in the proceeds models because it was used to identify firms more likely to be pursuing an IPO strategy but would not affect the proceeds of those firms that actually do go public.

12. We also tried Patents Granted with Application to Date/R&D Stock, but that variable was always dominated by the others, and so it is omitted in the discussion.

References

Aghion, Philippe, and Jean Tirole. 1994. "The Management of Innovation." *Quarterly Journal of Economics* 109(4), 1185–1209.

Banerjee, Abhijit V. 1992. "A Simple Model of Herd Behavior." *Quarterly Journal of Economics* 107(3), 797–818.

Bikhchandani, Sushil, David Hirshleifer, and Ivo Welch. 1992. "A Theory of Fads, Fashion, Custom, and Cultural Change as Informational Cascades." *Journal of Political Economy* 100(5), 992–1026.

Bioscan. 1989–1997.

Carter, Richard, Fredrick Dark, and Ajai Singh. 1998. "Underwriter Reputation, Initial Returns, and the Long-term Performance of IPO Stocks." *Journal of Finance* 53(2), 285–311.

Cetus Corporation. 1988. "Biotechnology Company Data Base." Emeryville, CA: Cetus.

Darby, Michael R., Qiao Liu, and Lynne G. Zucker. 2004. "High Stakes in High Technology: High-Tech Market Values as Options." *Economic Inquiry* 42(3), 351–369.

Darby, Michael R., and Lynne G. Zucker. 2003. "Growing by Leaps and Inches: Creative Destruction, Real Cost Reduction, and Inching Up." *Economic Inquiry* 41(1), 1–19.

Darby, Michael R., and Lynne G. Zucker. 2006. "Grilichesian Breakthroughs: Inventions of Methods of Inventing in Nanotechnology and Biotechnology." *Annales d'Economie et Statistique.*

Deeds, David L., Dona Decarolis, and Joseph E. Coombs. 1997. "The Impact of Firm-Specific Capabilities on the Amount of Capital Raised in an Initial Public Offering." *Journal of Business Venturing* 12, 31–46.

Genbank. 1990. Release 65.0 (machine-readable database). September. Palo Alto, Calif.: IntelliGentics.

GenBank. 1994. Release 81.0 (machine-readable database). Bethesda, Md.: National Center for Biotechnology Information.

Griliches, Zvi. 1990. "Patent Statistics as Economic Indicators: A Survey." *Journal of Economic Literature* 28, 1661–1707.

IPO Reporter. 1976–1992.

Jones, Lyle V., Lindzey, Gardner and Coggeshall, Porter E., eds. 1982. *An Assessment of Research-Doctorate Programs in the United States: Biological Sciences*. Washington, D.C.: National Academy Press.

Merton, Robert K. 1968. "The Matthew Effect in Science." *Science*, January 5, 56–63.

Romanelli, Elaine. 1991. "The Evolution of New Organizational Forms." *Annual Review of Sociology* 17, 79–103.

Securities Data Company. 1998a. *Global New Issues*. New York: Securities Data Company.

Securities Data Company. 1998b. *VentureXpert*. New York: Securities Data Company.

Stephan, Paula E., and Everhart, Steve. 1998. "The Changing Rewards to Science: The Case of Biotechnology." *Small Business Economics* 10(2), 141–151.

U.S. Department of Commerce. Economics and Statistics Administration. Bureau of Economic Analysis. 1992. *Regional Economic Information System, Version 1.3*. CD-ROM. Washington, D.C.: Bureau of Economic Analysis.

U.S. Department of Commerce. Patent and Trademark Office. 1993. *Patent Technology Set: Genetic Engineering.* CD-ROM. Washington, D.C.: U.S. Department of Commerce, Office of Information Systems.

Welch, Ivo. 1992. "Sequential Sales, Learning, and Cascades." *Journal of Finance* 47(2), 695–732.

Zucker, Lynne G., Marilynn B. Brewer, Amalya Oliver, and Julia Liebeskind. 1993. "Basic Science as Intellectual Capital in Firms: Information Dilemmas in rDNA Biotechnology Research." Working paper, UCLA Institute for Social Science Research.

Zucker, Lynne G., and Michael R. Darby. 1996. "Star Scientists and Institutional Transformation: Patterns of Invention and Innovation in the Formation of the Biotechnology Industry." *Proceedings of the National Academy of Sciences*, November 12, 12709–12716.

Zucker, Lynne G., Michael R. Darby, and Jeff Armstrong. 1998. "Geographically Localized Knowledge: Spillovers or Markets?" *Economic Inquiry* 36(1), 65–86.

Zucker, Lynne G., Michael R. Darby, and Jeff Armstrong. 2002. "Commercializing Knowledge: University Science, Knowledge Capture, and Firm Performance in Biotechnology." *Management Science* 48(1), 138–153.

Zucker, Lynne G., Michael R. Darby, and Marilynn B. Brewer. 1994. "Intellectual Capital and the Birth of U.S. Biotechnology Enterprises." Working paper No. 4653, NBER.

Zucker, Lynne G., Michael R. Darby, and Marilynn B. Brewer. 1998. "Intellectual Human Capital and the Birth of U.S. Biotechnology Enterprises." *American Economic Review* 88(1), 290–306.

Zucker, Lynne G., Michael R. Darby, and Máximo Torero. 2002. "Labor Mobility from Academe to Commerce." *Journal of Labor Economics* 20(3), 629–660.

12

Afterword

Naomi R. Lamoreaux and Kenneth L. Sokoloff

With globalization, innovation is today even more important to the U.S. economy than it was in the past. The country's comparative advantage in the generation of new ideas is increasingly evident in trade statistics, and the benefits we realize through our capacity to apply technologies conceived in one location to the actual production of goods and services elsewhere is reflected in our attention to strengthening the enforcement of intellectual property around the world. Regional specialization in invention is, as we have seen, nothing new, tracing back in American history to the differences across regions in the extent of markets in outputs and in technology present during the early stages of industrialization. But it is striking that substantial geographic variation in inventiveness persists, despite improvements in transportation and in the transmission of information, which have gone a long way toward equalizing access to trade in material products and knowledge of market conditions. What accounts for this pattern? Clearly part of the explanation for why the generation of new technologies remains so geographically concentrated across the regions of the United States and the rest of the world is differences in familiarity with the frontiers of technical knowledge. Just as hotbeds of innovation sometimes emerge out of pockets of expertise in cutting-edge technologies formed around clusters of leading universities, such as in Silicon Valley or in the Boston-Cambridge area, countries such as the United States, Japan, and Germany are advantaged in the race to make new discoveries by their relatively large ranks of people well educated and trained in technical fields.

Institutional supports for the market in technology, including those related to mobilizing finance, also seem likely to play a major role in accounting for the technological leadership of the United States, past, present, and future. These sorts of structures evolve unevenly and often have a geographic dimension, even within countries. Moreover, as the chapters in this book highlight, the availability of finance to support investments in technological development should not be presumed to materialize automatically on demand. The United States has been fortunate in that its institutions have proved remarkably flexible in adapting over time to changing circumstances in the market for technology. Indeed, the innovations made in the means of mobilizing capital to finance creative entrepreneurs are an important source for the comparative advantage of the United States in generating new technologies.

A prime example of the importance of institutional innovation in promoting technological advance is, of course, the enormous growth of the venture capital of financial markets that has occurred over the past several decades. Although investments that might be termed venture capital have long been with us, there is no doubt that the volume of funds allocated in that direction has exploded since a clarification in the Employee Retirement Income Security Act and other measures of financial deregulation in the late 1970s freed pension funds and other financial institutions to commit more of their resources to venture capital. Samuel Kortum and Josh Lerner (1998 and 2000) exploited this natural experiment through the use of instrumental variables techniques to show that increases in venture capital gave significant impetus to the pace of innovation, and it is likely that the effects over time have been even larger than they estimated because improvements in the efficiency of venture capital firms have undoubtedly followed through learning by doing and the establishment of complementary enterprises and institutions. The impact of the growth of venture capital can be gauged not only by the extraordinary surge in patenting since 1980, but also by the proliferation of start-ups focused on the development of new technologies over the same period.[1] It is easy to understand why the European Union, Israel, and other societies with labor forces similarly well qualified for R&D have sought to imitate the success of the United States by encouraging the formation and expansion of the venture capital segment of their financial markets.

Another institution that has figured prominently in the history of technological innovation in the United States has been its patent system. Quite revolutionary in design at inception, the U.S. patent system came to be much admired for providing broad access to property rights in new technological knowledge and for facilitating trade in patented technologies through strict examination of applications and enforcement. These features attracted the technologically creative, even those who lacked the capital to directly exploit their inventions, to devote their energies toward developing their ideas, and also fostered a division of labor between the conduct of inventive activity and the application of technical discoveries to actual production across individuals, firms, and regions. Especially after the Americans greatly impressed observers with their innovations at the Crystal Palace Exhibition of 1851, other countries began a long series of modifications of their patent regimes to make them more like the U.S. model. Making a tradable asset of technological knowledge proved to be a powerful institutional means of mobilizing capital for investment in inventive activity, and its continued effectiveness is vividly on display today in the business plans that those with new ideas to explore now routinely formulate and in the ways venture capital firms manage their investments.

Not so long ago, when large corporations such as IBM, Bell, General Motors, and Merck maintained in-house R&D laboratories to develop the bulk of the technologies they would exploit, it was common for many in industry and academe to question how useful patents were as a means of encouraging private parties to invest in inventive activity. Much emphasis was given to the possibility that information asymmetries inhibited trade in technological knowledge (whether patented or not), and it was suggested that alternative methods of extracting returns from inventions, such as through secrecy, were in practice much more important than patents. In principle it was recognized that patents might be helpful to independent inventors, but there was skepticism about how generally they could be put to effective use and something of a consensus that they were not central to the processes of technological innovation. The coincidence of the rise of venture capital with dramatic surges in patenting, in the formation of firms by entrepreneurial inventors, and in the pace of technical progress across a broad array of high-tech industries has, for the moment, chastened the critics of patent systems. These developments have led many to reconsider whether it might not be better for social welfare to allow both the researchers closest to the technological frontier as well as the market to exercise more influence in choosing which lines of R&D should be pursued. Indeed, the passage of the Bayh-Dole Act and the adoption of new policies promoting the patenting of discoveries in government laboratories such as the National Institutes of Health seem to reflect growing acceptance of the notion that patent systems can be extremely valuable for mobilizing finance and other entrepreneurial resources toward the practical application of knowledge. Moreover, it is well understood, and presumably approved of, that as these policies typically involve the patent rights to discoveries being shared by the researcher and the home institution, the de facto elevation of commercial considerations likely affects the kinds of research being carried out with public funds.

The criticisms of the patent system being leveled today have a very different character from those made in the past. They focus primarily on how the workings of what has long been a remarkably constructive institution may have been badly damaged by several recent reforms.[2] In 1982 the United States established a single specialized court for all appeals of patent cases, and in the early 1990s Congress mandated a new structure of fees and financing for the U.S. Patent Office with the aim of generating surpluses from the patent system that could be returned to the federal budget. It is argued that these changes have significantly eroded the effectiveness of the examination of applications, making it far too easy to obtain a patent, and led to an explosion of litigation. These are important concerns, and if the problems are not worked out, it is conceivable that they might seriously impair the effectiveness of this key institutional support to the market for technology. There are grounds for optimism, however. The United States has been distinguished throughout its history for the flexibility of its institutions, and it is not too late to remedy the missteps or unfortunate unforeseen consequences of the recent reforms.

At the beginning of the twenty-first century, the United States seems well positioned to maintain its leadership in the development of new technologies for some time to come. Although the country enjoys a comparative advantage in this activity on the basis of the quality of its universities, the human capital of its population, and its vast wealth and product markets, it is also greatly aided by institutions that are exceptionally well suited, by current world standards, to the mobilization of finance and other resources for investment in innovative activities. Globalization may soon deliver an integrated world product market, and the easy mobility of highly skilled labor may diminish the significance of the origins of technologically creative individuals, but institutions do not yet travel so well across national boundaries. That the spread across the world of intellectual property regimes with strict enforcement of patent rights has proceeded so slowly is but one outstanding example of this truth. As nations with formidable economic and technological potential such as China, India, and Brazil continue to catch up in level of development and confront problems and opportunities increasingly like our own, however, they will tend to adopt or redesign institutions that more closely resemble ours. It is then that the institutional edge the U.S. commands may be tested.[3] But this may take quite some time, and competition in this country has often inspired creative responses.

Notes

1. For more discussion of the dramatic changes in the organization of inventive activity that have followed accompanied the greater availability of venture capital, see chapter 10 by Josh Lerner and chapter 11 by Michael Darby and Lynne Zucker.

2. Perhaps the most eloquent and insightful exposition of these views is Jaffe and Lerner (2004).

3. A wonderful example of how quickly and powerfully a change in a key institution, the patent system, can have on the rate of invention in a developing country comes from the case of Taiwan. In his careful investigation of the strong patent system suddenly forced on the country by the U.S., Shih-Tse Lo (2004) offers persuasive evidence of a massive response in the financing and conduct of inventive activity, as reflected through expenditures on research and development, patenting in the United States by Taiwanese residents, and foreign direct investment in R&D-intensive sectors. Today, even with its rather small population, Taiwan ranks in the top five countries in the world in awards of U.S. patents. He reports similar findings for South Korea, which radically altered its patent system under circumstances much like those that influenced Taiwan.

References

Jaffe, Adam B., and Josh Lerner. 2004. *Innovation and Its Discontents.* Princeton, N.J.: Princeton University Press.

Kortum, Samuel, and Josh Lerner. 1998. "Stronger Protection or Technological Revolution: What Is behind the Recent Surge in Patenting." *Carnegie-Rochester Conference Series on Public Policy* 48, 247–304.

Kortum, Samuel, and Josh Lerner. 2000. "Assessing the Contribution of Venture Capital to Innovation." *RAND Journal of Economics* 31, 674–692.

Lo, Shih-Tse. 2004. "Strengthening Intellectual Property Rights: The Experience of the 1986 Taiwanese Patent Reforms." Ph.D. dissertation, University of California, Los Angeles.

Contributors

Ashish Arora
H. John Heinz III School of Public
Policy and Management
Carnegie Mellon University

Marco Ceccagnoli
College of Management
Georgia Tech

Wesley M. Cohen
Fuqua School of Business
Duke University

Michael R. Darby
Anderson School of Management
University of California, Los Angeles

Lance E. Davis
Division of Humanities and Social
Sciences
California Institute of Technology

Kira R. Fabrizio
Department of Organization and
Management
Goizueta Business School
Emory University

Margaret B. W. Graham
McGill University
Faculty of Management

William Janeway
Warburg Pincus LLC
New York, NY

Steven Klepper
Department of Social and Decision
Sciences
Carnegie Mellon University

Naomi R. Lamoreaux
Departments of Economics and
History
University of California, Los Angeles

Josh Lerner
Rock Center of Entreprenurship
Harvard Business School

Margaret Levenstein
Stephen M. Ross School of Business
University of Michigan

David C. Mowery
Haas School of Business
University of California, Berkeley

Larry Neal
Department of Economics
University of Illinois

Tom Nicholas
Entrepreneurial Management Group
Harvard Business School

Mary O'Sullivan
The Wharton School
University of Pennsylvania

Kenneth L. Sokoloff
Department of Economics
University of California, Los Angeles

Steven W. Usselman
School of History, Technology, and
Society
Georgia Institute of Technology

Lynne G. Zucker
Department of Sociology
University of California, Los Angeles

Index